Annals of Mathematics Studies

Number 103

SEMINAR ON MINIMAL SUBMANIFOLDS

EDITED BY

ENRICO BOMBIERI

PRINCETON UNIVERSITY PRESS

PRINCETON, NEW JERSEY

1983

ISBN 0-691-08324-X (cloth)
ISBN 0-691-08319-3 (paper)

Printed in the United States of America
by Princeton University Press, 41 William Street
Princeton, New Jersey

☆

Clothbound editions of Princeton University Press books are printed on acid-free paper, and binding materials are chosen for strength and durability. Paperbacks, while satisfactory for personal collections, are not usually suitable for library rebinding.

Library of Congress Cataloging in Publication data will be found
on the last printed page of this book

CONTENTS

INTRODUCTION

The present volume collects the papers which were presented in the academic year 1979-1980 at the Institute for Advanced Study, in the areas of closed geodesics and minimal surfaces, as part of the activities of a special year in differential geometry and differential equations. Starting with a survey lecture, they have been arranged according to dimension and approach, from classical to that of geometric measure theory.

We wish to extend our sincere thanks to all contributors, particularly for their collaboration in sending their texts and their revisions as well as for their patience in waiting for these notes to appear. Our thanks also to the National Science Foundation for supporting this special year at the Institute for Advanced Study.

ENRICO BOMBIERI

Seminar On
Minimal Submanifolds

SURVEY LECTURES ON MINIMAL SUBMANIFOLDS

Leon Simon*

Our aim here is to give a general (but necessarily brief) introduction to the theory of minimal submanifolds, including as many examples as possible, and including some discussion of the classical problems (Bernstein's, Plateau's) which have provided much of the motivation for the development of the theory.

Our first task is to discuss a notion of minimal "variety" in a sufficiently general sense to include the various classes of objects (e.g. algebraic varieties in $\mathbf{C}^n$, cones over smooth submanifolds of S^n, "soap film like" minimal surfaces in $\mathbf{R}^3$, branched minimal immersions, least area integral current representatives of homology classes) which arise naturally. This will be done in §2, after some classical introductory discussion of first and second variation in §1. In §3 we present some of the principal classes of examples of minimal varieties. §4 includes a discussion of some of the special properties of minimal submanifolds of $\mathbf{R}^n$. In §5 we give a brief survey of the known interior regularity theory, and in §6 we discuss the classical Bernstein and Plateau problems and the present state of knowledge concerning them. Finally, in §7, we discuss some selected applications of second variation of minimal submanifolds in geometry and topology.

*Research was partially supported by an N.S.F. grant at the Institute for Advanced Study, Princeton.

Seminar on Minimal Submanifolds
0-691-08324-X/83/003-50 $3.00/0 (cloth)
0-691-08319-3/83/003-50 $3.00/0 (paperback)

We would here like to recommend the survey articles [L1], [N1], [O1], [B], which cover topics only touched upon (or not mentioned at all) here. For other general reading we strongly recommend the works [FH1], [GE1], [L3], [L4], [N3], [O2].

In all that follows, N will denote an n-dimensional Riemannian manifold without boundary $(n \geq 2)$, and k is an integer with $1 \leq k < n$. U will always denote an open subset of N. We also often have occasion to consider a "smooth deformation" (i.e. a smooth isotopy) of N which holds everything outside a compact subset of U fixed. To be specific, let $(-1,1) \times N$ be equipped with the product metric, let $\phi : (-1,1) \times N \to N$ be a C^2 map, let $\phi_t : N \to N$ be defined by $\phi_t(x) = \phi(t,x)$ for $(t,x) \in (-1,1) \times N$, suppose ϕ_t is a diffeomorphism of each $t \in (-1,1)$, and suppose there is a compact $K \subset U$ such that

$$(0.1) \qquad \phi_0 = 1_N, \quad \phi_t|_{N\sim K} \equiv 1_{N\sim K}, \quad \phi_t(K) \subset K \qquad \forall t \in (-1,1).$$

X will denote the associated initial velocity tangent vector field on N, defined by

$$(0.2) \qquad X_x = \frac{\partial}{\partial t} \phi(t,x)\Big|_{t=0}.$$

Of course, given any C^1 vector field on N with compact support in U, there always exists ϕ as above such that (0.2) holds.

§1. *First and second variation*

We begin classically, with M denoting an embedded (but not necessarily properly embedded, oriented or complete) k-dimensional submanifold of N such that the k-dimensional volume $\mathcal{H}^k(M \cap K)$* of $M \cap K$ is finite for each compact $K \subset N$.

When U is such that $\mathcal{H}^k(M \cap U) < \infty$ (e.g. if $\bar{U}$ is compact), we say that M is *stationary* (or "minimal") in U if

*Here and subsequently, $\mathcal{H}^k$ denotes k-dimensional Hausdorff measure in N.

(1.1) $$\frac{d}{dt}|\phi_t(M\cap U)|_{t=0} = 0$$

for every ϕ as in (0.1). Here $|\phi_t(M\cap U)|$ denotes the k-dimensional volume of the submanifold $\phi_t(M\cap U)$; that is, $|\phi_t(M\cap U)| = \mathcal{H}^k(\phi_t(M\cap U))$.

M is said to be *stable* in U if it is stationary in U and if

(1.2) $$\frac{d^2}{dt^2}|\phi_t(M\cap U)|_{t=0} \geq 0$$

for each ϕ_t as in (0.1).

We say that M is stationary (respectively stable) in an arbitrary open set $W \subset N$ if it is stationary (respectively stable) in U for each open U with $\mu(M\cap U) < \infty$ and $U \subset W$.

The quantities appearing on the left of (1.1), (1.2) are called the first and second variation of M with respect to the deformation ϕ. We shall see that the first variation depends only on the initial velocity vector X of (0.2), and is otherwise independent of ϕ.

We now want to explicitly compute the first and second variation. To do this, we need the *area formula* ([FH1, §3.2]), which asserts that

(1.3) $$|\phi_t(M\cap U)| = \int_M J(t,x)\,d\mu,$$

where, for the moment, $\mu = \mathcal{H}^k$ (k-dimensional Hausdorff measure), and where $J(t,x)$ denotes the Jacobian of the restriction of ϕ_t to M. Thus

(1.4) $$J(t,x) = |d_x\phi_t(\tau_1)\wedge d_x\phi_t(\tau_2)\wedge\cdots\wedge d_x\phi_t(\tau_k)|,$$

where $d_x\phi_t$ denotes the linear map $T_xN \to T_{\phi_t(x)}N$ between tangent spaces induced by ϕ_t, and where $\tau_1,\cdots,\tau_k$ is any orthonormal basis for T_xM. (T_xM is of course equipped with the Riemannian inner product $\langle\ ,\ \rangle$ of T_xN.)

We thus have

$$\frac{d}{dt}\left|\phi_t(M\cap U)\right|_{t=0} = \int_M \frac{\partial}{\partial t} J(t,x)\Big|_{t=0} d\mu$$

and

$$\frac{d^2}{dt^2}\left|\phi_t(M\cap U)\right|_{t=0} = \int_M \frac{\partial^2}{\partial t^2} J(t,x)\Big|_{t=0} d\mu ,$$

and direct computation (using (1.4)) shows (see for example the appendix of [SL1] for details) that

$$\text{(1.5)} \qquad \frac{d}{dt}\left|\phi_t(M\cap U)\right|_{t=0} = \int_M \operatorname{div}_M X \, d\mu$$

$$\text{(1.6)} \qquad \frac{d^2}{dt^2}\left|\phi_t(M\cap U)\right|_{t=0} = \int_M \Big\{\operatorname{div}_M Z + (\operatorname{div}_M X)^2 + \sum_{i=1}^{k} |(\nabla_{\tau_i} X)^{\perp}|^2$$

$$- \sum_{i,j=1}^{k} <\tau_i, \nabla_{\tau_j} X><\tau_j, \nabla_{\tau_i} X> - \sum_{i=1}^{k} <R(\tau_i, X)X, \tau_i>\Big\} d\mu .$$

Here X, τ_i are as in (0.2) and (1.4) respectively, $Z_x = \ddot{\phi}_t(x)|_{t=0}$, $x \in M$, $\operatorname{div}_M X = \sum_{i=1}^{k} <\tau_i, \nabla_{\tau_i} X>$, ∇_{τ_i} denotes covariant differentiation in N with respect to τ_i, $Y^{\perp}$ (for any $Y \in T_xN$) denotes the orthogonal projection of Y onto $(T_xM)^{\perp}$, and R denotes the Riemannian curvature tensor of N, defined by $R(X,Y)Z = \nabla_X\nabla_Y Z - \nabla_Y\nabla_X Z - \nabla_{[X,Y]}Z$.

From (1.5), (1.6) we thus see that M is stationary in U (whether or not $\overline{U}$ is compact) if and only if

$$\text{(1.7)} \qquad \int_M \operatorname{div}_M X \, d\mu = 0$$

for every smooth vector field X on N with compact support in U, and M is stable in U if and only if (1.7) holds and

$$(1.8)\quad \int_M \Big\{(\mathrm{div}_M X)^2 + \sum_{i=1}^{k} |(\nabla_{\tau_i} X)^{\perp}|^2 - \sum_{i,j=1}^{k} <\tau_i, \nabla_{\tau_j} X><\tau_j, \nabla_{\tau_i} X>$$

$$- \sum_{i=1}^{k} <R(\tau_i, X)X, \tau_i>\Big\}\, d\mu \geq 0$$

for every such X.

To examine more closely the meaning of these definitions, at each point of M we write $X = X^T + X^{\perp}$, with $X^{\perp}$ denoting the part of X normal to M. Thus, letting $\nu^{k+1}, \cdots, \nu^n$ be a locally defined orthonormal set of vector fields normal to M, we have $X^{\perp} = \sum_{\alpha=k+1}^{n} <\nu^{\alpha}, X>\nu^{\alpha}$, and hence (since $<\tau_i, \nu^{\alpha}> = 0$),

$$(1.9)\qquad \mathrm{div}_M X^{\perp} = \sum_{\alpha=k+1}^{n} <\nu^{\alpha}, X> \mathrm{div}_M \nu^{\alpha}.$$

Now

$$\mathrm{div}_M \nu^{\alpha} = \sum_{i=1}^{k} <\tau_i, \nabla_{\tau_i} \nu^{\alpha}> = -<\nu^{\alpha}, \sum_{i=1}^{k} B(\tau_i, \tau_i)>,$$

where B denotes the second fundamental form of M, defined on $T_x M \times T_x M$ and taking values in $(T_x M)^{\perp}$ according to

$$(1.10)\qquad B(\xi, \eta) = -\sum_{\alpha=k+1}^{n} <\xi, \nabla_{\eta} \nu^{\alpha}>\nu^{\alpha}, \quad \xi, \eta \in T_x M$$

$$(= B(\eta, \xi)).$$

An alternative definition is

$$(1.10)' \qquad B(\xi, \eta) = (\nabla_{\xi} Y)^{\perp},$$

where Y is any smooth extension of η. The *trace* of B is called the *mean curvature* vector $\mathbf{H}$ of M; thus

$$\mathbf{H}(x) = \sum_{i=1}^{k} B(\tau_i, \tau_i) , \tag{1.11}$$

with τ_i as in (1.4).

It goes without saying (and is easily checked) that all these definitions are independent of the particular choice of $\tau_1, \cdots, \tau_k, \nu^{k+1}, \cdots, \nu^n$.

We now have from (1.9), (1.10), and (1.11) that

$$\begin{aligned} \operatorname{div}_M X^{\perp} &= -\sum_{\alpha=k+1}^{n} <\nu^\alpha, X><\nu^\alpha, \mathbf{H}> \\ &= -<X, \mathbf{H}> , \end{aligned} \tag{1.12}$$

and hence

$$\int_M \operatorname{div}_M X \, d\mu = \int_M \operatorname{div}_M X^T \, d\mu - \int_M <X, \mathbf{H}> d\mu .$$

To go further, we assume that M is actually compact with smooth boundary ∂M (possibly empty); then the classical divergence theorem tells us (since X^T is by definition a tangent vector field on M) that $\int_M \operatorname{div}_M X^T \, d\mu = -\int_{\partial M} <X, \eta> d\mathcal{H}^{k-1}$, where η denotes the co-normal of ∂M. Thus η is tangent to M, normal to ∂M, and points *into* M. Putting these facts together, we thus have

$$\int \operatorname{div}_M X \, d\mu = -\int_{\partial M} <X, \eta> d\mathcal{H}^{k-1} - \int_M <X, \mathbf{H}> d\mu , \tag{1.13}$$

and we see that (in case M is smooth, compact) that M is stationary in U if and only if

$$\partial M \cap U = \emptyset \quad \text{and} \quad H \equiv 0 \quad \text{on} \quad M \cap U . \tag{1.14}$$

This is of course a well-known classical characterization of stationary submanifolds. It also explains why the word "minimal" is used in the classical setting; one can in fact check that the following lemma holds.

LEMMA 1.1. *Suppose* M *is smooth, compact, and* $\xi \in M \sim \partial M$. *Then* $H \equiv 0$ *in a neighborhood of* ξ *implies that there is a (small) open* U *containing* ξ *such that*

$$|M \cap U| \leq |\phi_t(M \cap U)| \tag{1.15}$$

for all sufficiently small t, *whenever* ϕ *is as in* (0.1).

We emphasize that this is in general false if M is allowed to have singularities. (In §2 below we *shall* allow M to have singularities.)

For a brief sketch of the proof of (1.15), we first suppose that N is represented (locally, near ξ, via a coordinate chart) as a submanifold of R^n in such a way that the point ξ corresponds to $0 \in R^n$. Then, selecting suitable coordinate axes in R^n and supposing that U has been chosen sufficiently small, we can write

$$|\phi_t(M \cap U)| = \int_\Omega F(x, u_t, Du_t)\,dx , \tag{1.16}$$

for all sufficiently small t. Here $\Omega \subset R^k$, $u_t = (u_t^{k+1}, \cdots, u_t^n) : \Omega \to R^{n-k}$ is such that $Du_0(0) = 0$ and such that graph u_t represents $\phi_t(M \cap U)$, and $F(x, z, p)$ is the "non-parametric k-dimensional area integrand" associated with our local coordinate representation of N. $F(x, z, p)$ is a C^2 function of $x \in \Omega$, $z = (z^{k+1}, \cdots, z^n) \in R^{n-k}$, and $p = (p_i^\alpha)_{\substack{\alpha=k+1,\cdots,n \\ i=1,\cdots,k}} \in R^{k(n-k)}$, and it satisfies the uniform convexity condition

$$(1.17)\qquad \sum_{i,j=1}^{k} \sum_{\alpha,\beta=k+1}^{n} \frac{\partial^2 F}{\partial p_i^\alpha \partial p_j^\beta}(x,z,p)\xi_i^\alpha \xi_j^\beta \geq c \sum_{i=1}^{k} \sum_{\alpha=k+1}^{n} (\xi_i^\alpha)^2$$

for $|p|$, $|z|$ sufficiently small. (In the codimension 1 case, when $n-k=1$, $F(x,z,p)$ is in fact convex in p for all p, but this is not so in case $n-k \geq 2$.)

Now, by virtue of the fact that $\frac{d}{dt}|\phi_t(M\cap U)|_{t=0} = 0$, we have the identity

$$|\phi_t(M\cap U)| - |M\cap U| = \frac{1}{2}\frac{d^2}{dt^2}|\phi_t(M\cap U)|_{t=\theta}$$

for some θ between 0 and t. Direct computation now shows that $\frac{d^2}{dt^2}|\phi_t(M\cap U)| \geq 0$ for sufficiently small t (provided U has originally been chosen small enough). (In checking this we use differentiation under the integral in (1.16); one needs to use the convexity (1.17) and also the Poincaré inequality $\int_\Omega \psi^2 \leq c(\operatorname{diam}\Omega)^2 \int_\Omega |D\psi|^2$, $\psi \in C_0^1(\Omega)$.)

In case M is smooth, compact, stationary in U, and $\partial M \cap U = \emptyset$, we can obtain a somewhat more compact expression for the second variation of M. Indeed if $I(X)$ represents the expression on the right of (1.6) (with first term deleted by virtue of the fact that M is stationary in U), then

$$(1.18)\qquad I(X) = \int_M \Big(\sum_{i=1}^{k} |(\nabla_{\tau_i} X^\perp)^\perp|^2 - \sum_{i,j=1}^{k} \langle X, B(\tau_i,\tau_j)\rangle^2 - \sum_{i=1}^{k} \langle R(\tau_i, X^\perp)X^\perp, \tau_i\rangle\Big)\,d\mu\,.$$

In checking this we first show that $I(X) = I(X^\perp)$ and then use the fact that $\langle \tau_j, \nabla_{\tau_i} X^\perp\rangle = \sum_{\alpha=k+1}^{n} \langle \tau_j, \nabla_{\tau_i}\nu^\alpha\rangle\langle X, \nu^\alpha\rangle = -\langle B(\tau_i,\tau_j), X\rangle$ (ν^α, B as in (1.9), (1.10) above). For further details concerning this computation, see for example [SL1]. In the codimension 1 oriented case, we can write $X = \zeta\nu$, where ν is a smooth unit normal and ζ is a scalar function, and the expression for $I(X)$ becomes

$$(1.19)\qquad I(X) = \int_M \left\{|\nabla\zeta|^2 - \zeta^2\left(|B|^2 + \sum_{i=1}^{k} \langle R(\tau_i,\nu)\nu,\tau_i\rangle\right)\right\} d\mu ,$$

where $\nabla\zeta$ denotes the gradient (taken in M) of the scalar function ζ. Notice that $\sum_{i=1}^{k} \langle R(\tau_i,\nu)\nu,\tau_i\rangle$ is just $\mathrm{Ric}(\nu,\nu)$, where Ric is the Ricci curvature of N.

We should finally mention the meaning of the terms "stationary in U" and "stable in U" in case M is *immersed* rather than embedded. To do this we can suppose that M_0 is any compact k-dimensional Riemannian manifold (with or without boundary) and let $\psi : M_0 \to N$ be a smooth map (not necessarily an immersion), and let $J(\psi)$ be the Jacobian of ψ. That is, $J(\psi)(x) = \|\Lambda_k(d_x\psi)\|$, where $d_x\psi$ denotes the linear map $T_xM_0 \to T_{\psi(x)}N$ induced by ψ. The area associated with such a ψ is of course defined by

$$(1.20)\qquad A(\psi) = \int_{M_0} J(\psi)\, d\mathcal{H}^k .$$

By a *variation* of ψ in U_0 (U_0 open in M_0) we mean a 1-parameter family $\{\psi_t\}_{t\in(-1,1)}$ of smooth maps of M_0 into N, smoothly varying in t, such that $\psi_0 = \psi$ and such that, for some fixed compact $K \subset U_0$, $\psi_t(x) \equiv \psi(x)$ whenever $x \in M_0 \sim K$ and $|t| < 1$. Then ψ is said to be *stationary* in U_0 if $\frac{d}{dt} A(\psi_t)|_{t=0} = 0$ for every such variation of ψ, and ψ is said to be *stable* in U_0 if it is stationary in U_0 and if $\frac{d^2}{dt^2} A(\psi_t)|_{t=0} \geq 0$ for every such variation. If we define $X_\xi = \frac{\partial}{\partial t} \psi_t(\xi)|_{t=0}$ (so that X is defined on M_0 but takes values in TN) then one can derive expressions for first and second variations (i.e. for $\frac{d}{dt} A(\psi_t)|_{t=0}$ and $\frac{d^2}{dt^2} A(\psi_t)|_{t=0}$) which are the same as the right sides of (1.5), (1.6), but in which the various quantities must now be appropriately interpreted. (Near points $x_0 \in M_0$, where $J(\psi)(x_0) \neq 0$, ψ *embeds* a small neighborhood W of x_0 into N and one computes the formulae for first and second variation by computing the t-derivatives of the relevant Jacobian as before. On the other hand, the points where $J(\psi)(x_0) = 0$ of course contribute nothing to the formulae.) Of course from the previous discussion, it follows that if $U_0 \cap \partial M_0 = \emptyset$, then ψ is stationary in U_0 if and only if ψ locally embeds M_0 as a zero mean curvature submanifold near points where $J(\psi) \neq 0$.

§2. k-*varifolds*, k-*currents*

An examination of the discussion in §1 will show that the formulae (1.5), (1.6) (and their derivations) remain valid even in the presence of serious singularities in M. To make this statement precise, we first need to introduce some terminology.

We henceforth let M be a *countably* k-*rectifiable Borel set* in N. That is, M is Borel, $M \subset \bigcup_{j=1}^{\infty} M_j$, where M_j are (open) k-dimensional C^1 submanifolds of N (M_j not necessarily complete nor pairwise disjoint).(*)

We also allow the introduction of a *multiplicity function* of M; specifically, we let μ be the measure on N defined by $d\mu = \theta d\mathcal{H}^k$, where θ is a non-negative locally $\mathcal{H}^k$-summable function (called the *multiplicity function*) on N. (We take θ to be defined on all on N, rather than just on M, for reasons of purely technical convenience; we of course often have $\theta \equiv 0$ on $N \sim M$.)

(*)Modulo sets of $\mathcal{H}^k$-measure zero, this is equivalent to the requirement that M is contained in a countable union of images, under Lipschitz maps, of compact subsets of $\mathbf{R}^k$. (By [FH1, 3.1.16].)

The pair (M, μ) will then be called a k-varifold in N. (This corresponds exactly to the definition of k-dimensional rectifiable varifold as defined by Allard [AW1]; one sees this by virtue of [AW1, 3.5(1), 2.8(5)].)(*) (M, μ) is called an *integral* k-varifold if the multiplicity function θ is integer-valued.

The *support*, denoted $\operatorname{spt}(M, \mu)$, of the k-varifold (M, μ) is defined by

$$\operatorname{spt}(M, \mu) = N \sim \cup W \tag{2.1}$$

where the union is over all open W such that $\mu(M \cap W) = 0$. Notice that $\operatorname{spt}(M, \mu) \subset \overline{M}$ (closure of M taken in N), but equality may be far from true. In fact $\operatorname{spt}(M, \mu)$ is just the support (in the usual sense of measures [FH1, Ch. 2]) of the measure $\mu \llcorner M$. ($\mu \llcorner M$ is the (outer) measure on N defined by $\mu \llcorner M(A) = \mu(M \cap A)$.) $\mu(M)$ is called the *mass* of (M, μ).

The *regular set*, denoted $\operatorname{reg}(M, \mu)$, is defined to be the set of x in $\operatorname{spt}(M, \mu)$ such that $\operatorname{spt}(M, \mu) \cap W$ is a properly embedded k-dimensional C^1 submanifold of N for some open W containing x. The *singular set*, denoted $\operatorname{sing}(M, \mu)$, is defined by

$$\operatorname{sing}(M, \mu) = \operatorname{spt}(M, \mu) \sim \operatorname{reg}(M, \mu) . \tag{2.2}$$

Given a k-varifold (M, μ) in N, and a smooth map $\psi : N \to \tilde{N}$ ($\tilde{N}$ any other Riemannian manifold) such that $\psi | \operatorname{spt}(M, \mu)$ is proper,(†) we define the *image varifold* $\psi_{\#}(M, \mu)$ to be the k-varifold in $\tilde{N}$ given by

$$\psi_{\#}(M, \mu) = ((\psi(M))^*, \psi_{\#}\mu) , \tag{2.3}$$

where $(\psi(M))^* \subset \psi(M)$ is any countably k-rectifiable Borel set with $\mathcal{H}^k(\psi(M) \sim (\psi(M))^*) = 0$. Evidently such $(\psi(M))^*$ exists; in case ψ is a diffeomorphism we can clearly take $(\psi(M))^* = \psi(M)$. In (2.3), $\psi_{\#}\mu$ denotes the measure given by $d\psi_{\#}\mu = \tilde{\theta}\, d\mathcal{H}^k$, with multiplicity $\tilde{\theta}(y) \equiv \sum_{z \in \psi^{-1}(y) \cap M} \theta(z)$.

*For an alternative description of Allard's work, see [SL5].

†That is, $\psi^{-1}(K)$ is compact whenever $K \subset \tilde{N}$ is compact.

Notice that if ψ is a diffeomorphism then $\psi_{\#}(M, \mu)$ has the multiplicity function $\tilde{\theta} = \theta \circ \psi^{-1}$.

We then have (by virtue of the area formula for k-rectifiable sets – [FH1, §3.2]), that (with U open such that $\mu(M \cap U) < \infty$)

$$|\phi_{t\#}(M \cap U, \mu)| = \int_{M \cap U} J(t, x) d\mu$$

whenever ϕ is as in (0.1), where $|\phi_{t\#}(M \cap U, \mu)| \equiv (\phi_{t\#}\mu)(M \cap U)$ is the k-dimensional mass of the image varifold $\phi_{t\#}(M \cap U, \mu)$. Here $J(t, x)$ is precisely as in (1.4), where now $\tau_1, \cdots, \tau_k$ is any orthonormal basis for the *approximate* tangent space (still denoted T_xM) of (M, μ) at x. (A k-dimensional subspace $S \subset T_xN$ is called the approximate tangent space of (M, μ) at x if

$$(2.4) \qquad \lim_{t \downarrow 0} t^{-k} \int_{M \cap W} f(t^{-1}\psi(y)) d\mu(y) = \theta(x) \int_S f d\mathcal{H}^k, \quad \forall f \in C_c^0(T_xN),$$

where W is an open set containing x and $\psi : W \to T_xN$ a diffeomorphism with $\psi(x) = 0$ and $d_x\psi = 1_{T_xN}$ (the validity of (2.4) is independent of which particular ψ one chooses to use). Of course here $d_x\psi$ denotes the linear map $T_xN \to T_xN$ induced by ψ.

Using standard, measure-theoretic facts about densities ([FH1, 2.10.19(4)] and the fact that $\lim_{\rho \downarrow 0} \frac{\text{meas}(A \cap B_\rho^k(\xi))}{\text{meas}(B_\rho^k(\xi))} = 1$ for Lebesgue almost all $\xi \in A$, in case $A \subset \mathbf{R}^k$ is Lebesgue measurable), together with the definition of countably k-rectifiable set given above, one easily checks that such an approximate tangent space exists μ – a.e. in M.

We thus have, by exactly the computations of §1, that

(2.5) $$\frac{d}{dt}\,|\phi_{t\#}(M\cap U,\mu)|\Big|_{t=0} = \int_M \operatorname{div}_M X\, d\mu$$

(2.6) $$\frac{d^2}{dt^2}\,|\phi_{t\#}(M\cap U,\mu)|\Big|_{t=0} = \int_M \Big(\operatorname{div}_M Z + (\operatorname{div}_M X)^2 + \sum_{i=1}^{k} |(\nabla_{\tau_i} X)^{\perp}|^2$$

$$- \sum_{i,j=1}^{k} <\tau_i, \nabla_{\tau_j} X><\tau_j, \nabla_{\tau_i} X> - \sum_{i=1}^{k} <R(\tau_i, X)X, \tau_i>\Big)\, d\mu\,.$$

Analogously to §1, we define (M,μ) to be stationary in U (respectively stable in U) if the integral on the right of (2.5) vanishes (respectively if the right side of (2.5) vanishes and the right of (2.6) ≥ 0) for all smooth vector fields X which have compact support in U.

Notice also that again the first term on the right of (2.6) can always be deleted in case (M,μ) is stationary in U.

For later applications it is convenient here to introduce a slight generalization of the notion of "stationary". Namely, we say that (M,μ) has *generalized mean-curvature vector* $\mathbf{H}$ *in* U if $\mathbf{H} \in L^1(\mu, U)$ and if

$$\int_M \operatorname{div}_M X\, d\mu = -\int_M <X, \mathbf{H}>\, d\mu$$

whenever X is a smooth vector field with compact support in U. Evidently (M,μ) is stationary in U if and only if it has zero generalized mean curvature vector in U, and (by the classical formula (1.13)), in case M is a compact smooth manifold with $\partial M \cap U = \emptyset$ (and in case $\mu = \mathcal{H}^k \llcorner M$), the generalized mean curvature vector of $(M, \mathcal{H}^k \llcorner M)$ coincides with the classical mean curvature vector of M as a submanifold on N.

Our main reason for introducing this notion is the following remark.

REMARK 2.1. If N is (without loss of generality, by [NJ]) assumed to be isometrically embedded in $\mathbb{R}^p$, if $U \subset \mathbb{R}^p$ is open, and if (M, μ), considered as a k-varifold in N, is stationary in $U \cap N$, then, as a k-varifold in $\mathbb{R}^p$, (M, μ) has generalized mean curvature vector $\mathbf{H}$ which at each point x of $M \cap U$ satisfies

$$|\mathbf{H}| \le k|B_N| , \tag{2.7}$$

where $|B_N|$ denotes the length of the second fundamental form of N (as a submanifold of $\mathbb{R}^p$).

To see this, we note that $\int_M \operatorname{div}_M X \, d\mu = 0$ for any smooth vector field X on N with compact support in $U \cap N$. Thus if X is a smooth vector field on $\mathbb{R}^p$ with compact support in U, then

$$\int_M \operatorname{div}_M (X - X^\perp) d\mu = 0 , \tag{2.8}$$

where $X^\perp$ denotes the component of X normal to N. Thus, locally, $X^\perp = \sum_{\alpha=n+1}^{p} \langle \nu^\alpha, X \rangle \nu^\alpha$, where $\nu^{n+1}, \cdots, \nu^p$ is a smoothly varying orthonormal set of vectors in $\mathbb{R}^p$ normal to N. Then (since $\langle \tau_i, \nu^\alpha \rangle = 0$), we have

$$\begin{aligned} \operatorname{div}_M X^\perp &= \sum_{i=1}^{k} \sum_{\alpha=n+1}^{p} \langle X, \nu^\alpha \rangle \langle \tau_i, \nabla_{\tau_i} \nu^\alpha \rangle \\ &= - \sum_{i=1}^{k} \sum_{\alpha=n+1}^{p} \langle X, \nu^\alpha \rangle \nu^\alpha \cdot B_N(\tau_i, \tau_i) \\ &= - \langle X, \sum_{i=1}^{k} B_N(\tau_i, \tau_i) \rangle , \end{aligned}$$

where B_N is the second fundamental form on N as a submanifold of R^p. Thus (by (2.8))

$$\int \mathrm{div}_M X \, d\mu = -\int \langle X, \sum_{i=1}^{k} B_N(\tau_i, \tau_i) \rangle \, d\mu ,$$

and hence $\sum_{i=1}^{k} B_N(\tau_i, \tau_i)$ is the generalized mean curvature vector of (M, μ) (as a k-varifold in R^p) on U. (2.7) evidently follows.

We will also have occasion subsequently to consider k-*currents*. A k-current in N is simply an oriented *integral* k-varifold in N. That is, a k-varifold (M, μ), with integer-valued multiplicity function, together with an orientation ξ. Here, by an orientation ξ we mean that $\xi(x) = \pm \tau_1 \wedge \cdots \wedge \tau_k$ at μ-almost all points $x \in M$ (where $\tau_1, \cdots, \tau_k$ is any orthonormal basis for the approximate tangent space $T_x M$), and also that ξ is μ-measurable when considered locally (via coordinate charts of N) as a mapping from $W \cap M$ (W a coordinate neighborhood of N) into $\Lambda^k(R^n)$.

The notion of k-current coincides exactly with the notion of k-dimensional locally rectifiable current in the sense of Federer and Fleming [FF], and Federer [FH1]. This becomes clearer (it is formally proved from [FH1, 4.1.28]) when we note that, associated with each k-current (M, μ, ξ) in N, we have a linear functional T defined on the smooth k-forms ω with compact support in N by

$$T(\omega) = \int_M \langle \omega, \xi \rangle \, d\mu ; \tag{2.9}$$

here $\langle \omega, \xi \rangle(x) = \langle \omega(x), \xi(x) \rangle$ denotes the dual pairing between k-covectors and k-vectors on $T_x N$. We will often write $T = (M, \mu, \xi)$, thus identifying the k-current (M, μ, ξ) with the associated linear functional on k-forms. By analogy with Stokes' Theorem, we can define another functional (called the boundary of T) by

$$(\partial T)(\omega) = T(d\omega) \tag{2.10}$$

for all smooth (k–1)-forms ω with compact support in N. Notice that ∂T, so defined, may or may not represent a (k–1)-current on N. In fact there is a very useful theorem due to Federer and Fleming ([FH1, 4.2.16]) which asserts that if $T = (M, \mu, \xi)$ is a k-current and if the *mass* of ∂T is finite and if spt ∂T is compact (see below for the definition of mass and support), then ∂T does represent a (k–1)-current.

We wish finally to make some further remarks related to the fact that k-varifolds and k-currents generalize the notion of classical smooth unoriented and oriented k-dimensional submanifolds. In fact, given any unoriented submanifold M of N such that $\mathcal{H}^k(M \cap K) < \infty$ for each compact $K \subset N$, we will let $\mathbf{v}(M) = (M, \mathcal{H}^k \llcorner M)$ (†) denote the k-varifold naturally associated with M. (Notice that of course the multiplicity function is identically 1 on M.) If M is oriented in the classical sense (so that there is a k-vector-valued orientation ξ which is smoothly varying over M), then we let $\boldsymbol{\tau}(M)$ (or simply M) denote the k-current $(M, \mathcal{H}^k \llcorner M, \xi)$. In case M is compact with appropriately oriented smooth boundary ∂M, then Stokes' Theorem ensures that $\partial \boldsymbol{\tau}(M) = \boldsymbol{\tau}(\partial M)$, thus justifying the definition (2.10).

For any linear functional T defined on the smooth k-forms ω with compact support (regardless of whether or not T corresponds to a k-current in the above sense), we define the *support* spt T and *mass* $\mathbf{M}(T)$ of T by

$$\operatorname{spt} T = N \sim \cup W \tag{2.11}$$

$$\mathbf{M}(T) = \sup_{\|\omega\| \le 1} T(\omega) , \tag{2.12}$$

(†) Recall that if μ is a measure on N, then $\mu \llcorner M$ denotes the measure defined by $\mu \llcorner M(A) = \mu(M \cap A)$.

where the union is over all open W such that $T(\omega) = 0$ whenever ω has compact support in W, and where $\|\omega\|$ denotes the *co-mass* of a form ω with compact support in N, defined by $\|\omega\| = \sup_{x \in N} \|\omega(x)\|$, with

(2.13) $\|\omega(x)\| = \sup\{<\omega(x), v_1 \wedge \cdots \wedge v_k>: v_1, \cdots, v_k$ are orthonormal in $T_x N\}$.

Given a smooth map $\psi: N \to \tilde{N}$ ($\tilde{N}$ any other Riemannian manifold) such that ψ is proper, we define $\psi_\# T$ to be the linear functional on k-forms ω with compact support in $\tilde{N}$ by

(2.14) $$(\psi_\# T)(\omega) = T(\psi^\# \omega) (\psi^\# \omega = \text{pull-back of } \omega \text{ to } N).$$

(Notice that it is easy to extend this definition to the case when $\psi|_{\text{spt} T}$, rather than ψ, is proper.)

In case T is actually a k-current (M, μ, ξ), one easily checks that

(2.15) $$\mathbf{M}(T) = \mu(M), \quad \text{spt } T = \text{spt}(M, \mu) \subset \overline{M},$$

and (with ψ as above) $\psi_\# T$ is also a k-current, with the associated k-varifold being exactly $\psi_\#(M, \mu)$, as defined in (2.3).

For later reference we also introduce the notion of congruence mod ν between k-currents (where $\nu \geq 0$ is an integer). We say that two k-currents S, T (considered as linear functionals on forms with compact support as explained above) are *congruent mod* ν, denoted $S \equiv T \bmod \nu$, if there is a k-current R on N such that $S - T = \nu R$.

Also, we write $\partial S \equiv \partial T \bmod \nu$ if

$$\lim \mathbf{M}(\partial(S-T) - \nu(P_r + \partial Q_r) - (L_r + \partial N_r)) = 0$$

for some sequences $\{P_r\}$, $\{L_r\}$ of (k–1)-currents and for some sequences $\{Q_r\}$, $\{N_r\}$ of k-currents with $\lim(\mathbf{M}(L_r) + \mathbf{M}(N_r)) = 0$. (See [FH1, 4.2.26] for further details.)

We say that a k-current $T = (M, \mu, \xi)$ is *minimizing* in U (U open in N) if

(2.16) $$\mathbf{M}(T_K) \leq \mathbf{M}(S) (T_K = (M \cap K, \mu, \xi))$$

whenever K is a compact subset of U and whenever S is a k-current with $\partial S = \partial T_K$ and $\operatorname{spt}(S - T_K) \subset K$. We say that T is *homologically minimizing* in U if (2.16) holds whenever $T_K - S = \partial R$ for some (k+1)-current R with $\operatorname{spt} R \subset K$.

Evidently, for any k-current T in N,

(2.17) T minimizing in $U \Longrightarrow T$ homologically minimizing in U.

Also,

(2.18) T homologically minimizing in $U \Longrightarrow T$ stable in U.

The last implication is readily checked by using the homotopy formula for currents ([FH1, 4.1.9]) which guarantees that

$$\partial\phi_{\#}[\![0,t]\!] \times T_K = T_K - \phi_{t\#}T_K$$

whenever ϕ is as in (0.1). Here $[\![0,t]\!] \times T_K$ denotes the (k+1)-current $((0,t) \times M \cap K, \mu \times \mathcal{H}^1, \xi \wedge \frac{\partial}{\partial t})$ in $(-1,1) \times N$ ($(-1,1) \times N$ being equipped with the product metric).

The following *minimization principle* is important.

LEMMA 2.2. *Suppose* $T = (M, \mu, \xi)$ *is a* k*-current in* N *and suppose that there exists a smooth* k*-form* ω *on* U *such that*

(2.19) $d\omega \equiv 0$ *on* U, $\|\omega(x)\| \le 1$ *on* U, $\langle\omega(x), \xi(x)\rangle = 1$, μ-a.e. $x \in M \cap U$,

where $\|\omega(x)\|$ *is as in* (2.13).

Then T *is homologically minimizing in* U.

Proof. Let $K \subset U$ be compact, and let S be any k-current such that $S - T_K = \partial R$ for some R with $\operatorname{spt} R \subset K$. Then

$$(S - T_K)(\omega) = \partial R(\omega) = R(d\omega) = 0 ,$$

so that $T_K(\omega) = S(\omega)$. Since $\langle\omega(x), \xi(x)\rangle = 1$ μ - a.e., we have from (2.9) that $T_K(\omega) = \mathbf{M}(T_K)$, while from $\|\omega\| \le 1$ and the definition of $\mathbf{M}(S)$ we have $S(\omega) \le \mathbf{M}(S)$.

REMARK. Notice that equality in $S(\omega) \le M(S)$ evidently holds if and only if $<\omega(x), \tilde{\xi}(x)> = 1$ $\tilde{\mu}$ – a.e. on $\tilde{M}$, where $S = (\tilde{M}, \tilde{\mu}, \tilde{\xi})$.

The notion of k-current and k-varifold introduced here is sufficient for our present (expository) purposes, but of course we have not mentioned much of the general theory of currents and varifolds which may be vital in applications. For this the reader is referred to the standard references [FH1, Ch. 3, 4], [AW1, 2]. See also [SL5].

§3. *Some important classes of examples*

The division into classes here is convenient for the purposes of exposition, but is of course somewhat arbitrary and there is a good deal of overlap between some of the classes.

Class 1. Classical 2-dimensional minimal surfaces in R^3

The first examples in this class (other than the 2-planes) were discussed in 1776 by Meusnier:

(i) the catenoid $z = \cosh^{-1}\sqrt{x^2+y^2}$ (which is, up to isometry and change of scale, the only complete minimal surface of revolution in R^3); and

(ii) the right helicoid $z = \tan^{-1}(y/x)$ (which, again up to isometry and change of scale; is the only complete *ruled* minimal surface in R^3, other than the 2-planes).

Topologically the catenoid is the cylinder $S^1 \times R$ and the right helicoid is diffeomorphic to R^2.

In 1834 Scherk discovered a further example: Scherk's surface is given by $M = \bigcup_{\substack{\ell, k \text{ even} \\ k+\ell \equiv 0 (\text{mod } 4)}} \overline{M}_{k,\ell}$, where, for each $k, \ell = 0, \pm 2, \pm 4, \cdots, M_{k,\ell}$ is defined by

$$z = \log \frac{\cos \frac{\pi}{2}(y-\ell)}{\cos \frac{\pi}{2}(x-k)} \quad \text{for } |y-\ell| < 1,\ |x-k| < 1 .$$

This surface is complete (without boundary) and infinitely connected.

For sketches of these examples, see for example [SP].

It was not until much later that the following general procedure (due to Enneper and Weierstrass) for generating minimal surfaces in $\mathbf{R}^3$ was discovered:

Let D be either the open disc in $\mathbf{C}$ or $\mathbf{C}$ itself, and let f, g be any holomorphic functions on D with $f \neq 0$ on D. Then the mapping $\zeta \to x(\zeta) = x_1(\zeta), x_2(\zeta), x_3(\zeta) \in \mathbf{R}^3$, defined by

$$(3.1) \qquad x_k(\zeta) = \operatorname{Re}\left(\int_0^\zeta \phi_k(z)\,dz\right), \qquad k = 1, 2, 3,$$

where

$$\phi_1 = \frac{1}{2} f(1 - g^2), \quad \phi_2 = \frac{i}{2} f(1 + g^2), \quad \phi_3 = fg,$$

defines a minimal immersion of D into $\mathbf{R}^3$. (In checking this we make use of the general fact that an immersion from D into $\mathbf{R}^3$ is stationary if it is *harmonic* and *conformal*. In this particular case conformality holds because $\phi_1^2 + \phi_2^2 + \phi_3^2 \equiv 0$.) The particular example given by $D = \mathbf{C}$, $f(\zeta) \equiv 1$, $g(\zeta) \equiv \zeta$ is known as *Enneper's surface*. (See sketch in [SP].)

If we allow g to have poles and f corresponding zeros (of at least twice the order of the poles) then, up to translations, *any* minimal immersion from D into $\mathbf{R}^3$ can be represented as in (3.1) (see [O2] for a simple proof). (3.1) is called the *Weierstrass representation* for minimal immersion of D into $\mathbf{R}^3$. For further details, discussion and applications related to the Weierstrass representation, the reader is referred to [O2], [N1], [N3].

Class 2. Stationary 2-varifolds in $\mathbf{R}^3$ *which are not classical minimal surfaces.*

An example will serve to illustrate the kind of example we have in mind here. Suppose $M = \bigcup_{i=1}^{3} H_i$, where H_1, H_2, H_3 are half-planes with common boundary a line L along which the half-planes meet at equal angles of 120°. We claim that $(M, \mathcal{H}^2 \llcorner M)$ is stationary in all of R^3 (even though it is certainly not a classical surface). To see this, we let X be any smooth vector field with compact support in R^3. By applying the classical first variation formula (1.13) to each H_i, we obtain $\int_{H_i} \operatorname{div}_{H_i} X \, d\mathcal{H}^2 = -\int_L <X, \eta_i> d\mathcal{H}^1$, where η_i denotes the (inward pointing) co-normal for H_i on L. Adding these identities we then have

$$\int_M \operatorname{div}_M X \, d\mathcal{H}^2 = -\int_L <X, \sum_{i=1}^{3} \eta_i> d\mathcal{H}^1 = 0 ,$$

because $\sum_{i=1}^{3} \eta_i = 0$ by virtue of the fact that the H_i meet at equal angles of 120°.

This type of example plays an important role in modelling "soap-film" like minimal surfaces; see the further discussion in §5 (#2) below and also the paper [TJ].

Class 3. The minimal graphs in R^{k+1}

Suppose Ω is a domain in R^k, let $u \in C^2(\Omega)$, and let $M = \text{graph } u$. By direct computation (or see e.g. [GT, Ch. 15]) we see that M has zero mean curvature (and hence is stationary by the discussion of §1) if and only if u satisfied the minimal surface equation

$$\mathfrak{M}(u) = 0 \quad \text{on} \quad \Omega . \tag{3.2}$$

Here $\mathfrak{M}(u)$ is the non-linear operator (the "mean curvature operator") defined by

$$\mathfrak{M}(u) = (1 + |Du|^2)^{-1/2} \sum_{i,j=1}^{k} (\delta_{ij} - D_i u D_j u/(1 + |Du|^2)) D_{ij} u$$
$$\equiv \sum_{i=1}^{k} D_i \left(\frac{D_i u}{\sqrt{1 + |Du|^2}} \right) ,$$

where

$$D_i u = \partial u/\partial x_i , \qquad D_{ij} u = \partial^2 u/\partial x_i \partial x_j .$$

Such graphs are often called *non-parametric* minimal surfaces; this terminology (which is somewhat cumbersome) arose from the classical literature where two dimensional surfaces were usually represented as mappings from a fixed parameter domain, as in (3.1) above.

Notice that by the discussion of §1 know that a solution of the minimal surface equation provides at least a local minimum for the non-parametric area functional $I(u) = \int_\Omega \sqrt{1+|Du|^2}\, dx$. Indeed the equation $\mathfrak{M}(u) = 0$ is just the Euler-Lagrange equation for this functional.

We can actually show that, provided (3.2) holds, M (appropriately oriented) must in fact, as a k-current, be minimizing in $\Omega \times \mathbf{R}$ in the sense of the discussion of §2. To see this we let $\nu(x, x_{k+1}) = (-Du(x), 1)/\sqrt{1+|Du(x)|^2}$ be the unit normal function, which we note is constant on vertical lines and which coincides with the upward unit normal of M at a point $(x, u(x)) \in M$. Also let

$$\omega = \sum_{i=1}^{k+1} (-1)^{i-1} \nu_i \, \widehat{dx_i} ,$$

where $\widehat{dx_i} = dx_1 \wedge \cdots \wedge dx_{i-1} \wedge dx_{i+1} \wedge \cdots \wedge dx_{k+1}$. One readily checks that, provided M has the smooth orientation ξ such that $\xi \wedge \nu \equiv * \, dx_1 \wedge dx_2 \wedge \cdots \wedge dx_{k+1}$ on M, the conditions of the minimization principle (Lemma 2.2) hold. (The condition $d\omega = 0$ is easily seen to be equivalent to the minimal surface equation (3.2).)

Thus we deduce that M (as a k-current) is homologically minimizing in $\Omega \times \mathbf{R}$. The fact that it is actually minimizing follows from the fact that k^{th} singular homology group $H_k(\Omega \times \mathbf{R}) \approx H_k(\Omega) = 0$. (From the identification of singular homology and the homology groups associated with the chain complex of k-currents with compact support ([FH1, §4.4]) we deduce that if S, T are k-currents in $\mathbf{R}^{k+1}$ with $\mathrm{spt}(S-T)$ a compact subset of $\Omega \times \mathbf{R}$, and if $\partial S = \partial T$, then there is a (k+1)-current R such that R has compact support in $\Omega \times \mathbf{R}$ and $S - T = \partial R$.)

In recent times the minimal surface equation and the related non-parametric theory has been much studied, and many important results obtained. For an introductory discussion and some historical references, the reader is referred to [GT, Ch. 15]. Other references are [BDM], [TN], [GE2], [SL3].

Class 4. Complex submanifolds of Kähler manifolds

Suppose N is a Kähler manifold such that the real dimension of N is $n = 2q$, and let J be the associated almost complex structure.

LEMMA 3.1. *Let* (M, μ, ξ) *be any* 2k-*current in* N *with the property that, for* μ – a.e. $x \in M$, $\xi(x)$ *can be written* $\tau_1 \wedge J(\tau_1) \wedge \tau_2 \wedge J(\tau_2) \wedge \cdots \wedge \tau_k \wedge J(\tau_k)$ *for some orthonormal set* $\tau_1, \cdots, \tau_k \in T_x N$. *Then* T *is homologically minimizing in* $N \sim \mathrm{spt}\, \partial T$.

The proof (due to Federer) involves noting that, with ω equal to the kth exterior power of the Kähler form, appropriately normalized, all the conditions of the minimization principle (Lemma 2.2) hold. The condition $\|\omega\| \le 1$ is just *Wirtinger's inequality*. See [L3] for more details.

As a particular example of the above we deduce that any complex submanifold M of N must automatically be homologically minimizing in $N \sim \mathrm{spt}\, \partial M$. (Provided it is given the canonical orientation as described in [L3].)

Another particular application of this result is that each complex algebraic variety in $\mathbf{C}^n$ (represented as a k-current) is minimizing in $\mathbf{C}^n$.

Class 5. G *invariant submanifolds of* N

Here suppose G is a compact (not necessarily connected) Lie group of isometries of N. We say a k-varifold (M, μ) in N is G-*invariant* if $g_{\#}(M, \mu) = (M, \mu)$ for each $g \in G$. Also, taking ϕ as in (0.1), with U a

G-invariant open subset of N (i.e. $g(U) = U$), we say that ϕ is a *G-equivariant deformation* of N if

$$g^{-1} \circ \phi_t \circ g = \phi_t \qquad \forall t \in (-1,1) . \tag{3.3}$$

LEMMA 3.2. *Let* (M, μ) *be a G-invariant* k*-varifold in* N *with* $\mu(M) < \infty$*, and let* (M, μ) *be stationary in* U *with respect to G-equivariant deformations* ϕ *in the sense that the quantity on the left of (2.5) vanishes whenever* ϕ *is as in (0.1), (3.3). Also, suppose* $g(U) = U \;\; \forall g \in G$.

Then (M, μ) *is stationary in* U.

Proof. Let ϕ be arbitrary as in (0.1), let X be as in (0.2), and let $\psi : (-1,1) \times N \to N$ be defined by

$$\psi_t(x) = \int_G g \circ \phi_t \circ g^{-1}(x) d\sigma ,$$

where σ is the Haar measure on G. One readily checks that ψ satisfied the condition (0.1), (3.3). Hence $\frac{d}{dt} |\psi_{t\#}(M, \mu)| \big|_{t=0} = 0$. Letting $\tilde{X}_x = \frac{\partial \psi(t,x)}{\partial t}\Big|_{t=0}$, we have on the other hand by (2.5) that

$$\frac{d}{dt} |\psi_{t\#}(M, \mu)| = \int_M \operatorname{div}_M \tilde{X} \, d\mu .$$

We check by direct calculation that $\tilde{X}_x = \int_G d_y g(X_y) d\sigma$ where $d_x g : T_x N \to T_{g(x)} N$ denotes the linear map induced by g at x, and $y = g^{-1}(x)$. Hence

$$\int_M \operatorname{div}_M \tilde{X}\, d\mu = \int_G \int_M \langle \tau_i, \nabla_{\tau_i} d_y g(X_y) \rangle\, d\mu(x)\, d\sigma$$

$$= \int_G \int_M \langle \tau_i, d_y g(\nabla_{d_x g^{-1}(\tau_i)} X) \rangle\, d\mu(x)\, d\sigma \quad (y = g^{-1}(x))$$

$$= \int_G \int_M \langle d_x g^{-1}(\tau_i), \nabla_{d_x g^{-1}(\tau_i)} X \rangle\, d\mu(x)\, d\sigma\,,$$

where we have used the fact that g is an isometry at the last step. Since (M, μ) is G-invariant, this last expression equals $\int_G \int_M \langle \tau_i, \nabla_{\tau_i} X \rangle\, d\mu d\sigma$, which equals $\int_M \operatorname{div}_M X\, d\mu$, because $\sigma(G) = 1$. Thus we have proved $\int_M \operatorname{div} X\, d\mu = 0$.

Lemma 3.2 is essentially due to Hsiang and Lawson [HL], although they both state and prove their result in a more classical manner.

To discuss some particular consequences of Lemma 3.2, let G be connected, and let $G(x) = \{g(x) : g \in G\}$ denote the orbit of x. Two orbits $G(x)$, $G(y)$ are said to be of the same type if the subgroups $G_x = \{g \in G : g(x) = x\}$ and $G_y = \{g \in G : g(y) = y\}$ are conjugate in G. Evidently if ϕ_t is a G-equivariant variation of N, then $\phi_t(G(x)) = G(\phi_t(x))$ and $G_{\phi_t(x)} = G_x$, so that $\phi_t(G(x))$ is always an orbit of the same type as $G(x)$. Thus in particular we deduce from Lemma 3.2 that *if* $G(x)$ *(as a k-varifold in* N*) has maximum or minimum mass relative to (nearly) orbits of the same type, then it is stationary in* N. In particular any orbit which is isolated (in the sense that it has a tubular neighborhood which does not intersect any other orbit of the same type) is stationary in N. For further results, related discussion, and examples of minimal submanifolds of S^n which arise as particular examples when one takes $G \subset SO(n+1)$, the reader is referred to the paper of Hsiang and Lawson [HL] and its bibliography.

Here it is also appropriate to mention the result of W.-Y. Hsiang [HW] that every compact homogeneous space can be isometrically immersed in S^n (as a G-invariant stationary submanifold). Previously this had been established in the irreducible case by Takahashi [T]; in this irreducible case P. Li [LP] has shown that in fact the image set is a closed *embedded* minimal submanifold of S^n, with the immersion providing a covering map.

Class 6. Minimizing representatives of homology classes

If N is compact and the kth singular homology group $H_k(N) \neq 0$, each homology class in $H_k(N)$ is represented by a homologically minimizing k-current $T = (M, \mu, \xi)$. (See [FF], [FH1, 5.1.6].) Unfortunately the regularity properties of such representatives are not well understood as yet. In case $n = k+1$ (i.e. the codimension 1 case) it is known (see §4 below) that T has no singularities ($\text{sing}\, T = \emptyset$) for $k \leq 6$, isolated singularities for $k = 7$ and $\mathcal{H}^{\alpha}(\text{sing}\, T) = 0\ \forall \alpha > k-7$ in case $k > 7$. Recently F. J. Almgren, Jr. [A4] has announced that $\mathcal{H}^{\alpha}(\text{sing}\, T) = 0$ $\forall \alpha > k-2$ in arbitrary codimension. No other significant information about the structure of the singular set has so far been obtained.

Class 7. Embedded minimal hypersurfaces in N

Suppose N is compact and oriented. J. Pitts has proved in [P2] (making use, in part, of previous work of Almgren [A3] on the existence of stationary integral k-varifolds) that if $k \leq 5$ there exists a properly embedded compact minimal hypersurface $M \subset N$.

By using Pitts' arguments together with new regularity results for stable minimal hypersurfaces (extending [SSY]), Schoen and Simon [SS1] were able to extend these results for arbitrary k as follows: the result is unchanged in case $k = 6$, M has at most isolated singularities in case $k = 7$, and M has singular set S such that $\mathcal{H}^{\alpha}(S) = 0\ \forall \alpha > k-7$ in case $k > 7$.

Pitts can show (to appear) that the minimal hypersurfaces obtained by using his procedure are in general not stable, and hence must often be distinct from the examples discussed in 6 above when $H_k(N) \neq 0$.

Class 8. Incompressible immersed surfaces in N

Suppose N is compact, suppose Σ_g is the standard genus g sphere-with-g-handles, $g \geq 1$ and suppose that there is a smooth map $f : \Sigma_g \to N$ such that the induced homomorphism $f_* : \pi_1(\Sigma_g, *) \to \pi_1(N, f(*))$ is an injection. Then there is a branched minimal immersion $\psi : \Sigma_g \to N$ such that $\psi_* = f_*$ and such that ψ has least area $A(\psi)$ (defined as in (1.20)) relative to all smooth maps $h : \Sigma_g \to N$ with $h_* = f_*$. This result is due to Schoen and Yau [SY] and (independently) Sacks and Uhlenbeck [SU1]. By the Osserman-Gulliver result (see [O3], [GR]) on the non-existence of branch points, one can assert that ψ is an immersion in case dimension $N = 3$.

There always exists (see [SY]) an f as above whenever $\pi_1(N)$ has a subgroup which is isomorphic to $\pi_1(\Sigma_g)$, hence Schoen and Yau were in fact able to deduce the existence of a genus g stable immersed $M \subset N$ whenever $\pi_1(N)$ contains a subgroup isomorphic to $\pi_1(\Sigma_g)$ and dimension $N = 3$. For discussion of one of the consequences of this result, see §7 below.

Class 9. Immersed minimal spheres in N

Suppose again that N is compact without boundary. Under very mild topological restrictions (that the universal covering space of N is not contractible, which in particular is true if the second homotopy group $\pi_2(N)$ is non-zero), Sacks and Uhlenbeck have shown that there is a branched stationary immersion ψ of S^2 into N; in general this ψ is not stable (in the sense of the discussion at the end of §1). See [SU2].

Meeks and Yau have proved ([MY]) that, in case $\pi_2(N) \neq 0$, there is a *stable embedded* diffeomorph of S^2 in N; indeed as a consequence of their method, they are able to show that their surface minimizes within its homotopy class.

Class 10. Embedded surfaces representing minima in isotopy classes

We finally mention a general result, due to Meeks, Simon and Yau [MSY] concerning the existence of closed stable embedded surfaces in N in the case dimension of $N = 3$. To describe this result, suppose that M_0 is an arbitrary fixed closed surface in N, where N is compact, and let $\mathcal{I}(M_0)$ denote the isotopy class of M_0 in N. (That is, $\mathcal{I}(M_0)$ denotes the collection of surfaces M which can be expressed $M = \phi_t(M_0)$, $|t| < 1$, with $\phi : (-1,1) \times N \to N$ smooth and ϕ_t, defined by $\phi_t(x) = \phi(t,x)$, a diffeomorphism for each $|t| < 1$. Then we have:

THEOREM 3.3. *Let* $\alpha = \inf_{M \in \mathcal{I}(M_0)} \mathcal{H}^2(M) > 0$, *and let* $\{M_k\} \subset \mathcal{I}(M_0)$ *be any sequence in* $\mathcal{I}(M_0)$ *with* $\lim \mathcal{H}^2(M_k) = \alpha$. *Then there is a subsequence* $\{k'\} \subset \{k\}$, *positive integers* $\ell, n_1, \cdots, n_\ell$, *and a pairwise disjoint collection* $M^{(1)}, \cdots, M^{(\ell)}$ *of stable embedded oriented surfaces such that* $\lim M_{k'} = \sum_{j=1}^{\ell} n_j M^{(j)}$ *in the measure theoretic sense that* $\lim \int_{M_{k'}} f = \sum_{j=1}^{\ell} n_j \int_{M^{(j)}} f$ *for each* $f \in C^0(N)$.

Furthermore, if all $M^{(j)}$ *are two-sided in* N,

$$\sum_{j=1}^{\ell} \text{genus}\,(M^{(j)}) \le \text{genus}\, M_0$$

and in fact, up to isotopy, each $M^{(j)}$ *can be obtained from* M_0 *as one of the components obtained by cutting out disjoint annular neighborhoods of a suitable disjoint family of Jordan curves and sewing in pairs of discs.*

A modified statement holds if some of the $M^{(j)}$ are one-sided in N. There is also a modified result in case N is not compact. For further details and discussion of topological applications, the reader is referred to the paper [MSY].

§4. *Special properties of stationary* k-*varifolds in* $\mathbb{R}^n$

Here we want to briefly recall some of the nice "special" properties of minimal submanifolds in $\mathbb{R}^n$, working in the general setting of k-varifolds.

Property 1 (Harmonicity of coordinate functions)

LEMMA 4.1. *A k-varifold* (M, μ) *in* R^n *is stationary in* U *if and only if each of the coordinate functions* x_j *of* R^n *is harmonic on* (M, μ) *in* U.

We have first to explain our terminology here. In case M is smooth and $\mu = \mathcal{H}^k \llcorner M$, the statement that x_j is harmonic on (M, μ) is to be taken to mean that the restriction of x_j to M is harmonic on M in the usual sense of differential geometry. In general, we say that f defined in a neighborhood of $\operatorname{spt}(M, \mu) \cap U$ is harmonic on (M, μ) in U if

$$\int_M \operatorname{grad}_M \zeta \cdot \operatorname{grad}_M f \, d\mu = 0$$

$\forall \zeta \in C_0^1(U)$. Here $\operatorname{grad}_M \zeta(x) = \sum_{i=1}^{k} (D_{\tau_i} \zeta)(x) \tau_i$, where D_{τ_i} is ordinary directional differention in R^n, and $\tau_1, \cdots, \tau_k$ is an orthonormal basis for the approximate tangent space $T_x M$ of (M, μ) at x.

To prove Lemma 4.1, we let $f(x) \equiv x_j$ and note that then $\int_M \operatorname{grad}_M \zeta \cdot \operatorname{grad}_M f \, d\mu = \int_M e_j \cdot \operatorname{grad}_M \zeta \, d\mu$, where e_j is the jth standard basis vector for R^n. But

$$\begin{aligned} e_j \cdot \operatorname{grad}_M \zeta &= e_j \cdot \tau_i D_{\tau_i} \zeta \\ &= \operatorname{div}_M(\zeta e_j) \end{aligned}$$

on M (in the sense of §2). Thus we see that x_j is harmonic on (M, μ) in U if and only if $\int \operatorname{div}_M(\zeta e_j) d\mu = 0$ $\forall \zeta \in C_0^1(U)$. Thus, since e_j are on a basis for R^n, we deduce that each x_j is harmonic on (M, μ) in U if and only if $\int_M \operatorname{div}_M X \, d\mu = 0$ for every smooth vector field X with compact support in U; i.e. if and only if (M, μ) is stationary in U.

We should point out that one can define a Laplacian operator Δ_M for (M, μ) in U, provided (M, μ) is stationary in U. This satisfies

$$\int_M \psi\, \Delta_M \zeta \, d\mu = - \int_M \mathrm{grad}_M \zeta \cdot \mathrm{grad}_M \psi \, d\mu$$

whenever $\zeta, \psi \in C_0^2(U)$. Indeed one can readily check (by using the formula $\int \mathrm{div}_M X \, d\mu = 0$ with $X = \psi D\zeta$) that $\Delta_M \zeta \equiv \sum^n e^{ij} \frac{\partial^2}{\partial x_i \partial x_j} \zeta$, where $e^{ij}(x)$ denotes the matrix of the orthogonal projection of R^n onto $T_x M$. More generally, if (M, μ) has generalized mean curvature vector $\mathbf{H}$ in U (in the sense described in §2), then (M, μ) has a Laplacian operator Δ_M given by

$$\Delta_M \zeta = \sum_{i,j=1}^n e^{ij} \frac{\partial^2 \zeta}{\partial x_i \partial x_j} + (\mathbf{H} \cdot e_j) \frac{\partial \zeta}{\partial x_j}, \quad \zeta \in C^2(U).$$

Property 2. The monotonicity formula

LEMMA 4.2. *Suppose* (M, μ) *is stationary in* $U \subset R^n$, *and suppose* $\xi \in U$ *and* $B_R(\xi) \subset U$, $B_R(\xi)$ = *the open or closed ball in* R^n *with centre* ξ *and radius* R. *Then*

$$(4.1) \quad \sigma^{-k} \mu(B_\sigma(\xi) \cap M) - \rho^{-k} \mu(B_\rho(\xi) \cap M) + \int_{(B_\rho(\xi) \sim B_\sigma(\xi)) \cap M} \frac{|(x-\xi)^\perp|^2}{|x-\xi|^{k+2}} d\mu = 0$$

for each $0 < \sigma < \rho < R$, *where* $(x-\xi)^\perp$ *denotes the component of* $x - \xi$ *normal to* $T_x M$.

In particular (4.1) implies that $\sigma^{-k} \mu(B_\sigma(\xi) \cap M)$ is an increasing function of σ, and hence the density function

$$(4.2) \qquad \Theta^k(\xi) = \lim_{\sigma \downarrow 0} (\alpha_k \sigma^k)^{-1} \mu(B_\sigma(\xi) \cap M)$$

(α_k = volume of the unit k-ball in R^k) exists (and is real-valued) for all

$\xi \in U$. Furthermore Θ^k evidently (by the monotonicity of $\rho^{-k}(M \cap B_\rho)$) has the semicontinuity property

$$\Theta^k(y) \geq \limsup_{\xi \to y} \Theta^k(\xi) \tag{4.3}$$

for $y \in U$, and hence in particular is Borel measurable on U. One easily checks that it agrees $\mathcal{H}^k$ almost everywhere on $M \cap U$ with the multiplicity function $\theta \equiv \frac{d\mu}{d\mathcal{H}^k}$ of (M, μ).

Lemma 4.2 is easily proved by making appropriate choices of X in the first variation formula $\int_M \operatorname{div}_M X \, d\mu = 0$. (One can take, for example, $X_x = (|x-\xi|_\sigma^{-k} - \rho^{-k})_+(x-\xi)$, where $|x-\xi|_\sigma = \max\{|x-\xi|, \sigma\}$; this function is not smooth when $|x-\xi| = \sigma, \rho$ and hence an approximation argument is actually necessary to justify its use. See [AW1] or [MS] or [BJ] for similar computations.)

Actually if we hypothesize that (M, μ) has generalized mean curvature vector $\mathbf{H}$ in U, with $|\mathbf{H}| \leq \Lambda_0$ on $M \cap U$, then modification of the above argument gives that $\exp(-c\rho\Lambda_0) \cdot \rho^{-k} \mu(B_\rho(\xi) \cap M)$ is increasing for $0 < \rho < R$. Thus (4.2) and (4.3) remain valid in this case.

These facts are important in the regularity theory to be discussed below in §5.

Property 3. (The general convex hull property)

LEMMA 4.3. *If* (M, μ) *is a* k-*varifold which is stationary in* U, *if* $\mu(M) < \infty$, *and if* Γ *is a compact set such that* (M, μ) *is stationary in* $\mathbf{R}^n \sim \Gamma$, *then* $\operatorname{spt}(M, \mu) \subset$ *convex hull of* Γ.

Thus, modulo a set of $\mathcal{H}^k$-measure zero, we deduce that M is contained in the convex hull of Γ.

This lemma is rather easy to prove. To see it, we first note that it suffices to prove

$$\mu(M \cap \{x : x \cdot e_n > 0\}) = 0 \quad \text{if} \quad \Gamma \subset \{x : x \cdot e_n < 0\}, \tag{4.4}$$

$e_n = (0, \cdots, 0, 1)$, because the convex hull of Γ can be represented as the countable intersection of closed $\frac{1}{2}$-spaces which contain Γ in their interior, and any such $\frac{1}{2}$-spaces can be brought into coincidence with $\{x : x \cdot e_n \leq 0\}$ by an isometry of $\mathbf{R}^n$. On the other hand (4.4) is easily checked by making appropriate choices of X in the formula $\int_M \operatorname{div}_M X = 0$. For details the reader is referred to [AS] or [SL1]. The discussion in [AS] shows that the convex hull property holds for general stationary varifolds in $\mathbf{R}^n$ (without any rectifiability assumptions).

§5. *Interior regularity*

Here we want to briefly survey the main local interior regularity theorems presently known for stationary k-varifolds. The pioneering work in regularity theory was carried out by Reifenberg [RE], DeGiorgi [DG1], Fleming [F], and Almgren [A1], [A2].

The most general theorem for k-varifolds is due to W. K. Allard [AW1]. In stating this theorem we let (M, μ) denote a k-varifold in $\mathbf{R}^n$ with $0 \in \operatorname{spt} \mu$ and generalized mean curvature vector $\mathbf{H}$ in B^n_ρ (see the discussion of §2). Here B^n_ρ denotes the open k-ball of radius ρ and centre 0 in $\mathbf{R}^n$. We will also refer to the following hypotheses in which $\varepsilon > 0$, $\rho > 0$ are given:

$$
(5.1) \qquad \begin{cases} \operatorname{ess\,sup}_{M \cap B^n_\rho} \rho |\mathbf{H}| + ((\alpha_k \rho^k)^{-1} \mu(M \cap B^n_\rho) - 1) < \varepsilon \\ \Theta^k(x) \geq 1 \quad \mu - \text{a.e.} \quad x \in M \cap B^n_\rho . \end{cases}
$$

THEOREM 5.1. *There is an* $\varepsilon > 0$ *(depending only on* k, n *) such that (5.1) implies that, after redefinition of* M *on a set of* $\mathcal{H}^k$*-measure zero, there is an isometry* Q *of* $\mathbf{R}^n$ *with*

$$
Q(M \cap B^n_\rho \cap (B^k_{\rho/2} \times \mathbf{R}^{n-k})) = \text{graph } u
$$

$$
u = (u^{k+1}, \cdots, u^n) \in C^{1,\alpha}(B^k_{\rho/2})
$$

for each $\alpha \in (0,1)$. *Furthermore*

$$\sup_{B^k_{\rho/2}} |Du| + \sup_{\substack{x,y \in B^k_{\rho/2} \\ x \neq y}} (|x-y|/\rho)^{-\alpha} |Du(x) - Du(y)| \le c \,,$$

where c *depends only on* k, n *and* $\alpha \in (0,1)$.

For the proof of this theorem (in the more general case when H is merely in L^p, $p > n$) the reader is referred to the paper [AW1]. See also [SL5]. We do emphasize that the monotonicity formula (4.1) (or more correctly is analogue for the case $H \neq 0$) plays a decisive role in the proof.

REMARK 5.2. It clearly follows from the theorem above and from Remark 2.1, that if (M, μ) is an arbitrary k-varifold in N (which we assume is isometrically embedded in $\mathbb{R}^p$ for some p as in Remark 2.1), if (M, μ) is stationary in U and if $\xi \in \operatorname{spt}(M, \mu) \cap U$, then $\xi \in \operatorname{reg}(M, \mu)$ provided

$$0 < \Theta^k(\xi) \le \liminf_{\substack{x \to \xi \\ x \in M_0}} \Theta^k(x)$$

for some $M_0 \subset M$ with $\mathcal{H}^k(M \sim M_0) = 0$. Notice that, in view of (4.3), we see that this is equivalent to the requirement that $\Theta^k(\xi) \neq 0$ and $\Theta^k|_{M_0}$ is continuous at ξ for some $M_0 \subset M$ with $\mathcal{H}^k(M \sim M_0) = 0$.

In particular, if $\mu = \mathcal{H}^k$ on M (i.e. if the multiplicity function $\frac{d\mu}{d\mathcal{H}^k}$ of (M, μ) is identically 1 on M), then $\operatorname{sing}(M, \mu) \cap U$ has $\mathcal{H}^k$-measure zero.

Using Theorem 5.1 in combination with various other arguments, various regularity results can be proved for special classes of k-varifolds and k-currents. In particular, we have the following results:

1. *Regularity of codimension 1 minimizing* k*-currents*

Suppose $T = (M, \mu, \xi)$ is a homologically minimizing k-current in U (see §2 for the terminology) with $U \subset N$, U open, and suppose dimension $N = k+1$. Then

$$\text{sing } T \sim \text{spt } \partial T = \emptyset, \quad k \le 6$$
$$\text{sing } T \sim \text{spt } \partial T \text{ is isolated}, \quad k = 7$$
$$\mathcal{H}^{\alpha}(\text{sing } T \sim \text{spt } \partial T) = 0 \qquad \forall \alpha > k-7 \text{ in case } k > 7 .$$

(Here and subsequently, sing $T = \text{sing}(M, \mu)$.)

The proof of this theorem involves using Theorem 5.1 (actually in this case only a version for codimension-1 k-currents, originally proved by DeGiorgi [DG], is needed) together with the result of J. Simons [SJ] concerning the non-existence of stable cones in R^{k+1}, $k \le 6$, together with a dimension reducing argument of Federer [FH2]. For an elegant presentation of this work, the reader is referred to the notes of Giusti [GE1].

As an illustration of the fact that generic *qualitative* results like those above, when combined with the appropriate compactness theorem and a priori estimates lead to *quantitative* results, we mention the following curvature estimates:

If $T = (M, \mu, \xi)$ is a k-current in R^{k+1}, $k \le 6$, such that T is minimizing in U, if $y \in U \cap \text{spt } T$ and if $B_R(y) \subset U$, then

$$|B|(y) \le c/R , \tag{5.2}$$

where B is the second fundamental form of the (regular) hypersurface $U \cap \text{spt } T$, and c is a constant depending on k. For the simple argument needed to prove this, the reader is referred to [SL2]. In [SL2] it is shown that there is also an analogous result when $k = 7$ in the *non-parametric* case. (See #3 of §3 above.)

2. *Regularity of arbitrary codimension* k-*currents*

By considering the complex algebraic varieties in C^2 (see #4 of §3 above) one sees that, if dimension $N > k+1$, the best we could expect in general for minimizing k-currents T is that $\mathcal{H}^{k-2}(\text{sing } T \cap K) < \infty$ $\forall$ compact K with $K \cap \text{spt } \partial T = \emptyset$. F. J. Almgren, Jr. has recently announced ([A4]) that $\mathcal{H}^{\alpha}(\text{sing } T \cap K) = 0 \ \forall \alpha > k-2$. Prior to this it was only known that reg T was open and dense (which follows from Remark 5.2 above).

3. *Soap-film like* k-*varifolds in* N

We look at k-varifolds $(M, \mathcal{H}^k \llcorner M)$ in N which minimize in U in the following "soap-film-like" sense:

For each $x \in U \cap \text{spt}(\mathcal{H}^k \llcorner M)$, there is an open ball W containing x such that

$$\mathcal{H}^k(M \cap W) \le \mathcal{H}^k(\phi(M \cap W)) \tag{5.3}$$

whenever $\phi : \overline{W} \to \overline{W}$ is Lipschitz with $\phi|_{\partial W} = 1_{\partial W}$. Evidently this implies that $(M, \mathcal{H}^k \llcorner M)$ is stable in W in the sense of §2. However, if we take $k = 1$, $N = \mathbf{R}^3$, $M = \{(x,y) \in \mathbf{R}^2 : y = \pm x\}$ then one easily sees that $(M, \mathcal{H}^1 \llcorner M)$ is stable but not minimizing in the above sense. (On the other hand the example of #2 in §3 *is* minimizing in the above sense; see the discussion of [TJ].)

From the Allard Theorem (in particular, from Remark 5.2) any $(M, \mathcal{H}^k \llcorner M)$ satisfying (5.3) satisfies $\mathcal{H}^k(\text{sing } M \cap U) = \emptyset$. For some general results of this type, see [A5].

In the special case $N = \mathbf{R}^3$ and $k = 2$, Jean Taylor [TJ] has proved much more than this: she proves that sing $M \cap U$ consists of a family of closed $C^{1,\alpha}$ Jordan arcs along which 3 sheets of reg M meet at equal angles of 120°. (Near the endpoints of the arcs, $\text{spt}(M, \mathcal{H}^2 \llcorner M)$ looks like the standard tetrahedral soap film configuration; for details see [TJ].) Kinderlehrer, Nirenberg and Spruck [KNS] prove that the Jordan arcs of sing $M \cap U$ are in fact real analytic curves. (Previously they had been proved to be C^∞ by Nitsche [N4].)

4. *Regularity of* k-*currents mod* 2

Here let $\mathcal{C}_k^\nu$ denote the set of k-currents $T = (M, \mu, \xi)$ in N such that $\partial T_U \equiv 0 \bmod \nu$. (Here $T_U = (M \cap U, \mu, \xi)$, considered as a k-current in U rather than in N, and congruence of $\partial T_U \bmod \nu$ is to be interpreted in the sense of §2 with U in place of N.) Suppose now that $T \in \mathcal{C}_k^2$ and that

$$M(T) \le M(S) \tag{5.4}$$

for each $S \in \mathcal{C}_k^2$ with $\operatorname{spt}(S-T) \subset U$. (In this case we say that T is minimizing mod 2.) Then we have the following regularity results:

(i) (Almgren [A1]) In case $k = 2$, sing $T \cap U$ consists of at most isolated points (if $k = 2$ and $n = 3$, there are *no* singularities—Reifenberg [RE]),

(ii) (Federer [FH2]) $\forall k \ge 2$, $\mathcal{H}^{\alpha}(\operatorname{sing} T \cap U) = 0$ for each $\alpha > k-2$.

Prior to these results, Reifenberg [RE] had proved $\mathcal{H}^k(\operatorname{sing} T \cap U) = 0\ \forall k \ge 2$; this result actually follows directly from Remark 5.2, because, writing $T = (M, \mu, \xi)$, we see from (5.4) that the multiplicity function of (M, μ) must be 1 on $M \cap U$, and in addition, (M, μ) is stationary on U because $(\partial \phi_{t\#} T)_U = (\phi_{t\#} \partial T)_U \equiv 0 \bmod 2$ (for every ϕ_t as in (0.1)) from the fact that $(\partial T)_U \equiv 0 \bmod 2$.

5. k-*currents* mod 3

In case $T \in \mathcal{C}_k^3$ is minimizing mod 3 in U (that is, in case (5.4) holds under the same conditions as above, except that $\partial T_U \equiv 0 \bmod 3$ and $\partial S_U \equiv 0 \bmod 3$), then, writing $T = (M, \mu, \xi)$, the k-varifold (M, μ) must have multiplicity 1 on $M \cap U$ and (as in #4 above) it follows that from Remark 5.2 that $\mathcal{H}^k(\operatorname{sing} T \cap U) = 0$. Furthermore, in case $k = 2$ we know that $\operatorname{spt}(M, \mathcal{H}^k \llcorner M) \cap U$ is as described in #3 above. (Because one easily verifies that $(M, \mathcal{H}^k \llcorner M)$ minimizes in the "soap-film sense" of #3 above.)

6. k-*currents* mod 4

Finally we mention the case of k-currents which minimize mod 4. B. White [WB] has shown that these decompose into minimizing k-currents in the codimension-1 case. Hence #1 above can be applied to discuss sing T in this case. See [WB] for details.

§6. *Classical problems*

In this section we want to discuss the Bernstein problem and the Plateau problem, including a brief survey of the history of these problems and the current state of knowledge concerning them. This seems worthwhile, especially in view of the fact that attempts to extend the classical results related to these problems have directly or indirectly been responsible for much of the development of the theory of minimal submanifolds.

1. *The Bernstein problem*

In 1915 Bernstein proved that if $u = u(x, y)$, $(x, y) \in \mathbf{R}^2$, is a C^2 solution of the minimal surface equation (see #3 of §3 above) on all of $\mathbf{R}^2$, then u must be linear. (That is, $u(x, y) = ax + by + c$, a, b, c constants, are the only global solutions of the minimal surface equation on $\mathbf{R}^2$.)

Attempts to extend this result can be conveniently discussed in three main classifications:

(a) attempts to obtain a *local regularity theory* (for solutions of the minimal surface equation) which is strong enough to incorporate the original Bernstein result as a special case,

(b) attempts to generalize the result to higher dimensions,

(c) attempts to extend the result from the non-parametric to the parametric case.

The first result to be obtained under (a) above was due to Heinz [HE], who proved that if $u = (x, y)$ is a C^2 solution of the equation on the disc $B_R = (x, y) : x^2 + y^2 < R$, and if B_0 denotes the second fundamental form B of the graph of u (as a submanifold of $\mathbf{R}^3$) at the point $(0, u(0))$, then

$$|B_0| \leq c/R \tag{6.1}$$

where c is an absolute constant. Notice that in fact since the mean curvature (trace B_0) is zero, $-\frac{1}{2}|B_0|^2$ is just $\det(B_0)$ = Gauss curvature of the graph at $(0, u(0))$.

In particular if $u \in C^2(\mathbf{R}^2)$ is a solution over all of $\mathbf{R}^2$, then Heinz could let $R \to \infty$ in (6.1), thus obtaining $B_0 = 0$. Applying the same argument to translations of the graph, one thus deduces that $B \equiv 0$ on graph u; that is, graph u is a plane and hence u is linear.

Various improvements and extensions of the result (6.1) were obtained (see for example the discussion in [N2]; see also the estimates of [SSY], [SS1] which extend (6.1) to a higher dimensional parametric setting). For an extension in another direction (graphs with quasi-conformal Gauss map) see [JS], [SL3].

Under classification (b) there has been decisive progress in the last 20 years. The first break-through came in a paper of Fleming [F], in which he gave a new proof of the two-dimensional Bernstein result. DeGiorgi [DG2] was able to use Fleming's argument, in combination with an argument which strongly used the fact that the hypersurface under consideration was a graph (and hence had a unit normal $\nu = (\nu_1, \cdots, \nu_4)$ with $\nu_4 \geq 0$), in order to extend the result to the case of a 3-dimensional minimal graph in $\mathbf{R}^4$. Almgren then extended the result to $k = 4$, and J. Simons extended to $k \leq 7$. These latter results used the Fleming-DeGiorgi argument in combination with results on the non-existence of stable (codimension 1) minimal cones in low dimensions. For a discussion see [SJ] and [GE1]. Notice also that the Bernstein theorem can be proved by letting $R \to \infty$ in (5.2). See [SSY], [SL2] for discussion.)

Finally the Bernstein result was shown to be false in case $k \geq 8$ by Bombieri, DeGiorgi, and Giusti [BDG], who proved the existence of non-linear solutions of quadratic (and also higher) order growth. A nice conjecture (see [BG]) is that all solutions of the minimal surface equation of $\mathbf{R}^k$ satisfy $u(x) = 0(|x|^p)$ as $|x| \to \infty$ for some constant p. Notice that it follows from the gradient estimate of Bombieri DeGiorgi and Miranda [BDM] that u is linear (in all dimension $k \geq 2$) if u has such growth with $p = 1$.

Under classification (c) there have been a number of results (see [O2]) for complete surfaces in $\mathbf{R}^2$ under various restrictions on the *Gauss map* of the surface.

Perhaps the nicest result proved to date is due to Fischer-Colbrie and Schoen [FS], who prove that if M is complete, oriented and immersed in R^3, and if M is stable in subsets U with $\overline{U}$ compact (see §1), then M is a plane.

We also mention another recent result due to Schoen and Simon [SS2], which asserts that if M is an embedded, simply connected complete surface in R^3 such that the area growth of M satisfies $|M \cap B^3_R(0)| = 0(R^2)$ as $R \to \infty$, then M is a plane. (Notice that this is *false* for *immersed* M; Enneper's surface provides a counter example.)

2. *The Plateau problem*

The classical version of the Plateau problem is the following:

Given a smooth Jordan curve $\Gamma \subset R^n$, find a map $\psi : \overline{D} \to R^n (D = \{(x,y) \in R^2 : x^2+y^2 < 1\})$ such that $\psi|\partial D$ is a homeomorphism onto Γ, $\psi|D$ is a smooth immersion, and such that ψ has least area (defined as in (1.20) with $M_0 = \overline{D}$) relative to all maps $\phi : \overline{D} \to R^n$ where $\phi|D$ is smooth and $\phi|\partial D$ is a homeomorphism onto Γ.

In view of the existence of isothermal coordinates for any immersion of $\overline{D}$ into R^n, this problem can be reduced to the problem of minimizing the *Dirichlet integral* of $C^0(\overline{D}) \cap H^1(D)$ * maps $\phi : \overline{D} \to R^n$ such that $\phi|\partial D$ is a monotone mapping onto Γ which brings into correspondence a specified triple of points on ∂D with a specified triple on Γ. (For details of this reduction, see [L4] for example.)

This is essentially the approach adopted (independently) by Douglas [D] and Rado [RT]. They actually only managed to prove the existence of a *branched* minimizing immersion ψ. Much later Ossermann [O3] proved that $\psi|D$ is free of "true" branch points. That "false" branch points were also absent was proved by Alt [AH] and Gulliver [GR]. Courant [C] extended the existence theorem of Douglas and Rado to the higher genus case, and Morrey [M] extended it to a Riemannian setting.

---

*$H^1(D)$ here denotes the set of mappings $\psi = (\psi_1, \cdots, \psi_n) : D \to R^n$, such that each $\psi_j \in L^2(D)$ and each generalized partial derivative of ψ is in $L^2(D)$.

The natural question concerning boundary regularity of the Douglas-Rado solution (i.e. if the boundary Γ is smooth, is the solution comparably smooth?) was answered by Hildebrandt [H] (in a general P.D.E. setting, applying in particular to Morrey's solutions) and then (with simpler but less general proofs) by Kinderlehrer [K] and Nitsche [N2]. However the question of whether or not ψ is an immersion (i.e. locally a diffeomorphism) is still partially open: there may be isolated boundary branch points. Gulliver and Leslie [GL] showed that no such boundary branch points exist in case Γ is analytic. If Γ is extreme, then the convex hull property can be used to give an elementary proof of the absence of boundary branch points. The existence of boundary branch points when Γ is merely smooth is still an open question. We should also mention that Tomi [TF] is able to prove that there are at most *finitely many* solutions of this classical Plateau problem in case Γ is at least C^4 and in case there are no boundary branch points on Γ (which is the case if, for example, Γ is analytic by [GL].)

Different approaches to the Plateau problem, which can be generalized to higher dimensions, are the following:

(i) (unrestricted genus oriented problem). Find a 2-current T which minimizes mass amongst all 2-currents S with spt S compact and $\partial S = \Gamma$. (Here Γ is given an orientation, and identified with the 1-current $\tau(\Gamma)$ —see §3.)

(ii) (unrestricted genus unoriented problem). Find the 2-current T which minimizes mass relative to all 2-currents S with spt S compact and $\partial S \equiv \Gamma \bmod 2$. (Here Γ is interpreted as a 1-current as in (i), and congruence mod 2 is as discussed in §3.)

In case $n = 3$ (i.e. in the codimension 1 case) the nice fact is that solutions to *both* these problems are represented (in the sense of §3) by *compact, properly embedded surfaces* M with $\partial M = \Gamma$ in the classical differential geometric sense.

The actual *existence* of a minimizing 2-current for the above problems follows directly from standard compactness theorems ([FH1, 4.2.17, p. 432]) for 2-currents and 2-currents mod 2. In case $n = 3$, relevant regularity results needed to establish that these minimizing 2-currents are actually represented by compact surfaces-with-boundary (as described above) were established as follows:

For problem (i) interior regularity was proved by Fleming [F] (now one can of course use the general results of §5), and boundary regularity by Hardt and Simon [HS]. (The boundary regularity theory of [HS] actually applies locally to codimension-1 minimizing currents in all dimensions.)

For problem (ii) interior regularity was first proved by Reifenberg [RE] (see the discussion of §5) and boundary regularity by Allard [AW2].

The boundary regularity theory developed by Allard in [AW2] is in fact very general. One particular consequence is that if T is a minimizing k-current in R^n with spt T compact, and if $\partial T = \Gamma$, where Γ is a smooth (k–1)-dimensional submanifold, and if $\Gamma \subset \partial A$ for some uniformly convex subset A of R^n, then $W \cap \text{spt } T$ is a regular manifold-with-boundary Γ for some neighborhood W of ∂A.

§7. *Applications of second variation*

In this last section we want to discuss a few selected applications of the second variation formula (2.6) (or (1.18) in the case M is a classical smooth surface).

First we discuss some applications in which the stability inequality is used to obtain information about the relation between the geometry of the ambient space N and its topology.

The classical example of this type is due to Synge, who showed that if N is compact and even dimensional, and if the sectional curvatures of N are positive, then $\pi_1(N) = 0$. We demonstrate this in two dimensions (the proof in the arbitrary even dimensional case is only slightly more complicated—see [CE] for example for details). First note that $\pi_1(N) \neq 0$ implies that there exists a stable closed geodesic in N; i.e. a stable immersion of S^1 into N. (One sees this by taking a given closed curve Γ in N which is not hull-homotopic, and then taking a sequence $\{\Gamma_k\}$

of closed curves, parametrized by arc-length, which minimizes length within the class of curves homotopic to Γ.) Of course a subsequence of $\{\Gamma_k\}$ converges uniformly to a smooth geodesic Γ_*. Then (since $k = 1$ and $\dim N = 2$) we can use the stability inequality (1.19) with $\zeta \equiv 1$ in order to deduce that $\int_{\Gamma_*} (-K) \geq 0$, where K is the Gauss curvature of N. Thus we deduce that $K > 0$ everywhere in N is impossible.

J. Simons [SJ] was the first to notice a higher dimensional generalization of this. He proved that if N is compact, of dimension $k+1$, and if the Ricci curvature $\operatorname{Ric}(X, X)$ is positive for all X, then the kth singular homology group $H_k(N)$ of N must be zero. Indeed if $H_k(N) \neq 0$, then the discussion of §3 (#6) shows that there is a homologically minimizing k-current $T = (M, \mu, \xi)$ in N with $\mathcal{H}^{\alpha}(\operatorname{sing} T) = 0 \ \forall \alpha > k-7$. One readily checks that we can then use the formula (1.19) with $\zeta \equiv 1$. This actually requires some technical justification because sing T may be non-empty. However since $\mu(M \cap B_\rho^P(y)) \leq c\rho^k \ \forall \rho < 1$, $y \in \operatorname{spt} T$ (by the monotonicity results guaranteed by Remark 2.1 and #2 of §4), and since sing T can be covered by balls $\{B_{\rho_i}^P(x_i)\}_{i=1,\cdots,P}$ with $\sum_{i=1}^{P} \rho_i^{k-2} < \varepsilon$ (for any preassigned $\varepsilon < 0$, by virtue of the fact that $\mathcal{H}^{k-2}(\operatorname{sing} T) = 0$), the reader will see that it is possible to construct functions ζ_ε with $0 \leq \zeta_\varepsilon \leq 1$ on N, $\zeta_\varepsilon \equiv 0$ in some neighborhood of sing T, $\int_M |\nabla \zeta_\varepsilon|^2 d\mu \leq c\varepsilon$, and $\zeta_\varepsilon \to 1$ as $\varepsilon \to 0$, uniformly on compact subsets of $N \sim \operatorname{sing} T$. Thus we can justify using $\zeta \equiv 1$ in (1.19) by first plugging in ζ_ε, and then letting $\varepsilon \to 0$. We thus have $\int_M (-\operatorname{Ric}(\nu, \nu))\, d\mu \geq 0$, and hence we see that N cannot have positive Ricci curvature throughout N.

A more recent result in a similar spirit is due to Schoen and Yau [SY]. They prove that if $\dim N = 3$ and if $\pi_1(N)$ has a subgroup which is (abstractly) isomorphic to $\pi_1(\Sigma_g)$, where Σ_g is the sphere with g handles and $g \geq 1$, then N cannot have scalar curvature which is everywhere positive. Indeed (as discussed in #8 of §3) they proved under the above hypotheses that there is a stable, oriented immersed surface $M \subset N$. Taking (1.19), again with $\zeta \equiv 1$, this gives $-\int_M (|B|^2 + \mathrm{Ric}(\nu,\nu))d\mu \geq 0$, where ν is the unit normal for M. Using the Gauss curvature equations and the fact that B has zero trace (i.e. the fact that the mean curvature of M is zero), this gives $\int(-\frac{1}{2}|B|^2 + K - S)d\mu \geq 0$, where K is the intrinsic Gauss curvature of M, and S is the scalar curvature of N. Since $g \geq 1$ the Gauss Bonnet formula gives $\int_M K\,d\mu = 2-2g \leq 0$, and hence we have $-\int_M S\,d\mu \geq 0$. Thus S cannot be everywhere positive. (In fact by a result of Kazdan and Warner [KW], S cannot be everywhere non-negative unless N is flat.)

Next we briefly discuss an interesting result of Lawson and Simons, [LS] concerning the *non-existence* of k-varifolds in S^n which are stable in all of S^n. This is in fact checked by making special choices on the right of (2.6); let us call the right side of (2.6) I(X). Also let $e_1, \cdots, e_{n+1}$ be the standard orthonormal basis for R^{n+1} and let e_j^T denote the tangential component of e_j in S^n. Thus, for $x \in S^n$, $e_{jx}^T = e_j - \langle x, e_j\rangle x$.

Evidently, in the notation of (2.6), $\nabla_{\tau_i} e_j^T = (D_{\tau_i} e_j^T)^T = -\langle e_j, x\rangle \tau_i$, and in particular $\mathrm{div}_M e_j^T = -\langle e_j, x\rangle k$ and $(\nabla_{\tau_i} e_j^T)^\perp = 0$. We also note that since all the sectional curvatures of S^n are 1), we have

$$\begin{aligned}\langle R(\tau_i, e_j^T)e_j^T, \tau_i\rangle &= |e_j^T - \langle \tau_i, e_j^T\rangle \tau_i|^2 = |e_j^T|^2 - \langle \tau_i, e_j^T\rangle^2 \\ &= 1 - \langle e_j, x\rangle^2 - \langle \tau_i, e_j\rangle^2, \quad i = 1, \cdots, k, \quad j = 1, \cdots, n+1 .\end{aligned}$$

Substituting in the expression for $I(e_j^T)$, we then have

$$I(e_j^T) = \int_M ((k^2 - k)\langle e_j, x\rangle^2 - k(1 - \langle e_j, x\rangle^2) + \sum_{i=1}^k \langle \tau_i, e_j\rangle^2))d\mu ,$$

and summing over $j = 1, \cdots, n+1$ we then obtain

$$\sum_{j+1}^{n+1} I(e_j^T) = \int_M (k^2 - k) \dot{-} (nk + k) d\mu$$

$$= \int_M k(k-n) d\mu < 0 \quad \text{if} \quad 1 \le k < n .$$

Thus if $1 \le k < n$ we see that $I(e_j^T) < 0$ for some j, and hence (M, μ) is not stable in S^n.

Finally we want to make a couple of remarks concerning Jacobi fields on a smooth submanifold $M \subset N$. First, define the "index form" by

$$I(X, Y) = \int_M (\mathrm{div}_M X \, \mathrm{div}_M Y + \sum_{i=1}^{k} <(\nabla_{\tau_i} X)^{\perp}, (\nabla_{\tau_i} Y)^{\perp}>$$

$$- \sum_{i,j=1}^{k} <\tau_j, \nabla_{\tau_i} X> <\tau_i, \nabla_{\tau_j} Y> - \sum_{i=1}^{k} <R(\tau_i, X) Y, \tau_i>) d\mu ,$$

where the notation is as in (2.6). We note that $I(X, X)$ is just $I(X)$, the expression on the right of (2.6), in case M is stationary in U and X has compact support in U.

Suppose now M is compact with $\partial M \cap U = \phi$ and let $\phi_{t,u}$, $t, u \epsilon (-1,1)$, be a two-parameter family of diffeomorphisms of N which smoothly vary in t, u and which are such that $\phi_{0,0} = 1_N, \phi_{t,u}(K) \subset K, \phi_{t,u}|_{N \sim K} = 1_{N \sim K}$ for some compact $K \subset U$, and such that $\left.\frac{\partial \phi_{t,u}(x)}{\partial t}\right|_{(t,u)=(0,0)} = X_x$, $\left.\frac{\partial \phi_{t,u}(x)}{\partial u}\right|_{(t,u)} = Y_x$. Then one can show (see e.g. [SL1] for an indication of the proof)

$$\left.\frac{\partial^2 |\phi_{t,u}(M)|}{\partial t \, \partial u}\right|_{(t,u)=(0,0)} = I(X, Y) = I(Y, X) = I(X^{\perp}, Y^{\perp}) \tag{7.1}$$

provided M is stationary in U. Now suppose that ψ is any diffeomorphism $(-1,1)\times U \to N$, let $\psi_t(x) = \psi(t,x)$, define J on U by $J_x = \frac{\partial\psi(t,x)}{\partial t}\Big|_{t=0}$, and suppose that, for each W with $\overline{W} \subset U$, $\psi_t(M)$ *is stationary in* W for all sufficiently small t. (Thus $\{\psi_t(M)\}_{t\epsilon(-\varepsilon,\varepsilon)}$ is a 1-parameter family of minimal surfaces for some $\varepsilon < 0$.) Also let Y be an arbitrary smooth vector field with compact support in U and suppose that ζ_u is a 1-parameter family of diffeomorphisms of N such that $\frac{\partial}{\partial u}\zeta_u(x)|_{u=0} = Y_x$ and such that $\zeta_0 = 1_N$, $\zeta_u(K) \subset K$, $\zeta_u|_{N\sim K} = 1_{N\sim K}$ for some compact $K \subset U$. Then defining $\phi_{t,u} = \zeta_u \circ \psi_t$ and using the fact that $\psi_t(M)$ is stationary in a neighborhood of K (for all sufficiently small t), we deduce $\frac{\partial}{\partial u}\Big|\phi_{t,u}(M)|_{u=0} = 0$ for all sufficiently small t, and hence $\frac{\partial^2}{\partial t\,\partial u}\Big|\phi_{t,u}(M)|_{(t,u)=(0,0)} = 0$, so that

$$\text{(7.2)} \qquad I(J^{\perp}, Y^{\perp}) = 0 \ \forall Y \quad \text{with} \quad \operatorname{spt} Y \subset U .$$

$J^{\perp}$ is called a (normal) Jacobi field on $M \cap U$: the identity (7.2) is in fact the weak version of an elliptic system for $J^{\perp}$. In the special case when the codimension is 1 and $M \cap U$ is oriented, we can write $u = \nu.J$ where ν is a smooth unit normal for $M \cap U$. Then the relation (7.2) is equivalent to the equation $Lu = 0$, where L is the operator given by

$$\text{(7.3)} \qquad Lu = \Delta_M u + (|B|^2 + \operatorname{Ric}(\nu,\nu))u, \ u \,\epsilon\, C^2(M \cap U) .$$

(For further discussion of the case codimension > 1, see [SJ].)

The equation $Lu = 0$, L as in (7.3), has important applications. For example if $N = \mathbb{R}^{k+1}$ and if $\psi_t(x) \equiv x + te$ ($e \,\epsilon\, S^n$ fixed), so that ψ_t is just a translation of $\mathbb{R}^n$ and $\psi_t(M)$ is certainly stationary if M is. Thus, since in this case $J_x \equiv e$, the above discussion shows that

$$\Delta_M \nu.e + |B|^2 \nu.e = 0 \quad \text{on} \quad M$$

whenever M is a stationary smooth submanifold of R^{k+1}. This equation has important applications–see for example [BDM].

We finally make a remark concerning the eigenvalue problem

$$\begin{cases} Lu = \lambda u & \text{on} \quad M \cap U \\ u = 0 & \text{on} \quad \partial(M \cap U) \end{cases}$$

with L as in (7.3). As is well known, the minimum eigenvalue for this problem is given by

$$\lambda_1 = \inf \int_M (|\text{grad } u|^2 - (|B|^2 + \text{Ric}(\nu, \nu))\, u^2)\, d\mu \,,$$

where the inf is taken over all $u \in C_0^\infty(U)$ with $\int_M u^2 d\mu = 1$. Then evidently (in view of (1.19)) we have $\lambda_1 \geq 0$ *if and only if* M *is stable in* U. There are of course applications of this; for example it was used in [FS] to obtain the Bernstein result discussed in §6 above.

MATHEMATICS DEPARTMENT
RESEARCH SCHOOL OF PHYSICAL SCIENCES
AUSTRALIAN NATIONAL UNIVERSITY
CANBERRA, A.C.T. 2600 AUSTRALIA

REFERENCES

[A1] F. J. Almgren, Jr., *Some interior regularity theorems for minimal surfaces and an extension of Berstein's Theorem.* Annals of Math. 84 (1966), 277-292.

[A2] ________, *Existence and regularity almost everywhere of solutions to elliptic variational problems among surfaces of varying topological type and singularity structure,* Ann. of Math. 27 (1968), 321-391.

[A3] ________, The theory of varifolds, Princeton mimeographed notes, 1965.

[A4] ________, *Q-valued functions minimizing Dirichlet's integral and regularity of area minimizing currents up to codimension 2* (Research announcement to appear).

[A5] ________, Existence and regularity almost everywhere of solutions to elliptic variational problems with constraints, Memoirs A.M.S. #165 (1976).

[AH] H. W. Alt, *Verzweigungspunkte von H-Flächen* I, Math. Z. 127 (1972), 333-362.

[AW1] W. K. Allard, *On the first variation of a varifold,* Ann. of Math. 95 (1972), 417-491.

[AW2] W. K. Allard, *On the first variation of a varifold: boundary behaviour,* Ann. of Math. 101 (1975), 418-446.

[AW3] ________, *On boundary regularity for Plateau's problem,* Bull. A.M.S. 75 (1969), 522-523.

[AS] F. J. Almgren, Jr., L. Simon, *Existence of embedded solutions of Plateau's problem,* Ann. Scuola Norm. Sup. Pisa. 6 (1979), 447-495.

[B] E. Bombieri, *Recent progress in the theory of minimal surfaces,* L'Enseignement Math. 25 (1979), 1-8.

[BG] E. Bombieri, E. Giusti, *Harnack's inequality for elliptic differential equations on minimal surfaces,* Invent. Math. 15 (1972), 24-46.

[BDG] E. Bombieri, E. DeGiorgi, E. Guisti, *Minimal cones and the Bernstein problem,* Invent. Math. 7 (1969), 243-268.

[BDM] E. Bombieri, E. DeGiorgi, M. Miranda, *Una maggiorazione a priori relativa alle ipersuperfici minimali non-parametriche,* Arch. Rat. Mech. Anal. 32 (1969), 255-267.

[BJ] J. Brothers, *Existence and structure of tangent cones at the boundary of an area minimizing integral current,* Indiana Univ. Math. J. 26 (1977), 1027-1044.

[C] E. Calabi, *Minimal immersions of surfaces in Euclidean spheres,* J. Diff. Geometry 1 (1967), 111-125.

[CR] R. Courant, *Dirichlet's Principle, conformal mapping, minimal surfaces,* Springer-Verlag reprint (1970).

[CHS] L. Caffarelli, R. Hardt, L. Simon, (to appear).

[CE] J. Cheeger, D. Ebin, *Comparison theorems in Riemannian geometry,* North Holland (1975).

[DG1] E. DeGiorgi, *Frontiere orientate di misura minima,* Sem. Mat. Scuola Norm. Sup. Pisa (1961), 1-56.

[DG2] ________, *Una estensione del teorema di Bernstein,* Ann. Scuola Norm. Sup. Pisa 19 (1965), 79-85.

[D] J. Douglas, *Solution of the problem of Plateau,* Trans. A.M.S. 33 (1931), 263-321.

[F] W. Fleming, *On the oriented Plateau problem,* Rend. Circ. Mat. Palerino 11 (1962), 69-90.

[FF] H. Federer, W. Fleming, *Normal and integral currents,* Ann. of Math. (2) 72 (1960), 458-520.

[FH1] H. Federer, *Geometric Measure Theory,* Springer-Verlag, New York, 1969.

[FH2] ________, *The singular sets of area minimizing rectifiable currents with codimension one and of area minimizing flat chains modulo two with arbitrary codimension,* Bull. A.M.S. 76 (1970), 767-771.

[FS] D. Fischer-Colbrie, R. Schoen, *The structure of complete stable minimal surfaces in 3-manifolds of non-negative scalar curvature* (preprint).

[GE1] E. Giusti, *Minimal surfaces and functions of bounded variation,* Notes on Pure Mathematics, Australian National University No. 10, 1977.

[GE2] ———, *Superfici cartesiane di area minima,* Rend. Sem. Mat. Milano (1970), 3-21.

[GT] D. Gilbarg, N. Trudinger, *Elliptic partial differential equations of second order,* Springer-Verlag, New York, 1977.

[GR] R. Gulliver, *Regularity of minimizing surfaces of prescribed mean curvature,* Ann. of Math. 97 (1973), 275-305.

[GL] R. Gulliver, J. Leslie, *On boundary branch points of minimizing surfaces,* Arch. Rat. Mech. Anal. 52 (1973), 20-25.

[HL] W.-Y. Hsiang, H. B. Lawson, *Minimal submanifolds of low cohomogeneity,* J. Differential Geometry 5 (1970), 1-37.

[HW] W.-Y. Hsiang, *On compact, homogeneous minimal submanifolds,* Proc. Nat. Acad. Sci., U.S.A. 56 (1966), 5-6.

[HS] R. Hardt, L. Simon, *Boundary regularity and embedded solutions for the oriented Plateau problem,* Ann. of Math. 110 (1979), 439-486.

[H] S. Hildebrandt, *Boundary behaviour of minimal surfaces,* Arch. Rat. Mech. Anal. 35 (1969), 47-82.

[HE] E. Heinz, *Über die Lösungen der Minimal flächengleichung,* Nachr. Akad. Wiss. Gottingen II (1952), 51-56.

[JS] H. Jenkins, J. Serrin, *Variational problems of minimal surface type I,* Arch. Rat. Mech. Anal. 12 (1963), 185-212.

[K] D. Kinderlehrer, *The boundary regularity of minimal surfaces,* Ann. Scuola Norm. Sup. Pisa 23 (1969), 711-744.

[KW] J. Kazdan, F. Warner, *Prescribing curvatures,* Proc. Symp. Pure Math. A.M.S. 27 (1975), 309-319.

[KNS] D. Kinderlehrer, L. Nirenberg, J. Spruck, Journal D'Analyse 34 (1978), 86-119.

[LP] P. Li, *Minimal immersions of compact irreducible homogeneous Riemannian manifolds,* to appear.

[L1] H. B. Lawson, Jr., *Minimal Varieties,* Proc. Symp. Pure Math. 27 (1975), 143-175.

[L2] ———, *The equivariant Plateau problem and interior regularity,* Trans. A.M.S. 173 (1972), 231-250.

[L3] ———, *Minimal varieties in real and complex geometry,* Univ. of Montréal Press, Montréal, 1973.

[L4] H. B. Lawson, Jr., I.M.P.A. Volume, Reprint by Publish or Perish (1980).

[LS] H. B. Lawson, Jr., J. Simons, *On stable currents and their application to global problems in real and complex geometry*, Ann. of Math. (2) 98 (1973), 427-450.

[M] C. B. Morrey, Jr., *The problem of Plateau in a Riemannian manifold*, Ann. of Math. 49 (1948), 807-851.

[MM] M. Miranda, *Frontiere minimali con ostacoli*, Ann. Univ. Ferrara Sez. VII (N.S.) 16 (1971), 29-37.

[MS] J. H. Michael, L. Simon, *Sobolev and mean-value inequalities on generalized submanifolds of* R^n, Comm. Pure Appl. Math. 26 (1973), 361-379.

[MSY] W. Meeks, L. Simon, S.-T. Yau, Annals of Math. 116 (1982), 621-659.

[MY] W. Meeks, S.-T. Yau, *The classical Plateau problem and the topology of 3-manifolds*. To appear in Arch. for Rational Mech. Analysis.

[NJ] J. Nash, *The embedding problem for Riemannian manifolds*, Annals of Math. 63 (1956), 20-63.

[N1] J. C. C. Nitsche, *On new results in the theory of minimal surfaces*, Bull. A.M.S. 71 (1965), 195-270.

[N2] ———, Arch. Rat. Mech. Anal. 52 (1973), 319-329.

[N3] ———, Vorlesungen über Minimalflächen, Springer-Verlag (1975).

[N4] ———, *The higher regularity of liquid edges in aggregates of minimal surfaces* (1978).

[O1] R. Osserman, *Minimal Varieties*, Bull. A.M.S. 75 (1969), 1092-1120.

[O2] ———, *A survey of minimal surfaces*, Van Nostrand, New York, 1969.

[O3] ———, *A proof of the regularity everywhere of the classical solution of Plateau's problem*, Ann. of Math. 91 (1970).

[P1] J. Pitts, Bull. A.M.S. 82 (1976), 503-504.

[P2] ———, *Existence and regularity of minimal surfaces in Riemannian manifolds* (preprint, 1979).

[RT] T. Rado, *On Plateau's problem*, Ann. of Math. (2) 31 (1930), 457-469.

[RE1] E. Reifenberg, *Solution of the Plateau problem for m-dimensional surfaces of varying topological type*, Acta. Math. 104 (1960), 1-21.

[RE2] ———, *An epiperimetric inequality related to the analiticity of minimal surfaces. On the analiticity of minimal surfaces.* Ann. of Math. 80 (1964), 1-21.

[SJ] J. Simons, *Minimal varieties in Riemannian manifolds*, Ann. of Math. (2) 88 (1968), 62-105.

[SP] M. Spivak, *A comprehensive introduction to differential geometry Vol. III*, Publish or Perish.

[SU1] J. Sacks, K. Uhlenbeck, *Minimal immersions of compact Riemannian surfaces*, preprint.

[SU2] ———, *The existence of minimal immersions of 2-spheres*, Ann. of Math. 113 (1981), 1-24.

[SL1] L. Simon, *First and second variation in geometry and topology*, (preprint, Melbourne University, 1979).

[SL2] ———, *Remarks on curvature estimates for minimal hypersurfaces*, Duke Math. J. 43 (1976), 545-553.

[SL3] ———, *Boundary regularity for solutions of the non-parametric least area problem.* Annals of Math. 103 (1976), 429-455.

[SL4] ———, *A Hölder estimate for maps between surfaces in Euclidean space*, Acta. Math. 139 (1977), 19-51.

[SL5] ———, *Lectures on Geometric Measure Theory*, Proceedings of Centre for Mathematical Analysis (to appear 1983), Australian National University.

[SSY] R. Schoen, L. Simon, S.-T. Yau, *Curvature estimates for minimal hypersurfaces*, Acta Math. 134 (1975), 276-288.

[SY] R. Schoen, S.-T. Yau, *Existence of incompressible minimal surfaces and the topology of 3-manifolds with non-negative scalar curvature*, Annals of Math. 110 (1979), 127-142.

[SS1] R. Schoen, L. Simon, *Regularity of Stable minimal hypersurfaces*, Comm. Pure Appl. Math. 34 (1981), 741-797.

[SS2] ———, *Regularity of simply connected surfaces with quasiconformal Gauss map.* These proceedings.

[TJ] J. Taylor, *The structure of singularities of soap-bubble-like and soap-film-like minimal surfaces*, Ann. of Math. 103 (1976), 489-539.

[TT] F. Tomi, A. Tromba, *Extreme curves bound an embedded minimal surface of the type of the disc*, Math. Z. 158 (1978), 137-145.

[TF] F. Tomi, Arch. for Rational Mech. Analy. 52 (1973), 312-318.

[T] T. Takahashi, *Minimal immersions of Riemannian manifolds*, J. Math. Soc. Japan, 18 (1966), 380-385.

[TN] N. Trudinger, *A new proof of the interior gradient bound for the minimal surface equation in* n *dimensions*, Proc. Nat. Acad. Sci. U.S.A. 69 (1972), 821-823.

ON THE EXISTENCE OF SHORT CLOSED GEODESICS AND THEIR STABILITY PROPERTIES

W. Ballmann,* G. Thorbergsson,* W. Ziller

In 1905 Poincaré [16] suggested the problem of finding elliptic closed geodesics for a Riemannian metric on S^2 and claimed that any convex surface has at least one such closed geodesic without self-intersections. But his proof contains large gaps and recent results of Grjuntal [8] show that his theorem is actually false. In 1979 Thorbergsson [17] showed that any metric on S^2 whose Gaussian curvature satisfies $\frac{1}{4} \leq K \leq 1$ has at least one closed geodesic of elliptic-parabolic type. We will summarize some recent results of the authors which show that any Riemannian metric on S^n satisfying certain pinching assumptions has at least two such closed geodesics.

In Section I we discuss the existence of closed geodesics which, in a certain geometric sense, can be considered short. Theorems of this type were first proved by Alber [3], [4] and Klingenberg [11], but both proofs rely on topological results of Alber [2] which recently turned out to be false. We will indicate some topological results which nevertheless enable us to prove theorems of the same type.

In Section II we examine the stability properties of the closed geodesics found in Section I. We consider conditions on the Riemannian

*This work was done under the program Sonderforschungsbereich "Theoretische Mathematik" SFB 40 at the University of Bonn.

Seminar on Minimal Submanifolds
0-691-08324-X/83/053-11 $1.05/0 (cloth)
0-691-08319-3/83/053-11 $1.05/0 (paperback)
For copying information, see copyright page.

metric which imply that some of these geodesics are of elliptic-parabolic type.

I. *On the existence of short closed geodesics*

In 1927 Birkhoff [6] showed that any metric on S^n has at least one closed geodesic. This closed geodesic can be considered short, since it is obtained by shortening the family of closed curves consisting of a great circle and all small circles parallel to it.

To obtain the existence of more than one closed geodesic one uses Lusternik-Schnirelmann theory. The most beautiful result was obtained by Lusternik-Schnirelmann in 1929 [13], who showed that any metric on S^2 has at least three closed geodesics without self-intersections. Actually their proof contains some gaps which have been filled in only recently by Ballmann [5]. But their methods do not seem to apply to S^n, $n > 2$.

Let g be a Riemannian metric on S^n. Lusternik-Schnirelmann theory applies to the energy functional

$$E : \Pi \to R ; \quad c \to \frac{1}{2} \int g(\dot{c}, \dot{c}) ,$$

where $\Pi = \Lambda/O(2)$. Here Λ is the space of piecewise smooth curves $S^1 \to S^n$, and $O(2)$ acts by reparametrizations on S^1. The closed geodesics are the critical points of positive energy of E.

Let Π^κ (resp. $\Pi^{\kappa-}$) be the space of curves of energy $\leq \kappa$ (resp. $< \kappa$). For each nonvanishing homology class $x \in H_*(\Pi^{\kappa-}, \Pi^0)$ one defines:

$$\kappa_x = \inf_{z \in x} \max_{c \in z} E(c) .$$

It is easy to show that $\kappa_x > 0$ and that there exists a critical point of energy κ_x. However, two different homology classes x and y may give rise to the same critical point. One says that x is subordinate to y if there exists a cohomology class ξ of dimension > 0 such that $\xi \cap y = x$. It is the main result of Lusternik-Schnirelmann theory that $\kappa_y \geq \kappa_x$ if x

is subordinate to y, and that $\kappa_y = \kappa_x$ implies the existence of infinitely many critical points of energy κ_x. Therefore, if there exists a chain of r homology classes $x_1, \cdots, x_r$ such that x_i is subordinate to x_{i+1}, then E has at least r critical points.

The main difficulty in the application of Lusternik-Schnirelmann theory is to distinguish between prime and multiple closed geodesics. This difficulty arises because for each closed geodesic $c(t)$ the iterates $c^q(t) = c(qt)$, $q \in N$, are also critical points to the energy functional. Under the assumptions on the Riemannian metric discussed below, we are able to prove that certain subordinate homology classes give rise to different prime closed geodesics.

Lusternik and Alber considered the inclusion $i:(\Gamma, \Gamma^0) \to (\Pi, \Pi^0)$ where Γ denotes the space of great and small circles on S^n and Γ^0 the space of point circles. Their idea was to try to find chains of subordinate homology classes in $i_* H_*(\Gamma, \Gamma^0)$ corresponding to such chains in $H_*(\Gamma, \Gamma^0)$. In fact, for the standard metric the only homology classes that give rise to prime closed geodesics are those in $i_* H_*(\Gamma, \Gamma^0)$. Notice that the closed geodesics guaranteed by Lusternik-Schnirelmann theory for a nonvanishing homology class in the image of i_* can be considered short.

In the following we restrict ourselves to homology with Z_2-coefficients.

The Thom isomorphism implies $H_*(\Gamma, \Gamma^0) \simeq H_{*-(n-1)}(G_{2,n-1})$, where $G_{2,n-1}$ is the Grassmannian of unoriented 2-planes in $(n+1)$-space. Alber [1] computed that $G_{2,n-1}$ has $g(n)$ subordinate homology classes, where $g(n) = 2n-s-1$, $n-s = 2^k$, $s < 2^k$. Thus $n+1 \leq g(n) \leq 2n-1$ and $g(n) \geq (3n-1)/2$. Hence $H_*(\Gamma, \Gamma^0)$ has $g(n)$ subordinate homology classes with respect to cohomology classes in $H^*(\Gamma - \Gamma^0) \simeq H^*(G_{2,n-1})$.

Let $j: \Gamma - \Gamma^0 \to \Pi - \Pi^0$ be the inclusion. In [2] Alber published the theorem that i_* is injective and j^* is surjective. As a consequence, i_* would carry subordinate homology classes into subordinate homology classes. But Ballmann found a mistake in his proof, and we can now show that j^* is not surjective for $n > 2$. This contradicts the subordination of certain of the homology classes. The reason behind this is that $O(2)$ does not act freely on Λ.

We can prove however:

THEOREM 1. (Π, Π^0) *contains* n *subordinate homology classes which lie in the image of* i_*.

Note that a homology class of (Π, Π^0) does not determine a unique critical point of the energy functional but a unique critical level κ and a unique compact nonvoid set G of critical points of energy κ. Let $x_1, \cdots, x_n$ be the homology classes of Theorem 1, numerated such that x_{i-1} is subordinate to x_i. We then get n critical levels $0 < \kappa_1 \leq \cdots \leq \kappa_n$ and sets of closed geodesics $G_1, \cdots, G_n$. Moreover $\kappa_{i-1} = \kappa_i$ implies that G_i contains infinitely many closed geodesics. We emphasize again that all these closed geodesics could be iterates of one prime closed geodesic if $\kappa_1 < \cdots < \kappa_n$.

We can show that (Π, Π^0) does not contain $g(n)$ subordinate homology classes in the image of i_* for $n > 3$. The surprising result is that under certain geometric assumptions one can define a space which depends on the metric but does contain $g(n)$ subordinate homology classes.

By g_0 we denote the standard metric on S^n with sectional curvature $K = 1$.

THEOREM 2. *If the injectivity radius of* g *is* $\geq \pi$ *and* $g < 4g_0$, *then* $(\Pi^{8\pi^2-}, \Pi^0)$ *contains* $g(n)$ *subordinate homology classes.*

Note that the injectivity radius is $\geq \pi$ if $\frac{1}{4} < K \leq 1$, or if $0 < K \leq 1$ and n is even.

The subordinate homology classes in Theorem 2 are obtained as follows: $g < 4g_0$ implies that $i(\Gamma)$ lies in $\Pi^{8\pi^2-}$. The injectivity radius estimate ensures that any closed geodesic has length at least 2π. Thus $\Pi^{8\pi^2-}$ does not contain any multiple closed geodesics, which makes the action of O(2) on $(\Lambda^{8\pi^2-}, \Lambda^0)$ free up to topological equivalence. This is the essential reason why the g(n) subordinate homology classes in (Γ, Γ^0) are carried over into subordinate homology classes in $(\Pi^{8\pi^2-}, \Pi^0)$.

The above theorems can be used to replace Alber's topological result [2] in the proofs of the following two theorems due to Klingenberg [11] and Alber [3], [4]:

THEOREM 3 (Klingenberg). *If the metric satisfies* $\frac{1}{4} < K \leq 1$, *then it has at least* n *closed geodesics without self-intersections, whose lengths lie in the interval* $[2\pi, 4\pi)$.

The proof consists in showing that the critical levels of the classes in Theorem 1 satisfy $\kappa_i < 8\pi^2$. By the curvature assumptions every geodesic loop has length at least 2π. Therefore the sets G_i consist of prime closed geodesics without self-intersections and lengths in $[2\pi, 4\pi)$.

The homology classes in Theorem 2 give rise to critical sets $H_1, \cdots, H_{g(n)}$ with critical levels $2\pi^2 \leq \kappa_1 \leq \cdots \leq \kappa_{g(n)} < 8\pi^2$. This proves:

THEOREM 4 (Alber). *If the injectivity radius is* $\geq \pi$ *and* $g < 4g_0$, *then the metric has* g(n) *closed geodesics without self-intersections and lengths in* $[2\pi, 4\pi)$.

Alber actually had the stronger assumption $g_0 \leq g < 4g_0$ in addition to the injectivity radius estimate. This is the so-called Morse condition which played an important role in the earlier literature.

It is conceivable that the curvature assumption $\frac{1}{4} < K \leq 1$ implies the existence of a diffeomorphism h such that $h^*g < 4g_0$, which would prove Theorem 4 under the assumption $\frac{1}{4} < K \leq 1$. Using the proof of the pinching theorem of Grove-Karcher-Ruh [9] we obtain at least:

THEOREM 5. *If the metric satisfies* $0.83 \leq K \leq 1$, *then it has* g(n) *closed geodesics without self-intersections and lengths in* $[2\pi, 2.2\pi)$.

Under the hypothesis of Theorem 4 or 5 and under the generic assumption that all closed geodesics are nondegenerate critical points, one gets $\frac{n(n+1)}{2}$ short closed geodesics. This is optimal, as shown by an n-dimensional ellipsoid with pairwise different principal axes close to 1. For a general metric, even arbitrarily close to g_0, it is not known whether $g(n)$ is the optimal number. This is related to the following open problem: does there exist a function on $G_{2,n-1}$ with only $g(n)$ critical points? In fact, not even the category of $G_{2,n-1}$ seems to be known.

For a general manifold M one can use a nontrivial homotopy class $f: S^k \to M$ to map the space Γ of great and small circles of S^k into M. One obtains the following result:

THEOREM 6. *If* M *is even dimensional,* $\frac{1}{16} < K \leq 1$, *and* M *is not a* Z_2*-homology sphere, then the metric has* $\alpha(M) = \min\{i > 0 | H_i(M, Z_2) \neq 0\}$ *closed geodesics without self-intersections and lengths in* $[2\pi, 4\pi)$.

II. *On the Poincaré map of the short closed geodesics*

Let c be a closed geodesic.

Assume that all eigenvalues of the linearized Poincaré map of c lie on the unit circle. Generically such a closed geodesic is of twist type [12], which is expressed by a certain open condition on the 3-jet of the Poincaré map. If c is of twist type, then by the fixed point theorem of Birkhoff-Lewis there are in any neighborhood of c in T_1M infinitely many periodic orbits of the geodesic flow with periods tending to infinity. Note that periodic orbits of the geodesic flow correspond to closed geodesics on M. But even more, the theorem of Kolmogorov-Arnold-Moser implies that in any neighborhood of $\dot{c}$ in T_1M there exist tori T^n which are invariant under the geodesic flow, on which the flow is quasiperiodic. The measure of these tori is positive, and furthermore the measure of the closure of the periodic orbits is positive in any neighborhood of $\dot{c}$. For $n = 2$ this implies that c is stable. For $n > 2$ this implies that the geodesic flow is not ergodic.

We call c elliptic if all eigenvalues z have modulus 1, $z \neq \pm 1$, and the Poincaré map consists of 2-dimensional rotations. c is called parabolic if ± 1 are the only eigenvalues. We say that c is of elliptic-parabolic type if the Poincaré map decomposes into an elliptic and a parabolic part. c is called hyperbolic if no eigenvalue lies on the unit circle.

If c is nonhyperbolic, all of the above is true on a center manifold, if the Poincaré map is of twist type there. This is again a generic condition for nonhyperbolic closed geodesics. See [14] for a discussion of these theorems.

We now examine the stability properties of the closed geodesics found in Section I. Let G_1 and $H_{g(n)}$ be the sets of closed geodesics defined above. They are the shortest and longest geodesics found in Section I. Note that G_1 exists for any metric on S^n, whereas $H_{g(n)}$ is defined only under the conditions of Theorems 4 or 5.

THEOREM 7. *Let* $c \in G_1$. *If the metric satisfies* $\delta \le K \le 1$, *then*

(i) $\delta > \frac{1}{4}$ *implies that* c *is nonhyperbolic.*

(ii) $\delta > \frac{9}{16}$ *implies that all eigenvalues of the linearized Poincaré map* P *of* c *lie on the unit circle. Furthermore* P *splits into* 2×2 *blocks* $\begin{pmatrix} \cos\phi & \sin\phi \\ -\sin\phi & \cos\phi \end{pmatrix}$ *and* $\begin{pmatrix} 1 & -1 \\ 0 & 1 \end{pmatrix}$ *where* $-\pi < \phi \le 0$.

(iii) $\delta > \left(\frac{2p-1}{2p}\right)^2$ *for some integer* $p \ge 2$ *implies in addition that* $-\frac{2\pi}{p} < \phi \le 0$.

THEOREM 8. *Let* $c \in H_{g(n)}$. *If the metric satisfies* $\delta \le K \le 1$, *then*

(i) $\delta > \frac{4}{9}$ *implies that* c *is nonhyperbolic.*

(ii) $\delta > \frac{16}{25}$ *implies that all eigenvalues of the linearized Poincaré map* P *of* c *lie on the unit circle. If* 1 *is not an eigenvalue, then* P *splits into* 2×2 *blocks* $\begin{pmatrix} \cos\phi & \sin\phi \\ -\sin\phi & \cos\phi \end{pmatrix}$ *and* $\begin{pmatrix} -1 & 1 \\ 0 & -1 \end{pmatrix}$ *where* $0 < \phi \le \pi$.

(iii) $\delta > \left(\frac{2p}{2p+1}\right)^2$ *for some integer* $p \geq 2$ *implies that all eigenvalues* $e^{i\phi}$ *of* P *satisfy* $0 \leq \phi \leq \frac{2\pi}{p}$.

Note that Theorem 5 is probably not optimal and that the set $H_{g(n)}$ might be defined for $\frac{1}{4}$-pinched metrics, to which Theorem 8 would then apply. We can do this at least for $n = 2$.

Part (i) of Theorem 7 for $n = 2$ and part (i) of Theorem 8 has been proved in [17].

One can draw the following conclusions:

(1) In the set of metrics satisfying $\frac{9}{16} < K \leq 1$ there exists an open and dense set of metrics for which the geodesic flow is not ergodic.

(2) In the set of metrics satisfying $0.83 < K \leq 1$ there exists an open and dense set of metrics with two short elliptic closed geodesics of twist type.

(3) In the set of metrics satisfying $\frac{1}{4} < K \leq 1$ there exists an open and dense set of metrics with infinitely many closed geodesics.

(4) In the set of metrics on S^2 satisfying $\frac{1}{4} < K \leq 1$ there exists an open and dense set of metrics with a stable closed geodesic.

For the other closed geodesics of Section I one can estimate, under appropriate pinching assumptions, how many eigenvalues lie on the unit circle. In particular:

THEOREM 9. *There exists a pinching constant* $\delta_n < 1$ *such that any metric on* S^n *satisfying* $\delta_n < K \leq 1$ *has at least* $g(n)-1 \geq n$ *non-hyperbolic short closed geodesics.*

These results are obtained by looking at the index of the iterates $c^q(t) = c(qt)$. On the one hand, the pinching assumptions imply a certain growth of the index of c^q by the Morse-Schoenberg comparison theorem. On the other hand, Bott [7] showed that the linearized Poincaré map determines the growth of the index of c^q.

Most of the above results generalize to Finsler metrics. This is of interest since the problem of finding periodic orbits of prescribed energy E for a Hamiltonian system H on T^*S^n can be reduced to Finsler metrics on S^n if, for every $q \in S^n$, $H^{-1}(E) \cap T_q^*S^n$ is a compact convex hypersurface containing the origin in its interior. This includes Hamiltonians C^2-close to g_0.

We obtain:

(1) If the Finsler metric satisfies $\frac{9}{16} < K \leq 1$, then there exists a short closed geodesic of elliptic-parabolic type. Hence there exists an open and dense set of such metrics for which the geodesic flow is not ergodic. Notice however, that Katok [10] constructed special nonsymmetric Finsler metrics, arbitrarily close to g_0, for which the geodesic flow is ergodic.

(2) Any Finsler metric F on S^n satisfying $\frac{16}{25} < K \leq 1$ and $F^2 < \frac{25}{16} g_0$ has at least two closed geodesics of elliptic-parabolic type.

As Weinstein pointed out to us, one can use [15] or [18] to prove that any Hamiltonian on T^*S^n sufficiently C^2-close to the kinetic energy of g_0 has at least two closed geodesics of elliptic-parabolic type. Using these perturbation techniques one can prove that the conclusions of all of the above theorems are satisfied for Hamiltonians sufficiently C^2-close to g_0. But our results are global in nature, since they are obtained under explicit conditions on the Hamiltonian.

Notice also that it does not follow from these perturbation results, together with the present pinching theorems, that there exists a $\delta < 1$, so that any δ-pinched metric satisfies the conclusions of Theorems 7-9. The reason is that the pinching theorems only show that sufficiently δ-pinched metrics are C^1-close to g_0.

The example of Grjuntal, mentioned in the introduction, shows that there exist convex metrics on S^2 such that any closed geodesic without self-intersections is hyperbolic. Thus some pinching assumptions are necessary for the above theorems. One can also construct Finsler metrics, arbitrarily close to g_0, which have only two closed geodesics of elliptic-parabolic type among their short closed geodesics.

W. BALLMANN AND G. THORBERGSSON
UNIVERSITY OF BONN
5300 BONN 1
WEST GERMANY

W. ZILLER
DEPARTMENT OF MATHEMATICS
UNIVERSITY OF PENNSYLVANIA
PHILADELPHIA, PENNSYLVANIA 19104

BIBLIOGRAPHY

[1] S.I. Alber, Homologies of a Space of Planes and their Application to the Calculus of Variations. Doklady Akad. Nauk SSSR (N.S.) 91 (1953), 1237-1240 [Russian].

[2] ———, On Periodicity Problems in the Calculus of Variations in the Large. Uspehi Mat. Nauk (N.S.) 12 (1957), 57-124 [Russian]; Amer. Math. Soc. Transl. (2) 14 (1960), 107-172.

[3] ———, Topology of Function Spaces. Doklady Akad. Nauk SSSR (N.S.) 168 (1966), 727-730 [Russian]; Soviet Mathematics 7 (1966), 700-704.

[4] ———, The Topology of Functional Manifolds and the Calculus of Variations in the Large. Uspehi Mat. Nauk (N.S.) 25 (1970), 57-122 [Russian]; Russ. Math. Surveys 25 (1970), 51-117.

[5] W. Ballmann, Der Satz von Lusternik und Schnirelmann. Bonner Math. Schriften 102 (1978), 1-25.

[6] G.D. Birkhoff, Dynamical Systems. Amer. Math. Soc. Colloq. Publ. 9, Amer. Math. Soc., New York 1927.

[7] R. Bott, On the Iteration of Closed Geodesics and the Sturm Intersection Theory. Comm. Pure Appl. Math. 9(1956), 171-206.

[8] A.I. Grjuntal, The Existence of Convex Spherical Metrics all of whose Closed Geodesics without Self-Intersections are Hyperbolic. Izvestija Akad. Nauk SSSR 43 (1979), 3-18 [Russian].

[9] K. Grove, H. Karcher, E. Ruh, Jacobi Fields and Finsler Metrics on Compact Lie Groups with an Application to Differentiable Pinching Problems. Math. Ann 211 (1974), 7-21.

[10] A.B. Katok, Ergodic Perturbations of Degenerate Integrable Hamiltonian Systems, Izvestija Akad. Nauk SSSR 37 (1973), 539-576 [Russian]; Math. USSR Izvestija 7 (1973), 535-571.

[11] W. Klingenberg, Simple Closed Geodesics on Pinched Spheres. J. Diff. Geom. 2 (1968), 225-232.

[12] W. Klingenberg, F. Takens, Generic Properties of Geodesic Flows. Math. Ann. 197 (1972), 323-334.

[13] L. Lusternik, L. Schnirelmann, Sur le problème de trois géodésiques fermées sur les surfaces de genre O. C. R. Acad. Sci. Paris 189 (1929), 269-271.

[14] J. Moser, Stable and Random Motions in Dynamical Systems. Ann. of Math. Studies 77, Princeton University Press, Princeton 1973.

[15] ———, Periodic Orbits near an Equilibrium and a Theorem by Alan Weinstein. Comm. Pure Appl. Math. 29 (1976), 727-747.

[16] H. Poincaré, Sur les lignes géodésiques des surfaces convexes. Trans. Amer. Math. Soc. 6 (1905), 237-274.

[17] G. Thorbergsson, Non-Hyperbolic Closed Geodesics. Math. Scand. 44 (1979), 135-148.

[18] A. Weinstein, Bifurcations and Hamilton's Principle. Math. Z. 159 (1978), 235-248.

EXISTENCE OF PERIODIC MOTIONS OF CONSERVATIVE SYSTEMS

Herman Gluck and Wolfgang Ziller

Given a classical conservative system with finitely many degrees of freedom, we prove there exists a periodic motion at each energy level E for which the set M_E of points at potential levels $\leq E$ is compact.

In 1917 G. D. Birkhoff [B_1] proved this in the special case that M_E is a sphere. In 1951 Fet and Lyusternik [F-L] proved this when M_E is a closed manifold. In 1947 Seifert [S] proved this when M_E is homeomorphic to an n-cell.

Our proof generalizes Seifert's in the same spirit that Fet-Lyusternik's proof generalized Birkhoff's.

Furthermore, if M_E is homeomorphic to an n-cell, we show there exist at least n periodic motions, provided the Hamiltonian satisfies a certain nonresonance condition.

This is analogous to results of Ekeland and Lasery [E-L] on the existence of n periodic orbits for a convex Hamiltonian system on R^{2n}, and also to results (see e.g. [B-T-Z]) on the existence of n closed geodesics for any Finsler metric on S^n, both under certain nonresonance conditions.

CONTENTS

Seminar on Minimal Submanifolds
0-691-08324-X/83/065-34 \$2.20/0 (cloth)
0-691-08319-3/83/065-34 \$2.20/0 (paperback)

1. *Introduction*

We begin by giving precise statements of the above results, first in the setting of Lagrangian mechanics, and then in the setting of Hamiltonian mechanics.

We model a conservative system with n degrees of freedom as the motion of a single particle on a smooth manifold M^n, called *configuration space*. This space is given a Riemannian metric, in terms of which the *kinetic energy* T of the particle is one-half the square of the speed. A *potential energy* function $U : M^n \to R$ is also given.

The motion takes place along paths which give stationary values to the integral of $T-U$. The total energy $E = T+U$ remains constant during such motions, which are thus constrained to run within the subset $M_E = \{q \in M : U(q) \leq E\}$ of configuration space, since $T \geq 0$.

Note that such systems are *reversible*: any solution traversed backwards is also a solution. In particular, if ∂M_E is nonempty, then periodicity can appear as to-and-fro motion along a *brake orbit*, that is, one which connects boundary points of M_E but otherwise runs through the interior.

THEOREM A. *Given a classical conservative system as above, and an energy level* E *for which* M_E *is compact and nonempty, there exists a periodic motion of energy* E. *Furthermore, if* ∂M_E *is nonempty, a brake orbit is guaranteed.*

The requirement that M_E be compact is in general necessary: just consider a ball rolling along an infinite inclined plane. This hypothesis can sometimes be replaced by other topological or geometric assumptions on M_E, as in [T], [B].

If we recast Theorem A in the language of Hamiltonian mechanics, we gain a broadening of scope.

THEOREM A′. *Let* M *be a smooth manifold and* $H : T^*M \to R$ *a smooth Hamiltonian function on the cotangent bundle of* M. *Assume that* H *is of "classical type", that is, convex and even on each fibre* T^*M_q.

Let E *be a regular value of* H *such that* $H^{-1}(E)$ *is compact and nonempty. Then there exists a periodic solution of Hamilton's equations having energy* E.

REMARKS. 1) Theorem A′ is stronger than Theorem A in that we do not assume H to be quadratic on each fibre. This is exactly the broadening of Seifert's theorem [S] provided by Weinstein [W_2]. It is essentially achieved by using Finsler metrics instead of Riemannian metrics.

2) If $H^{-1}(E)$ projects to all of M, then the periodic solution can be guaranteed without assuming evenness. If $H^{-1}(E)$ projects to part of M, existence of a brake orbit is guaranteed.

3) If M_E is an n-cell, there is a much stronger theorem of Rabinowitz [Ra]: *If* $H^{-1}(E)$ *projects to an* n*-cell in* M, *and if* H *is star-shaped on each fibre, then there exists a periodic motion of energy* E. In particular, Rabinowitz does not need the evenness of H, but the periodic motion obtained may not be a brake orbit. In fact, we give in Section 7 an example of an irreversible Hamiltonian system with no periodic brake orbits, even though $H^{-1}(E)$ projects to an n-cell for all $E > 0$.

4) When $H^{-1}(E)$ projects to part of M, our proof depends critically on the assumption of evenness to guarantee that any solution traversed backwards is also a solution. For applications it would be of primary importance to drop this assumption from Theorem A′.

The proof of Theorems A and A′ depends on the following geometrical results, which may have some appeal in their own right.

THEOREM B. *A compact Riemannian (or more generally, Finsler) manifold-with-boundary has a geodesic chord.*

Here a *geodesic chord* is simply a geodesic whose endpoints lie on the boundary but which otherwise runs through the interior of the manifold.

It would be more ambitious to seek a geodesic chord which meets the boundary of the manifold orthogonally. In general, no such *orthogonal geodesic chord* exists, as shown clearly by the following example of Bos [Bo

Figure 1

However, we do prove

THEOREM C. *A compact Riemannian (or more generally, Finsler) manifold-with-boundary which is locally convex has an orthogonal geodesic chord.*

The definition of *locally convex* is given in Section 4.

REMARK. Note that neither Theorem B nor C requires the Finsler metric to be symmetric. It is only in applying these results to prove Theorems A and A′ that we rely on symmetry.

We first prove Theorem C. This is done by using a curve-shortening process with free boundary points, guided principally by the ideas of G.D. Birkhoff [B_1], [B_2] and Seifert [S]. The new catalyst is an application of the relative Hurewicz theorem in order to carry out Birkhoff's minimax method.

We then prove Theorem B by using Seifert's trick of adding a neck to a manifold-with-boundary to make it locally convex, and invoking Theorem C.

Finally, we use the Principle of Least Action of Euler-Maupertuis-Jacobi [A] to interpret Theorems A and A′ as promising a closed geodesic or geodesic chord on M_E in a certain *Jacobi metric*. When M_E is all of M (large values of E), the closed geodesic is guaranteed by the theorem of Birkhoff-Fet-Lyusternik [B_1], [F-L]. When M_E is a manifold-with-boundary (small values of E), the geodesic chord is produced by coupling Theorem B with a limiting procedure due to Seifert. This limiting procedure is required because the Jacobi metric vanishes on the boundary of M_E, and hence Theorem B cannot be applied directly.

Now suppose that E is an energy level for which M_E is topologically an n-cell. Seifert suggested in [S] the possibility of using Lyusternik-Schnirelmann theory [L-S] to guarantee the existence of n periodic orbits of energy E. We prove this in Theorem D with the aid of a very restrictive nonresonance condition.

NONRESONANCE CONDITION. *Take any parametrization of* M_E *as an n-cell, and consider the "straight line segments" connecting boundary points of* M_E. *We ask that the maximum length of all such line segments should be strictly smaller than twice the length of the shortest brake orbit, all measurements to be made in the Jacobi metric.*

Some remarks about this condition are given at the end of Section 9.

THEOREM D. *Given a classical conservative system (either as in Theorem A or A′) and an energy level* E *for which* M_E *is homeomorphic to an n-cell and for which the above nonresonance condition is satisfied. Then there exist at least* n *periodic orbits of energy* E.

To put Theorem D into perspective, recall the celebrated theorem of Weinstein [W_1]: *If* $H = \frac{1}{2}\sum \alpha_i^{-2}(p_i^2 + q_i^2)$, *then any Hamiltonian sufficiently close to* H *has* n *periodic orbits of energy* E. The conditions of Theorem D are much more restrictive: they only apply if $\max \alpha_i < 2 \min \alpha_i$ and if in addition the perturbation of H is even in q. But Theorem D seems to have two payoffs:

1) The n periodic orbits obtained are all brake orbits.

2) One should be able to compute explicitly how much of a perturbation of H is allowed.

For more details on this, see the remarks at the end of Section 9.

To prove Theorem D, we use

BOS' THEOREM. *A compact Riemannian (or symmetric Finsler) manifold-with-boundary which is locally convex and homeomorphic to an* n*-cell has at least* n *orthogonal geodesic chords.*

This result generalizes that of Lyusternik and Schnirelmann, who showed in [L-S] that a smooth n-cell in R^n with convex boundary has at least n orthogonal geodesic chords.

Note that the nonresonance condition does not appear at the level of Bos' theorem. It only appears when we use it to prove Theorem D. Note also that, unlike Theorems B and C, Bos' theorem needs the symmetry of the metric.

REMARKS ON DIFFERENTIABILITY. We formulate all our results for smooth (i.e., C^∞) manifolds and maps. The only place where a specific level of differentiability is needed is to insure that we can locally connect by unique geodesics which depend continuously on their endpoints. This can be done for C^3 Riemannian and C^4 Finsler metrics. But all results also hold for C^2 Riemannian and C^2 Finsler metrics (see the end of Section 6).

2. *Survey of the major ideas*

(2.1) *The Principle of Least Action (Euler-Maupertuis-Jacobi, 1747)*

One form of this principle states:

Consider a conservative system as described in the Introduction, and a fixed energy level E. *In the region of configuration space* M *where* $U < E$, *define a new metric (the Jacobi metric) by*

$$d\bar{s}^2 = (E - U(q))\, ds^2 ,$$

in which ds^2 *represents the original metric on* M. *Then the orbits of the system with total energy* E *meet this region precisely along all the geodesics in the Jacobi metric.*

The value of this principle is that it encourages a geometric point of view. For example, if $E > \max_M U$ then $M_E = M$ and periodic motions of energy E correspond to closed geodesics on M in the Jacobi metric. But if E is a regular value of U with $\min_M U < E < \max_M U$, then M_E is a manifold with boundary, and the Jacobi metric degenerates to zero on the boundary. In that case, periodic motions of energy E may also correspond to curves which connect boundary points of M_E, but otherwise run through the interior, where they are geodesics in the Jacobi metric. Motion along such *brake orbits* will then have a to-and-fro character, oscillating between momentary rest points at either end.

For further information, see the book [A] by Arnol'd.

(2.2) *The ideas of G. D. Birkhoff (1917)*

The basic ideas used in the search for closed geodesics and periodic motions were developed by Birkhoff. They appeared in the paper [B_1], which was awarded the first Bocher Prize in 1923. See also [B_2].

Looking for periodic motions of conservative systems, Birkhoff took the geometric point of view suggested by the Principle of Least Action. He saw, as did many others, that the existence of closed geodesics was often clear on intuitive grounds. For example, to search for a closed geodesic on a torus, one places an elastic string in the position of a non-nullhomotopic closed curve. Trying to shorten itself, the string finally comes to rest along a closed geodesic. This was first made precise by Hilbert [H] in 1900.

But it was Birkhoff in 1917 who first interpreted the elastic contractions of a closed curve as the iteration of a specific mathematical process. Applied on a closed Riemannian (or more generally, Finsler) manifold, this *curve shortening process* homotopes a closed curve to an inscribed geodesic polygon.

Iterating this procedure, one obtains a sequence of shorter and shorter closed curves, such that a subsequence converges either to a point or to a closed geodesic. If the original curve is not nullhomotopic, then a closed geodesic appears in the limit.

The curve shortening process is continuous with respect to variations of the beginning curves. Capitalizing on this feature of continuity, Birkhoff applied his process to whole families of closed curves at the same time. Working on a sphere with an arbitrary Riemannian metric, he selected a family of closed curves covering the sphere simply and then considered the longest curve in the family after each step of the shortening process. Some subsequence of these converged to a closed geodesic.

Describing the imagery behind his *minimax* method, Birkhoff wrote:

> "There is a minimum length of closed string, constrained to lie in a given closed surface of genus 0, which may be slipped over that surface; in some intermediate position the string will be taut and will then coincide with a closed geodesic."

(2.3) *The idea of Fet and Lyusternik (1951)*

Fet and Lyusternik [F-L] observed that Birkhoff's minimax method could be used to prove the existence of a closed geodesic on any closed Riemannian (or Finsler) manifold. Simply pick the first integer k for which the homotopy group $\pi_k(M)$ is nonzero. The existence of such an integer (at most equal to the dimension of M) is guaranteed by the Hurewicz theorem. Then an essential map $f: S^k \to M$ can be used to transfer to M a family of closed curves covering S^k. Applying the curve shortening process to this family leads one to the desired closed geodesic.

Fet and Lyusternik observed that their argument works equally well for Finsler manifolds, even without symmetry.

So as a corollary one gets Theorems A and A′ for large values of the energy (that is, for M compact and $E > \max_M U$).

(2.4) *The idea of Seifert (1947)*

Seifert [S] demonstrated the existence of periodic motions of a classical conservative system at energy level E whenever M_E is a cell, thus proving Theorem A in this case. He did this by adapting Birkhoff's shortening process to apply to curves which run between boundary points of M_E, but otherwise lie in the interior. Shortening the family of diameters on the n-cell, he deduced the existence of a brake orbit. The main difficulty he faced was that the Jacobi metric vanished on the boundary of M_E. To overcome this, he used a limiting procedure in which he carefully controlled the behavior of geodesics near the boundary of M_E.

(2.5) *The idea of Weinstein (1978)*

Weinstein [W_2] observed that a possible generalization of Seifert's theorem is to assume that in the Hamiltonian formulation, the smooth function $H: T^*M \to R$ is convex and even on each fibre. He proved that whenever M_E is an n-cell, there exists a brake orbit and therefore a periodic orbit on $H^{-1}(E)$, thus giving Theorem A′ in this case. The place of the Jacobi Riemannian metric on the interior of M_E is now taken by a Jacobi-Finsler metric, and the discussion of the behavior of geodesics near the boundary of M_E is replaced by a discussion of the behavior of solutions of Hamilton's equations there, using the convexity of H.

(2.6) We add to these ideas by observing that the relative Hurewicz theorem ought to be used to generalize the arguments of Seifert and Weinstein, just as the absolute Hurewicz theorem was used by Fet and Lyusternik to generalize Birkhoff's ideas. Some reorganization of their work is required in order to carry this out.

3. *Finsler metrics*

We give a quick summary of the basic facts about Finsler metrics.

Let M be a smooth manifold and $F: TM \to R$ a continuous function. If $q = q(t)$ represents a smooth curve on M, then we may integrate the function F along the curve:

$$I = \int_C F\left(q(t), \frac{dq}{dt}(t)\right) dt .$$

The simplest problem in the calculus of variations is to determine those paths between fixed endpoints which yield local extrema (more generally, stationary values) of the above integral.

DEFINITION. A function $F : TM \to R$ which is continuous everywhere and smooth off the zero section is called a *Finsler metric* if it satisfies the following conditions:

1) $F(q, \dot{q})$ is positively homogeneous of degree 1 in $\dot{q}$, that is,

$$F(q, \lambda\dot{q}) = \lambda F(q, \dot{q}) \qquad \text{for } \lambda > 0 .$$

Note that this implies $F = 0$ on the zero-section of TM. Condition 1 is necessary and sufficient for the above integral to be independent of the choice of parameter t along the curve.

2) $F > 0$ off the zero-section of TM.

This condition suggests thinking of the integral $I = \int_C F$ as giving the "length" of the curve C.

3) On the complement of the zero-section of TM, the quadratic form $D^2 F^2$, with matrix

$$\left(\frac{\partial^2 F^2}{\partial \dot{q}_i \, \partial \dot{q}_j}(q, \dot{q})\right)$$

in local coordinates, should be positive definite. This is equivalent to the Legendre condition of the calculus of variations.

A Finsler metric is said to be *symmetric* if in addition it satisfies:

4) $F(q, -\dot{q}) = F(q, \dot{q})$.

We use the following properties of Finsler metrics:

5) $F(q, \dot{q})$ is convex in $\dot{q}$, that is,

$$F\left(q, \frac{\dot{q}_1 + \dot{q}_2}{2}\right) \leq \frac{F(q, \dot{q}_1) + F(q, \dot{q}_2)}{2} .$$

This follows from Condition 3 above. But caution: the two are not equivalent. Example:

$$F(q, \dot{q}) = (\Sigma\, \dot{q}_i^{\,4})^{1/4} .$$

6) A symmetric Finsler metric $F : TM \to R$ is a norm on each fibre TM_q .

7) Orthogonality: write $v \perp w$ in TM_q if, taking the differential of F with respect to $\dot{q}$, we get

$$D_{\dot{q}} F(q, v)(w) = 0 .$$

This holds if and only if w is tangent to the hypersurface of vectors of fixed length $|v| = F(q, v)$ at the point (q, v) of TM_q. Caution: this orthogonality relation is in general not symmetric, that is, $v \perp w$ does not imply $w \perp v$. Also, $v \perp w$ implies $v \perp -w$, but does not necessarily imply $-v \perp w$. Caution: symmetry of the Finsler metric does *not* imply symmetry of orthogonality.

8) Conversely, given a hyperplane $W = \{w\}$ in TM_q, there exist precisely two rays:

$$V_1 = \{\lambda v_1 : \lambda \geq 0\} \quad \text{and} \quad V_2 = \{\lambda v_2 : \lambda \geq 0\}$$

in TM_q such that $v \in V_i$ implies $v \perp w$ for all $w \in W$.
This property is guaranteed by the strict convexity of the constant distance spheres (Condition 3).

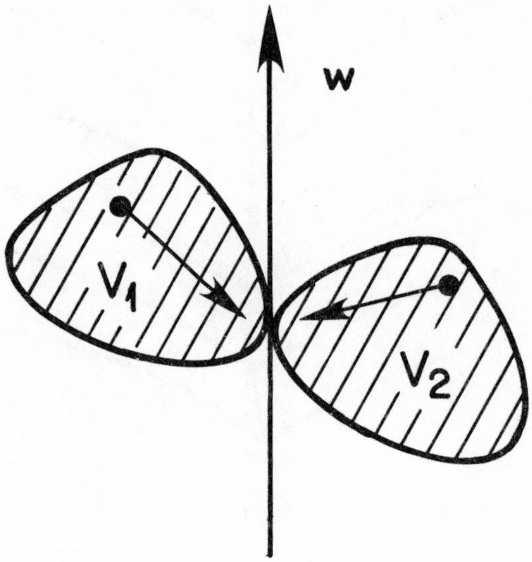

Figure 2

Note: the two rays V_1 and V_2 will unite to form a line if the Finsler metric is symmetric, but not in general. If we think of these rays as functions of q and W, then these functions are smooth.

9) By analogy with Riemannian geometry, define *geodesics* to be curves on M which locally minimize arc length (that is, the integral of F). Given a point $(q, \dot{q})$ of TM, there exists a geodesic of M passing through q with velocity $\dot{q}$. This geodesic is unique up to translation of its time parameter.

10) Given a compact set K in a Finsler manifold M, there exists a real number $d > 0$ such that any two points of K at distance $< d$ from each other can be joined by a unique geodesic arc of length $< d$, depending continuously on the two points [F].

11) Given a two-sided hypersurface S in a Finsler manifold M, pick a point $q \in S$ and consider the hyperplane $TS_q \subset TM_q$. Let $V_1(q)$ and

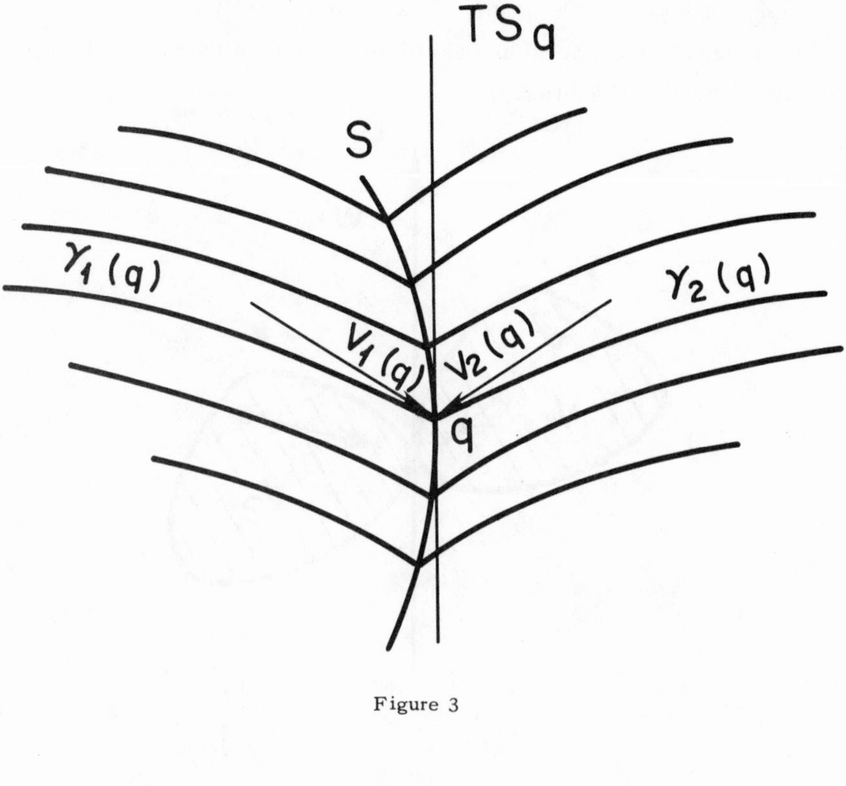

Figure 3

$V_2(q)$ be the two rays orthogonal to TS_q in TM_q, as promised by Property 8 above. Let $\gamma_1(q)$ and $\gamma_2(q)$ be geodesics terminating at q with nonzero tangent vectors in $V_1(q)$ and $V_2(q)$. If we let q vary, the geodesics $\gamma_1(q)$ fill out, nonsingularly for a short distance, a small one-sided neighborhood of S in M. Similarly for the $\gamma_2(q)$ on the other side of S. Furthermore (by First Variation of Arc Length, which is true for Finsler metrics just as for Riemannian ones), these geodesics are everywhere orthogonal to the hypersurfaces at constance distance from S.

12) It follows easily from 11 that if M is compact, then there exists a real number $d > 0$ such that any point of M within distance d of ∂M can be joined by a unique shortest geodesic of length $< d$ to the boundary. Furthermore, this shortest connection depends continuously on the point.

4. *Proof of the existence of one orthogonal geodesic chord*

(4.1) *The curve shortening process with free boundary*

Let N be a Riemannian (or more generally, Finsler) manifold-with-boundary. It is not necessary to assume the Finsler metric is symmetric. To do the curve shortening process on N, we must assume that N is *locally convex.*

DEFINITION. Enlarge N to a Riemannian (or Finsler) manifold N′. Let q be a point of ∂N. We say that N is *locally convex at* q if for sufficiently small geodesically convex neighborhoods U of q in N′, we have $N \cap U$ convex. It is easy to see that this definition is independent of the choice of N′. We say that N is *locally convex* if it is locally convex at each of its boundary points.

It is not hard to see that if N is a compact, locally convex manifold-with-boundary, then there exists a real number $d > 0$ such that any two points in the interior of N, at distance $< d$, can be joined by a unique shortest geodesic arc which also lies in the interior of N and which depends continuously on the two points. We can also choose d small enough so that any point in the interior of N, whose distance to ∂N is $< d$, has a unique shortest geodesic to ∂N which meets the boundary orthogonally.

Let c be a curve on N joining two points on ∂N. All curves will be assumed to be parametrized proportional to arc length. Let L denote the length of c and let r be an integer such that $L/r < d$.

Divide the curve c into r equal segments, each of length L/r, by the division points $q_0, q_1, \cdots, q_{r-1}, q_r$. Replace each interior segment $\widehat{q_i q_{i+1}}$ of c by the unique geodesic arc $\overline{q_i q_{i+1}}$ of length $< d$. Replace the initial segment $\widehat{q_0 q_1}$ by the geodesic arc $\overline{q_0' q_1}$ which joins q_1 orthogonally to ∂N at q_0'. Similarly replace the terminal segment $\widehat{q_{r-1} q_r}$ by $\overline{q_{r-1} q_r'}$. This replaces c by the r-sided geodesic polygon

$$c' = \overline{q_0' q_1} \cup \overline{q_1 q_2} \cup \cdots \cup \overline{q_{r-2} q_{r-1}} \cup \overline{q_{r-1} q_r'}.$$

which meets ∂N orthogonally. The length of c' will be strictly smaller than the length of c, except when $c' = c$.

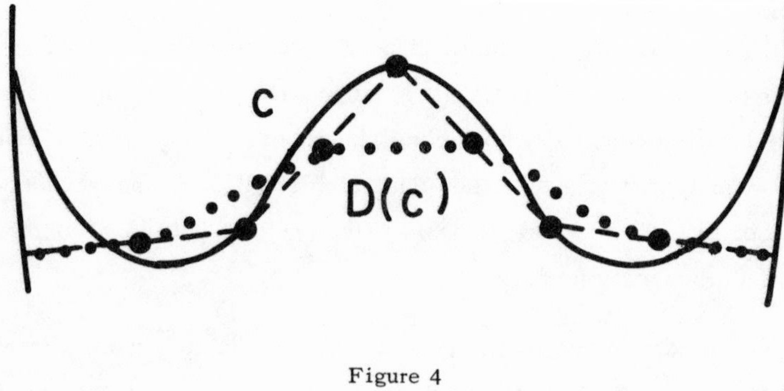

Figure 4

Now take the r midpoints of the segments of c'. Each pair of successive midpoints are at distance $< d$ apart, so may be connected by a unique shortest geodesic arc. Doing so produces a $r-1$ sided geodesic polygon. Augmenting this by half the first and last segments of c', we form the $r+1$ sided geodesic polygon c''.

Denote c'' by $D(c)$, so that the symbol D stands for the curve shortening process. Note that D can be defined simultaneously on all curves of length $\leq L$ which connect boundary points of N.

This curve shortening process with "sliding boundary points" enjoys the following properties:

1) The length of $D(c)$ is less than the length of c, unless c is already an orthogonal geodesic chord, in which case $D(c) = c$.
2) Once the integer r is fixed, $D(c)$ depends continuously on c in the C^1 topology.
3) The curve c is homotopic to $D(c)$, with endpoints constrained to slide along ∂N. Furthermore, this homotopy depends continuously on c.

Note that the same integer r used for c can also be used for $D(c)$. Thus the curve shortening process can be iterated to yield a sequence of curves:

$$c_0 = c, \; c_1 = D(c), \cdots, c_i = D^i(c), \cdots .$$

The length of c_{i+1} is strictly less than that of c_i, unless c_i is already an orthogonal geodesic chord on N, in which case $c_i = c_{i+1} = c_{i+2} = \cdots$. Each curve is homotopic to the next in a natural way.

The following lemma was first proved by Birkhoff [B_1]; we give his proof as modified by Alber [Al].

LEMMA. *The sequence* $c_0, c_1, c_2, \cdots$ *either converges to a point of* ∂N, *or else some subsequence converges to an orthogonal geodesic chord.*

Suppose first that the decreasing sequence of lengths $L_0, L_1, L_2, \cdots$ of these curves converges to 0. There exists an $\varepsilon < d$ such that the ε-neighborhood of each point of N is convex. Pick j so large that $L_j < \varepsilon$. Then each c_i, $i \geq j$, remains in the ε-neighborhood of an endpoint of c_j. It follows that the sequence of curves $c_0, c_1, c_2, \cdots$ converges to some point of ∂N.

Suppose next that the decreasing sequence of lengths $L_0, L_1, L_2, \cdots$ converges to $L^* > 0$. Choose a subsequence (c_{n_i}) of the broken geodesics $c_1, c_2, \cdots$ whose vertices converge. The subsequence then automatically converges to a broken geodesic c^* of length L^*, which meets ∂N orthogonally at its endpoints.

Note that

$$\begin{aligned} \text{length}(D(c^*)) &= \text{length}(D(\lim c_{n_i})) = \text{length}(\lim(Dc_{n_i})) \\ &= \text{length}(\lim c_{n_i+1}) = \lim \text{length}(c_{n_i+1}) \\ &= L^* = \text{length}(c^*) . \end{aligned}$$

But then c^* must be an orthogonal geodesic chord, according to Property 1.

(4.2) *Proof of Theorem C*

Let N^n be a compact, locally convex Riemannian or Finsler manifold-with-boundary. We want to produce an orthogonal geodesic chord. This is where the relative Hurewicz theorem comes in.

If $\pi_1(N, \partial N) \neq 0$, we pick a curve on N which connects two points of ∂N but which cannot be deformed into ∂N rel ∂N. Applying the curve shortening process of the previous section then yields an orthogonal geodesic chord, indeed, at least one in each nontrivial relative homotopy class.

If $\pi_1(N, \partial N) = 0$, pick the first integer k for which the relative homotopy group $\pi_k(N, \partial N) \neq 0$. The existence of such an integer $\leq n$ is guaranteed by the relative Hurewicz theorem. Then consider an essential map $f_0 : (B^k, S^{k-1}) \to (N, \partial N)$. Let C be a family of parallel straight line segments, filling B^k. Applying f_0, we get a family $C_0 = f_0(C)$ of curves on N, each one running between points of ∂N.

Let L be the maximum length of a curve in the family C_0, and let r be an integer such that $L/r < d$. Apply the curve shortening process of the previous section to all the curves in the family C_0 simultaneously. Doing so deforms the family C_0 into a family C_1, and by sending points on C_0 into corresponding points on C_1 we obtain a deformed map $f_1:(B^k, S^{k-1}) \to (N, \partial N)$ which is homotopic to f_0 and satisfies $C_1 = f_1(C)$. Iterating, we get the maps f_i and the families $C_i = f_i(C)$ of curves on N.

Suppose for each curve c in C, there is an integer $j(c)$ such that the curve $f_{j(c)}(c)$ has length less than d. Then all further iterates $f_i(c)$, $i \geq j(c)$, also have length less than d. By compactness of the family C, there will then be a single integer j for which $f_j(c)$ has length less than d for *all* c in C. But then the map f_j can clearly be homotoped to a map into ∂N, contrary to the essentiality of f_0.

It follows that for some curve c in C, each iterate $f_i(c)$ has length $\geq d > 0$. By the Lemma, some subsequence of these curves converges to an orthogonal geodesic chord, completing the proof of Theorem C.

5. *Proof of the existence of one geodesic chord*

Let N be a compact Riemannian or Finsler manifold-with-boundary. In this section we show that N has a geodesic chord. Of course, in most cases such chords obviously exist, e.g., if ∂N is convex at some point. But if ∂N is concave everywhere, then we rely on a trick due to Seifert [S]: add a *neck* to N to make it locally convex, then use Theorem C. We give a simplified version of Seifert's construction which only works in the Riemannian case; for the Finsler case see [Rz].

Begin by viewing N within a slightly larger Riemannian manifold. Use the field of geodesics orthogonal to ∂N to parametrize a neighborhood of ∂N by $\partial N \times [-\varepsilon, \varepsilon]$ so that $(q, 0)$ corresponds to q and so that $q \times [-\varepsilon, \varepsilon]$ is the geodesic segment orthogonal to ∂N at q, parametrized by arc length, with negative parameters indicating points within N.

The parallel surfaces $\partial N \times t$ are orthogonal to these geodesics, by First Variation of Arc Length. Hence the metric on this neighborhood of N may be written as $g_t + dt^2$, where g_t is a varying metric on N.

Now let $\phi(t)$ be a C^∞ function as shown:

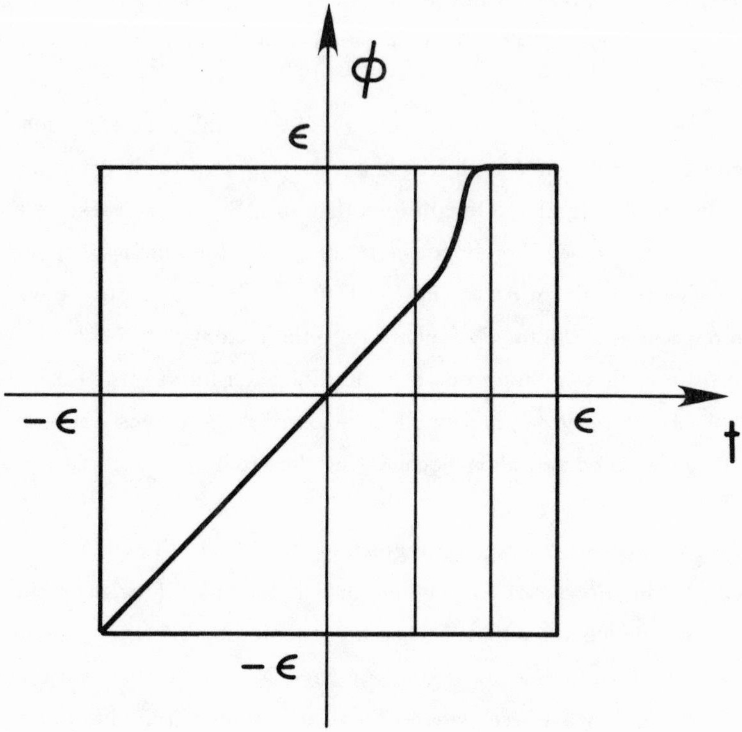

Figure 5

Then $g_{\phi(t)} + dt^2$ is an alternative metric on this neighborhood, agreeing with the original metric on N. Adding this neighborhood, with this alternative metric, to N produces the manifold-with-boundary N′. Since N′ is cylindrical near its boundary, it is certainly locally convex, and the whole process can be viewed as adding a neck to N.

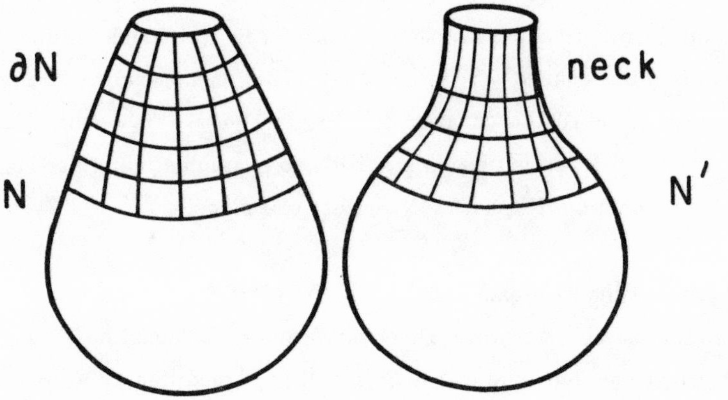

Figure 6

Notice that the segments $q \times [-\varepsilon, \varepsilon]$ are still geodesics in N'. Hence we can take an orthogonal geodesic chord for N', guaranteed by Theorem C, and follow it from $\partial N'$ down the neck until it first crosses ∂N into N. If we follow it further through N until it again meets ∂N, that further portion will be a geodesic chord on N, and we have Theorem B.

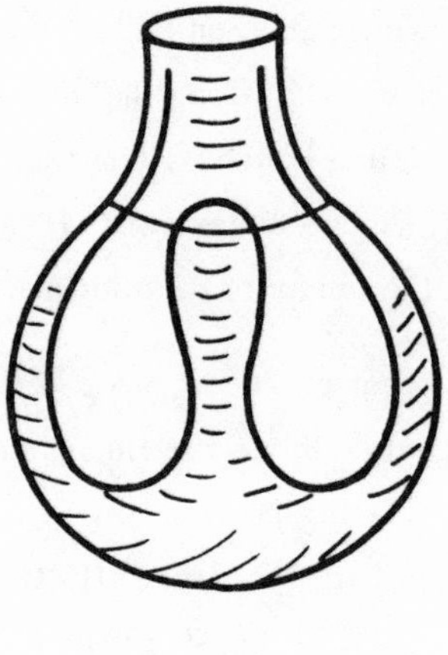

Figure 7

REMARK. Note that the orthogonal geodesic chord of N' may meet N in several pieces, as shown in the figure above, so that we do not in general get an orthogonal geodesic chord for N (although we do get one which is orthogonal to ∂N at one end). This phenomenon is the crux of the difficulty in finding n periodic orbits when M_E is a cell. The non-resonance condition is chosen precisely to guarantee that an orthogonal geodesic chord on N' meets N in just one piece.

6. *Proof of the existence of one periodic orbit*

In this section we prove Theorems A and A′ by coupling Theorem C with a limiting procedure due to Seifert [S] and modified by Weinstein [W_2].

(6.1) Note first that Theorem A′ implies Theorem A. For suppose we start with configuration space M^n, its Riemannian metric ds^2, kinetic energy $T = \frac{1}{2} ds^2$ and potential energy $U : M \to R$. To switch from Lagrangian to Hamiltonian mechanics, we take the Legendre transform of $L = \frac{1}{2} ds^2 - U$ to get the Hamiltonian $H = \frac{1}{2} ds^{*2} + U$, where ds^{*2} is the dual Riemannian metric defined on T^*M. Clearly H is convex and even on each fibre T^*M_q. Thus we are in the situation of Theorem A′, which then delivers the promised periodic orbit.

Notice also that the same construction will not work if the kinetic energy is given by a Finsler metric on M^n, since a Finsler metric which is not Riemannian cannot be C^2 along the zero section of TM. But then H cannot be smooth and convex along the zero section of T^*M. And this ruins the proof, since convexity of H at the zero section is essential in discussing the behavior of solutions near ∂M_E.

(6.2) We now begin the proof of Theorem A′. We first have to define the potential function $U : M \to R$. Since the Hamiltonian $H(q,p)$ is convex and even in p, it takes its minimum at $p = 0$ for each fixed $q \in M$. Thus define U by $U(q) = H(q,0)$. Since $H - U \geq 0$, any motion of total energy E must run within $M_E = \{q \in M : U(q) \leq E\}$.

Consider first the case $E > \max_M U$, so that $M_E = M$. Consider the unique function $K : T^*M \to R$ satisfying

i) $K^{-1}(1) = H^{-1}(E)$

ii) $K(q, p)$ is homogeneous of degree two in p.

The Hamiltonian system defined by K has the same orbits on $K^{-1}(1) = H^{-1}(E)$ as does H, up to reparametrization, since dH and dK are proportional there. Hence it is sufficient to seek a periodic orbit for K.

Now the Legendre transform $L : TM \to R$ of K will also be convex, even and homogeneous of degree two on each fibre. Hence $F = (2L)^{1/2}$ is a Finsler metric on M, called the *Jacobi-Finsler metric.* Now periodic orbits of the Hamiltonian system K correspond to periodic orbits of the Lagrangian system L, and hence by the Principle of Least Action correspond to closed geodesics in the Jacobi-Finsler metric. These in turn are guaranteed by Birkhoff-Fet-Lyusternik, so we have Theorem A′ in this case. Since the theorem of Birkhoff-Fet-Lyusternik does not need the symmetry of the Finsler metric, we do not need the evenness in Theorem A′ for this case.

(6.3) Now consider the case when E is a regular value of H, $\min_M U < E < \max_M U$. It is easy to see that E is also a regular value of U, hence M_E is a manifold-with-boundary.

In this case, the alternative Hamiltonian K can only be defined on the portion of T^*M over Int M_E. The corresponding Jacobi-Finsler metric F, given above, is also defined just on Int M_E. It vanishes as one approaches ∂M_E. If one starts in the circumstances of Theorem A with the Riemannian metric ds^2 on M, switches to Hamiltonian formulation via the Legendre transform, and then back again to Lagrangian formulation with the Jacobi-Finsler metric F, we have

$$F^2 = (E - U(q))\, ds^2 .$$

Our problem is now to find a brake orbit on M_E. Such a brake orbit can be thought of as a geodesic chord in the Jacobi metric.

If M_E is an n-cell, such a brake orbit was found by Seifert in the Riemannian case and by Weinstein in the Finsler case, thus giving Theorem A′ in this situation. We must remove the hypothesis that M_E is an n-cell.

(6.4) To begin, fix a regular value E of U, $\min U < E < \max U$, and assume that M_E is compact. For sufficiently small $\delta \geq 0$, $E-\delta$ will also be a regular value of U, and hence

$$M_{E-\delta} = \{q \in M : U(q) \leq E-\delta\}$$

will be a manifold-with-boundary, a slightly shrunken version of M_E. The Jacobi-Finsler metric F, though it vanishes on ∂M_E, is a genuine Finsler metric on $M_{E-\delta}$.

Applying Theorem B, we get, for each small $\delta > 0$, a geodesic chord c_δ on $M_{E-\delta}$. The plan is to choose a sequence of values for δ, converging to 0, and show that the corresponding chords c_δ converge to a brake orbit on M_E.

To carry this out, three difficulties must be avoided:

1) The curves c_δ might converge to a point curve on ∂M_E. It turns out this can't happen.
2) The curves c_δ might become arbitrarily long as $\delta \to 0$. This can happen, and will be avoided by imposing a certain uniformity on the choice of the c_δ.
3) Even if the curves c_δ have bounded length, they might converge to a curve which spirals towards ∂M_E. It turns out this also can't happen.

(6.5) LEMMA. *The geodesic chords* c_δ *on* $M_{E-\delta}$ *can be selected so that their lengths in the Jacobi metric* F *are bounded from above, independent of* δ.

Simply consider how the geodesic chord c_δ is obtained. For each small $\delta > 0$, we take the manifold $M_{E-\delta}$ and add a neck to it, producing the manifold $M'_{E-\delta}$ which is locally convex. We pick an integer $k \le n$ for which $\pi_k(M'_{E-\delta}, \partial M'_{E-\delta}) \neq 0$, clearly independent of δ, and select an essential map

$$f_0^\delta : (B^k, S^{k-1}) \to (M'_{E-\delta}, \partial M'_{E-\delta}) .$$

We take a family C of parallel straight line segments filling B^k and apply the map f_0^δ to get the family C_0^δ of curves on $M'_{E-\delta}$. Applying the curve shortening process, a subsequence of the iterates of one of the curves in such a family converges to an orthogonal geodesic chord on $M'_{E-\delta}$, and a portion of this yields the required geodesic chord c_δ on $M_{E-\delta}$.

All that is necessary is to choose the maps f_0^δ in a uniform way, *including* $\delta = 0$, and then appeal to compactness to find a uniform upper bound L for the lengths of the curves in the families C_0^δ. This same number L will then automatically bound the lengths of the geodesic chords c_δ.

The uniform choice of the maps f_0^δ is summarized in the following diagram:

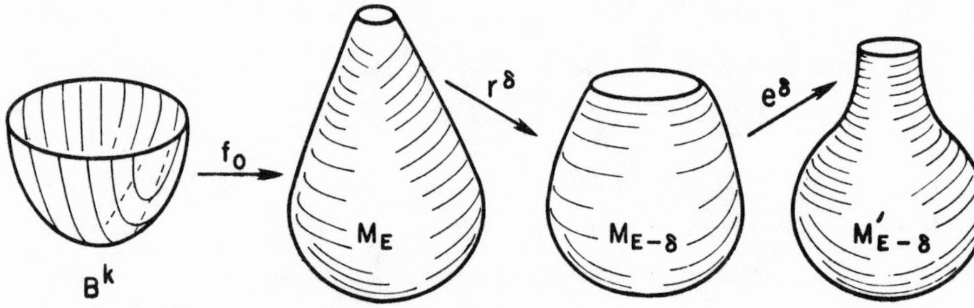

Figure 8

We select an essential map

$$f_0 : (B^k, S^{k-1}) \to (M_E, \partial M_E) .$$

For small $\delta \geq 0$, let $r^\delta : M_E \to M_{E-\delta}$ be a diffeomorphism which restricts to the identity on $M_{E-2\delta}$, varying continuously with δ. For each such δ, pick the real number ε_δ (the length of the neck) as in the proof of Theorem B, varying continuously with δ and such that $\varepsilon_0 = 0$. Let the function ϕ in that proof also vary continuously with δ. Let $e^\delta : M_{E-\delta} \to M'_{E-\delta}$ be a diffeomorphism varying continuously with δ, and such that e^0 is the identity map of M_E. Then let

$$f_0^\delta = e^\delta r^\delta f_0 : (B^k, S^{k-1}) \to (M'_{E-\delta}, \partial M'_{E-\delta}) .$$

If we do all this for $0 \leq \delta \leq \delta_0$, then by compactness of $B^k \times [0, \delta_0]$, there will exist a uniform upper bound L for the lengths of the curves in the families $f_0^\delta(C) = C_0^\delta$, and the lemma is proved.

The geodesic chord c_δ on $M_{E-\delta}$ in the Jacobi metric corresponds, by the Principle of Least Action, to a solution $q_\delta(t)$, $0 \leq t \leq T_\delta$, of the equations of motion. The endpoints $q_\delta(0)$ and $q_\delta(T_\delta)$ lie on $M_{E-\delta}$. As $\delta \to 0$, these endpoints get closer and closer to ∂M_E.

Lemma 6.5 overcomes the second difficulty listed. The first and third are handled by the following

(6.6) LEMMA (Weinstein). *There exist constants* $0 < \tau_1 < \tau_2$ *such that* $\tau_1 \leq T_\delta \leq \tau_2$ *for all sufficiently small* δ.

The proof given by Weinstein on pp. 516-517 of $[W_2]$ uses the result of Lemma 6.5 and works without change in the present circumstances, since no use is made there of the assumption that M_E is an n-cell. The proof involves a careful study of the behavior of solutions near ∂M_E, and uses the convexity of the Hamiltonian.

To get the required brake orbit now, we take a sequence $\delta_j \to 0$ such that $q_{\delta_j}(0)$ converges to a point $\overline{q}$ on ∂M_E and such that $T_{\delta_j} \to T_0$, with $\tau_1 \leq T_0 \leq \tau_2$. Let $q_0(t)$ be the orbit of energy E which begins at $\overline{q}$ at time $t = 0$ with zero velocity. Then

$$U(q_0(T_0)) = \lim U(q_{\delta_j}(T_{\delta_j})) = \lim E - \delta_j = E .$$

Hence this orbit returns at time T_0 to ∂M_E, where it again has zero velocity. Since H is even, the motion continues back and forth on this brake orbit, and we have the periodic motion promised by Theorem A′.

REMARK ON DIFFERENTIABILITY. To prove Theorem A and A′ for C^2 Riemannian metrics and Hamiltonians, we use an approximation technique. Let H be a C^2 Hamiltonian and choose a sequence H_j of C^∞ Hamiltonians that converge to H in the C^2 topology. By Theorem A′ H_j has a brake orbit c_j of total energy E and period T_j. Since in Lemma (6.5) we can clearly find a bound independent of j, the proof of (6.6) also implies that $0 < r_1 \leq T_j \leq r_2$ independent of j. Hence a subsequence of c_j converges to a periodic orbit of H.

7. *Nonsymmetric Finsler metric with no periodic brake orbit*

In the nonsymmetric case a periodic orbit could either lie completely in the interior of M_E or be a closed brake orbit that bounces off ∂M_E several times. The following example shows that the search for periodic orbits cannot be limited to such closed brake orbits. It is in the same spirit as the example of Katok [K] of a Finsler metric on S^2 with only two closed geodesics. See also Ziller [Z].

Consider the Hamiltonian

$$H_0 = \frac{1}{2}(p_1{}^2 + p_2{}^2) + \frac{1}{2}(q_1{}^2 + q_2{}^2) \qquad \text{on } R^4 .$$

It represents a pair of uncoupled harmonic oscillators. The corresponding Hamiltonian vector field is

$$\xi_{H_0} = p_1 \frac{\partial}{\partial q_1} + p_2 \frac{\partial}{\partial q_2} - q_1 \frac{\partial}{\partial p_1} - q_2 \frac{\partial}{\partial p_2} .$$

The regular constant energy surfaces are spheres; on each of these the Hamiltonian flow is the Hopf flow of period 2π. The orbits project to ellipses in configuration space. The extreme cases are circles (equal sharing of energy by the two oscillators) and brake orbits (oscillations in phase).

Consider next the Hamiltonian

$$K = -q_2 p_1 + q_1 p_2 \qquad \text{on } R^4 .$$

The function K represents the angular momentum about the origin of the point $(q_1(t), q_2(t))$. The corresponding Hamiltonian vector field is

$$\xi_K = -q_2 \frac{\partial}{\partial q_1} + q_1 \frac{\partial}{\partial q_2} - p_2 \frac{\partial}{\partial p_1} + p_1 \frac{\partial}{\partial p_2} .$$

The orbits are all closed, of period 2π. The corresponding Hamiltonian flow leaves configuration space invariant, and is simply a rotation about the origin there at unit angular velocity.

Finally, consider the Hamiltonian

$$H_\alpha = H_0 + \alpha K ,$$

which we think of as a slight perturbation of H_0.

CLAIM. 1) If $0 \leq \alpha < 1$, the equal energy surfaces of H_α are spheres.

2) If furthermore α is irrational, then on each equal energy sphere there are precisely two periodic orbits, neither a brake orbit.

Proof of 1). We can rewrite

$$H_\alpha = \frac{1-\alpha}{2}(p_1{}^2 + p_2{}^2 + q_1{}^2 + q_2{}^2) + \frac{\alpha}{2}(q_1 + p_2)^2 + \frac{\alpha}{2}(q_2 - p_1)^2 .$$

Thus $H_\alpha \geq 0$, and equals 0 only at the origin. Furthermore

$$dH_\alpha = (q_1 + \alpha p_2)dq_1 + (q_2 - \alpha p_1)dq_2 + (p_1 - \alpha q_2)dp_1 + (p_2 + \alpha q_1)dp_2 ,$$

which vanishes only at the origin. Hence each $E > 0$ is a regular value of H_α and so $H_\alpha^{-1}(E)$ is diffeomorphic to a three-sphere.

Proof of 2). Noting that $\xi_{\alpha K} = \alpha \xi_K$ and using the above formulas, we compute $[\xi_{H_0}, \xi_{\alpha K}]$ and find that it vanishes. Hence the flow of $\xi_{H_\alpha} = \xi_{H_0} + \xi_{\alpha_K}$ is just the flow of ξ_{H_0} composed with the flow of $\xi_{\alpha K}$. But this means that the Hamiltonian system H_α may be interpreted as viewing the Hamilton system H_0 from a coordinate system in the (q_1,q_2)-plane which rotates with constant angular velocity α.

In this rotating coordinate system, only the circular orbits of H_0 will appear to be closed, since α is irrational. We get two at each energy level, one going in each direction but at different speeds over the same circular path in configuration space. In particular, there are no brake orbits.

8. *Proof of Bos' theorem*

Let N^n be a locally convex Riemannian or Finsler manifold-with-boundary. Let $B(N)$ be the family of all piecewise C^1 curves $c:[0,1] \to N$ with $c(0)$ and $c(1)$ lying on ∂N. Put the C^1 topology on $B(N)$, and let $B_0(N)$ denote the subset of point curves on ∂N.

For each homology class Γ in $H_*(B(N), B_0(N))$, we define its *minimax*

$$\ell_\Gamma = \inf_{C \epsilon \Gamma} \sup_{c \epsilon C} \ell(c) ,$$

where $\ell(c)$ denotes the length of the curve c.

Using the curve shortening process with free boundary described in (4.1), it is not hard to see that there exists an orthogonal geodesic chord of length ℓ_Γ. Lyusternik-Schnirelmann theory tells us that if two homology classes Γ_1 and Γ_2 are subordinated (that is, there exists a cohomology class ξ in $H^*(B(N))$ with $\xi \cap \Gamma_1 = \Gamma_2$), then $\ell_{\Gamma_1} \geq \ell_{\Gamma_2}$, with equality implying that there are infinitely many orthogonal geodesic chords of that length.

If the Finsler metric is symmetric, then every chord $c(t)$ defines another one by the formula $\bar{c}(t) = c(1-t)$, and we consider these two chords as geometrically the same. So we work instead in the quotient space $(B/Z_2, B_0/Z_2)$, where we have divided out by the Z_2-action $c \to \bar{c}$.

To prove Bos' theorem, we assume in addition that N^n is homeomorphic to an n-cell, and seek n subordinated homology classes of dimensions ≥ 1 in $H_*(B/Z_2, B_0/Z_2)$.

To obtain these, first parametrize N^n by an n-cell D^n and note that $B(N^n)$ is homotopy equivalent to the set $L(D^n)$ of oriented straight line segments on D^n which connect points on ∂D^n. This is simply because any curve c in B can be deformed into the straight line connecting the same endpoints. Then $(B/Z_2, B_0/Z_2)$ is homotopy equivalent to the corresponding pair $(L/Z_2, L_0/Z_2)$.

Now it is easy to see that $L/Z_2 - L_0/Z_2$ is homeomorphic to the open unit disk bundle of the canonical $n-1$ dimensional vector bundle over RP^{n-1}. Hence the pair $(L/Z_2, L_0/Z_2)$ is homeomorphic to the closed disk bundle modulo its unit sphere bundle.

The Thom isomorphism with Z_2 coefficients then gives n homology classes for $(L/Z_2, L_0/Z_2)$ in dimensions $n-1, n, n+1, \cdots, 2n-2$, which are subordinated with respect to the unique one-dimensional cohomology class in $L/Z_2 - L_0/Z_2$. The same is therefore true of $(B/Z_2, B_0/Z_2)$.

Applying Lyusternik-Schnirelmann theory as described above, we then get at least n distinct orthogonal geodesic chords, completing the proof of Bos' theorem.

REMARKS. 1) The theorem is optimal, since the flat metric on an ellipsoid has precisely n orthogonal geodesic chords if the axes all have different lengths.

2) If N^n is a locally convex Riemannian (or symmetric Finsler) manifold-with-boundary, not necessarily homeomorphic to an n-cell, one can easily decide how many orthogonal geodesic chords the above method gives.

For example, if N^n is homeomorphic to $S^{n-1} \times [0,1]$, then the corresponding $(B/Z_2, B_0/Z_2)$ has two components. One of these, containing the curves which begin and end at the same boundary component of N, is homotopically trivial. The other, containing the curves connecting the two different boundary components, has the homotopy type of S^{n-1}. We get two subordinated homology classes, and hence at least two orthogonal geodesic chords (shortest and longest) connecting these boundary components to one another.

3) If N^n is homeomorphic to an n-cell but the Finsler metric is not symmetric, we can only guarantee two orthogonal geodesic chords (not n), since (B, B_0) only has homology in dimensions $n-1$ and $2n-2$.

9. *Proof of the existence of* n *periodic orbits*

We are given a classical conservative system, either as in Theorem A or as in Theorem A′, with configuration space M^n and potential function $U: M \to R$. We are also given a regular value E of U such that M_E is topologically an n-cell. Finally, we have the

NONRESONANCE CONDITION. Take any parametrization of M_E as an n-cell, and consider the "straight line segments" connecting boundary points of M_E. We ask that the maximum length of all such line segments should be strictly smaller than twice the length of the shortest brake orbit, all measurements to be made in the Jacobi metric.

We want to find n distinct periodic motions of energy E, each a brake orbit.

Consider the motion of energy E which begins at time $t = 0$ at some point of ∂M_E with zero velocity. As time increases, the given point moves into the interior of M_E. This orbit is, up to reparametrization, a geodesic in the Jacobi metric on M_E. Let s denote the arc length parameter in the Jacobi metric along this geodesic, beginning with $s = 0$ when $t = 0$. Seifert showed in [S] that we can introduce a regular coordinate system on a neighborhood of ∂M_E in M_E in which all these curves appear as coordinate curves, and on which $s^{2/3}$ is a regular parameter. See also Ruiz [Rz].

The sets of constant geodesic distance δ from ∂M_E are coordinate hypersurfaces for sufficiently small $\delta \geq 0$. Let $M_{E-\delta}$ denote the closed region bounded by such a hypersurface. It is also an n-cell, approximating M_E from within. Note that our current definition of $M_{E-\delta}$ differs from the one used in Section 6. Note also that the family of special geodesics starting from ∂M_E is orthogonal to each hypersurface $\partial M_{E-\delta}$, as follows immediately from the Gauss Lemma and the fact that the Jacobi metric vanishes on ∂M_E. (For a Gauss Lemma for Finsler metrics, see Warner [Wa].)

Fix a small value of $\delta > 0$. We next add a neck to the cell $M_{E-\delta}$ to produce the locally convex cell $M'_{E-\delta}$, exactly as in Section 5. Applying Bos' theorem to $M'_{E-\delta}$, we get n orthogonal geodesic chords. We will use the nonresonance condition to establish the following

CLAIM. *For sufficiently small $\delta > 0$, each orthogonal geodesic chord on $M'_{E-\delta}$ meets $M_{E-\delta}$ in a single piece, which is an orthogonal geodesic chord there.*

Suppose the condition of the claim is violated for a sequence $\delta_k \to 0$, and corresponding orthogonal geodesic chords c_k on $M'_{E-\delta_k}$. Let d_k and e_k denote the two extreme pieces of the intersection of c_k with $M_{E-\delta_k}$. Note that each is orthogonal to $\partial M_{E-\delta_k}$ at one of its endpoints.

Now replace the δ_k by a subsequence so that d_k and e_k converge to brake orbits as $k \to \infty$. The corresponding arguments from Section 6 suffice for this. Hence

$$\limsup \text{length}(c_k) \geq 2(\text{length of shortest brake orbit}) .$$

But c_k is obtained by applying the curve shortening process to some "straight line segment" across $M'_{E-\delta_k}$. Taking limits of subsequences

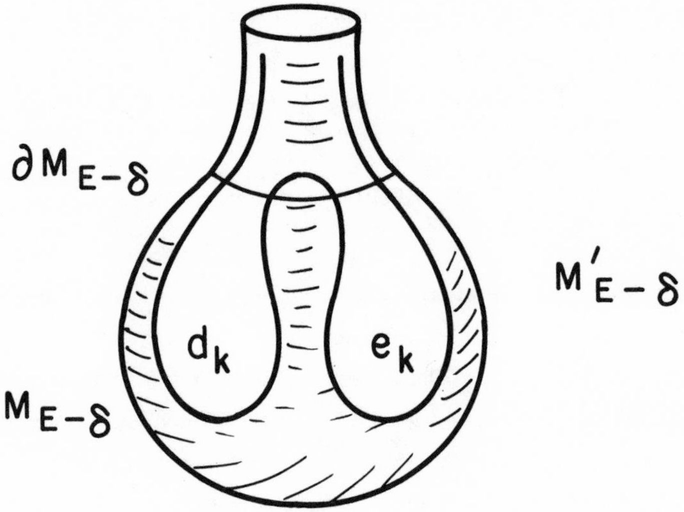

Figure 9

again, some "straight line segment" across M_E has length at least twice that of the shortest brake orbit, contradicting the nonresonance condition. The claim follows.

Now fix $\delta > 0$ so small that the claim is satisfied. Taking the n distinct orthogonal geodesic chords on $M'_{E-\delta}$ promised by Bos' theorem and intersecting them with $M_{E-\delta}$, we get n orthogonal geodesic chords there. Each of these extends immediately to a brake orbit on M_E by using the special geodesics which run between $\partial M_{E-\delta}$ and ∂M_E. Thus we get n brake orbits on M_E, and the proof of Theorem D is complete.

REMARKS. 1) Suppose we start with a specific Hamiltonian, such as

$$H(q, p) = \frac{1}{2} \Sigma (p_i^2 + q_i^2)$$

on R^{2n}, which describes the motion of n uncoupled harmonic oscillators. For any value of $E > 0$, M_E is an n-cell and H satisfies the nonresonance condition. Clearly any Hamiltonian C^2 close to H also satisfies the nonresonance condition. It should be possible to compute explicitly how much we can perturb H in the C^2 topology without destroying the condition.

2) Suppose instead that we start with the Hamiltonian

$$H(q,p) = \frac{1}{2} \Sigma \frac{p_i^2 + q_i^2}{\alpha_i^2} .$$

Again, for any value of $E > 0$, M_E is an n-cell. When the ith oscillator has all the energy, we get a brake orbit whose length in the Jacobi metric is proportional to α_i. The largest α_i gives the maximum length of the "straight line segments" referred to in the nonresonance condition. The smallest α_i is similarly proportional to the length of the shortest brake orbit. Hence the nonresonance condition in this case reads:

$$\max \alpha_i < 2 \min \alpha_i .$$

If this is satisfied, one should again be able to compute explicitly how much of a perturbation of H can still be guaranteed to satisfy the nonresonance condition.

3) The nonresonance condition we use is similar to one used by Ekeland and Lasery [E-L]. Given a convex Hamiltonian on R^{2n}, they proved the existence of n periodic motions of energy E provided $H^{-1}(E)$ lies between a Euclidean sphere of radius r and one of radius R, such that $R < 2r$. One should be able to recover their theorem from Theorem D by using the doubling trick in Weinstein's paper [W_2].

4) It is natural to try to avoid the nonresonance condition by seeking a curve shortening process which takes place directly on M_E in the Jacobi metric. But nearby ∂M_E one does not have unique shortest connections for neighboring points. Perhaps some modification of the traditional curve shortening process might work there.

H. GLUCK AND W. ZILLER
DEPARTMENT OF MATHEMATICS
UNIVERSITY OF PENNSYLVANIA
PHILADELPHIA, PENNSYLVANIA 19104

REFERENCES

[Al] S.I. Alber, *On periodicity problems in the calculus of variations in the large*, Uspehi Mat. Nauk 12 No. 4 (76) (1957), 57-124 (Russian); Amer. Math. Soc. Transl. (2) 14 (1960).

[A] V.I. Arnol'd, MATHEMATICAL METHODS OF CLASSICAL MECHANICS, Springer-Verlag (1978).

[B-T-Z] W. Ballmann, G. Thorbergsson, W. Ziller, *On the existence of short closed geodesics and their stability properties*, this volume p. 53.

[Ba] V. Bangert, *Closed geodesics on complete surfaces*, Math. Ann. 251 (1980), 83-96.

[B_1] G.D. Birkhoff, *Dynamical systems with two degrees of freedom*, Trans. Amer. Math. Soc. 18 (1917), 199-300.

[B_2] ________, DYNAMICAL SYSTEMS, Colloq. Publ. 9, Amer. Math. Soc. (1927).

[Bo] W. Bos, *Kritische Sehnen auf Riemannschen Elementarraumstücken*, Math. Ann. 151 (1963), 431-451.

[E-L] I. Ekeland, J.M. Lasery, *Nombre de solutions périodiques des equations de hamilton*, C.R. Acad. Sci. Paris 288 (1979).

[F] A.I. Fet, *Variational problems on closed manifolds*, Amer. Math. Soc. Transl. 90 (1953).

[F-L] A.I. Fet, L.A. Lyusternik, *Variational problems on closed manifolds*, Dokl. Akad. Nauk. SSSR 81 (1951), 17-18.

[G] K. Grove, *Condition (C) for the energy integral on certain path spaces and applications to the theory of geodesics*, J. Diff. Geom. 8 (1973), 207-223.

[K] A.B. Katok, *Ergodic properties of degenerate integrable Hamiltonian systems*, Izv. Akad. Nauk. SSSR 37 No. 3 (1973), (Russian); Math. USSR-Izv. 7 (1973), 535-571.

[L-S] L. Lyusternik, L. Schnirelmann, MÉTHODES TOPOLOGIQUES DANS LES PROBLÈMES VARIATIONNELS, Moscow (1930), (Russian); Hermann, Paris (1934).

[M] F. Mercuri, *The critical point theory for the closed geodesic problem*, Math. Z. 156 (1977), 231-245.

[Ra] P.H. Rabinowitz, *Periodic solutions of a hamiltonian system on a prescribed energy surface*, preprint, Univ. of Wisconsin (1978).

[Rz] O.R. Ruiz, *Existence of brake orbits in Finsler mechanical systems*, Springer Lecture Notes in Math. 597 (1977), 542-567.

[Rn] H. Rund, THE DIFFERENTIAL GEOMETRY OF FINSLER SPACES, Grundl. Math. Wiss. 101, Springer, Berlin (1959).

[S] H. Seifert, *Periodische Bewegungen mechanischer Systeme*, Math. Z. 51 (1948), 197-216.

[T] G. Thorbergsson, *Closed geodesics on noncompact Riemannian manifolds*, Math. Z. 159 (1978), 249-258.

[Wa] F. Warner, *The conjugate locus of a Riemannian manifold*, Amer. J. Math. 87 (1965), 575-604.

[W_1] A. Weinstein, *Normal modes for nonlinear hamiltonian systems*, Inv. Math. 20 (1973), 47-57.

[W_2] ———, *Periodic orbits for convex hamiltonian systems*, Ann. of Math. 108 (1978), 507-518.

[Z] W. Ziller, *Geometry of the Katok example*, to appear in Journal of Ergodic Theory and Dynamical Systems.

ARE HARMONICALLY IMMERSED SURFACES AT ALL LIKE MINIMALLY IMMERSED SURFACES?

Tilla Klotz Milnor

1. *Introduction*

For a long time I have wondered just how much of the standard textbook material on minimally immersed surfaces carries over in an interesting way to the larger class of harmonically immersed surfaces. Of course, minimal surfaces are enormously special. Not all of their properties will have reasonable counterparts. Still, things are closer to being the same than I once expected. Indeed, the classification theorem in §6 indicates that some essential facts about minimal surfaces may be best understood as special cases of observations about harmonically mapped surfaces.

2. *Basic notions*

We work with maps $X : R \to M^n$ of a Riemann surface R in a Riemannian manifold M^n of dimension $n \geq 3$. (Though less is needed, C^∞ smoothness is assumed throughout.) The map X is called *conformal* in case the angle measurement associated with the induced metric I coincides with angle measurement on R wherever I is non-degenerate. Associated with any immersion X is the second fundamental form B (as defined in [8]) and the mean curvature vector field

$$\mathcal{H} = \mathrm{tr}_I B \ .$$

Seminar on Minimal Submanifolds
0-691-08324-X/83/099-12 $1.10/0 (cloth)
0-691-08319-3/83/099-12 $1.10/0 (paperback)

One calls a map X *minimal* in case it is conformal and extremal with respect to the ordinary area integral

$$A = \int dA_I .$$

A conformal immersion X is minimal if and only if $\mathcal{H} \equiv 0$.

When a Riemannian metric g is prescribed on R, we associate with a map $X : (R, g) \to M^n$ the energy function

$$e = e(X) = \frac{1}{2} \operatorname{tr}_g I$$

and the g-mean curvature vector field

$$\mathcal{H}^g = \operatorname{tr}_g B .$$

In particular, $\mathcal{H}^I \equiv \mathcal{H}$ wherever I is nondegenerate. Since X need not be conformal, g need not be proportional to I. We assume throughout, however, that g is a *conformal metric* on R. This means that $g = \lambda dz d\bar{z}$ for some function λ, where $z = x + iy$ here and below denotes a conformal parameter on R.

DEFINITION 1. Associated with any immersion $X : (R, g) \to M^n$ with energy function e is the *energy 1 metric* $\Gamma = eg$, so called because among all conformal metrics on R, Γ is the only one for which $X : (R, \Gamma) \to M^n$ has energy function $\equiv 1$. One can compute Γ without reference to g as follows. If $I = E dx^2 + 2F dx dy + G dy^2$ for $z = x + iy$ on R, then

(1) $$2\Gamma = (E+G) dz d\bar{z} = (E+G)(dx^2 + dy^2) .$$

Thus, it makes sense to speak of the energy 1 metric Γ of any map $X : R \to M^n$. However, Γ may fail to be Riemannian at points where I is degenerate. Note that $\Gamma = I$ if and only if X is conformal.

In terms of the energy 1 metric Γ, the usual definition of a harmonic map $X: R \to M^n$ (see [6]) takes on a form closer to the standard definition of a minimal immersion.

DEFINITION 2. A map $X: R \to M^n$ is *harmonic* if and only if it is extremal for the Γ-area integral

$$A_\Gamma = \int dA_\Gamma .$$

A map is minimal if and only if it is harmonic and conformal, that is, harmonic with $\Gamma \equiv I$.

The following characterization of harmonic immersions $X: R \to M^n$ gives still another picture of the special place of minimal immersions among harmonic immersions. (See [11] and [12].) The notation $\mathrm{Cod}(\Gamma, I)$ means that I satisfies the classical Codazzi-Mainardi equations with respect to Γ as metric, or, in other words, that I is a Codazzi tensor for the Riemannian metric Γ.

Fact 1. An immersion $X: R \to M^n$ is harmonic if and only if $\mathrm{Cod}(I, \Gamma)$ and $\mathcal{H}^\Gamma \equiv 0$.

For those familiar with harmonic mappings, we note that $\mathrm{Cod}(I, \Gamma)$ forces the tangent component of the tension field of X to vanish, while $\mathcal{H}^\Gamma \equiv 0$ forces the normal component to vanish. When X is conformal, $\Gamma \equiv I$, so that $\mathrm{Cod}(I, I)$ is automatic, and the condition $\mathcal{H}^\Gamma \equiv 0$ becomes the condition $\mathcal{H}^I \equiv \mathcal{H} \equiv 0$.

There is a quadratic differential $\Omega = h dz^2$ on R associated with any map $X: R \to M^n$. It is defined by taking

$$h = E + G - 2iF \tag{2}$$

if $I = Edx^2 + 2Fdxdy + Gdy^2$ for any $z = x + iy$ on R. Note that $\Omega \equiv 0$ when X is conformal. One calls Ω holomorphic in case h is complex analytic in z for any z on R. A harmonic map $X: R \to M^n$ always has

Ω holomorphic (see [5] or [11]), but this necessary condition is not sufficient, as one checks by comparing Facts 1 and 2. (Also, see [12].)

Fact 2. If Γ is the energy 1 metric for an immersion $X: R \to M^n$, then

(i) Γ is complete if I is,

(ii) Ω is holomorphic if and only if $\mathrm{Cod}(\Gamma, I)$, and

(iii) if $\mathrm{Cod}(\Gamma, I)$, the intrinsic curvatures $K(\Gamma)$ and $K(I)$ are related by $K(\Gamma) \leq \mu K(I)$ where $0 \leq \mu \equiv \det I/\det \Gamma \leq 1$.

The properties of the energy 1 metric make it especially useful in the study of harmonic immersions. Indeed, the results discussed below are often obtained by substituting Γ for I at strategic points in some statement or argument about minimal immersions. Another device used is the substitution of Ω or its coefficient function h in place of 0. Finally, we suggest in §4 that a mapping Φ be used in place of the Gauss spherical image mapping when looking at harmonic immersions $X: R \to E^n$. (When X is minimal, Φ coincides with the Gauss mapping.)

3. *A Weierstrass representation*

In this section we describe a representation (see [10]) for harmonic maps $X: R \to E^3$ which specializes to the classical Weierstrass representation (see [15], for example) in case X is minimal. The initial discussion applies to any map $X: R \to E^n$ where $n \geq 3$. We take R to be a region in the x,y-plane, with $z = x + iy$ a particular conformal parameter on R. It is well known that $X: R \to E^n$ is harmonic if and only if

$$\Delta X \equiv \partial^2 X/\partial z\, \partial \bar{z} \equiv 0,$$

where $2\partial/\partial z = \partial/\partial x - i\partial/\partial y$ and $2\partial/\partial \bar{z} = \partial/\partial x + i\partial/\partial y$. Thus X is harmonic if and only if

$$\phi \equiv 2\partial X/\partial z \equiv (\phi_k)$$

is holomorphic. It follows that a harmonic immersion X has the expression

(3) $$X = \mathrm{Re} \int_{P_0}^{P} \phi dz + \text{constant}$$

with

$$I = |\mathrm{Re}\phi|^2 dx^2 - 2 < \mathrm{Re}\phi, \mathrm{Im}\phi > dxdy + |\mathrm{Im}\phi|^2 dy^2$$

(4) $$4\Omega = hdz^2 , \qquad h = \sum_1^n \phi_k{}^2 = \phi^2$$

$$2\Gamma = |\phi|^2 dz d\bar{z} , \qquad |\phi|^2 = \sum_1^n |\phi_k|^2 .$$

The immersion X is minimal if and only if $h = \phi^2 \equiv 0$.

REMARK 1. Formula (3) provides a way to generate all harmonic maps of any R in E^n. Specifically, given n holomorphic differentials $\phi_k dz$ on R with no real periods, the map X given by (3) for $\phi = (\phi_k)$ is harmonic with I, Γ and Ω given by (4). The map X is an immersion if and only if $\mathrm{Re}\phi$ and $\mathrm{Im}\phi$ are linearly independent. The map X is minimal if and only if $h = \phi^2 \equiv 0$.

REMARK 2. If R is simply-connected and $X : R \to E^n$ is harmonic, the *associate harmonic maps* $X_\theta : R \to E^n$ are defined by

$$X_\theta = \mathrm{Re} \int_{P_0}^{P} e^{i\theta} \phi dz + \text{constant}$$

for a global conformal parameter z on R. When X is minimal, the X_θ all share the same induced metric I. If X is only harmonic, the X_θ are not isometric, but they do share the same energy 1 metric Γ as $X = X_0$.

Because $h = \phi_1{}^2 + \phi_2{}^2 + \phi_3{}^2 \equiv 0$ for a minimal $X : R \to E^3$, it is no surprise that two functions f and g suffice to describe such an X. The pleasant fact is that one can use the same functions f and g together with the coefficient $h = \phi_1^2 + \phi_2^2 + \phi_3^2$ of Ω to describe any harmonic $X : R \to E^3$. Put another way, one can start with the pair f, g which represent a minimal immersion, and by suitable variation of the complex analytic function h, obtain a family of harmonic maps with $\Omega = hdz^2$ stacked vertically over the original minimal surface.

LEMMA. *Suppose* f, h, fg *and* $\sqrt{f^2g+h}$ *are analytic and* g *is meromorphic on a region* R *in the plane. Then the functions*

$$(5) \qquad \phi_1 = \frac{f}{2}(1-g), \qquad \phi_2 = \frac{if}{2}(1+g), \qquad \phi_3 = \sqrt{f^2g+h}$$

are analytic, and satisfy

$$(6) \qquad h = \sum_1^3 \phi_k^2 .$$

Moreover, every triple of analytic functions ϕ_k *on* R *can be represented in the form (5) with* h *given by (6) unless* $\phi_1 \equiv i\phi_2$.

THEOREM 1. *Suppose* $X : R \to E^3$ *is given by (3) for a holomorphic vector field* $\phi = (\phi_1, \phi_2, \phi_3)$. *If* $\phi_1 \equiv i\phi_2 \equiv 0$, x_1 *and* x_2 *are constant, so that* X(R) *lies along a vertical line. If* $\phi_1 \equiv i\phi_2 \not\equiv 0$, *then* $x_1 - ix_2$ *is analytic, so that* X(R) *can be locally represented as the graph of a harmonic function* $x_2 = x_3(x_1, x_2)$. *If* $\phi_1 \not\equiv i\phi_2$, *the representation of the lemma applies. The map* X *is then an immersion if and only if* g *has a pole of order* m *at every zero of* f *of order* m, *while* $|g| \neq 1$ *wherever* $\bar{g}(g+h/f^2) \leq 0$. *In particular, any harmonic immersion* $X : R \to E^3$ *has (at least locally) a representation in terms of the functions* f, g *and* h *of the lemma, where* $\Omega = hdz^2$ *is given by (2).*

REMARK 3. Theorem 1 and the lemma give the classical Wierstrass representation of a minimal $X : R \to E^3$ in case $h \equiv 0$. However, we use the symbol g to name the square of the function historically given that name, so that *one must replace* g *by* g^2 *when* $h \equiv 0$ *to get the standard formulas.*

REMARK 4. Suppose the harmonic map $X : R \to E^3$ is represented by f, g and h, with $\sqrt{g}$ meromorphic. Then one can associate with X the minimal surface $\tilde{X} : R \to E^3$ represented by f, g and $h \equiv 0$. In this situation, $\sqrt{g}$ is equal to $\tilde{X}$ followed by its Gauss spherical image map followed by stereographic projection to the extended plane. We are led to the following application of Bernstein's Lemma.

COROLLARY TO THEOREM 1. *If* $X : R \to E^3$ *is harmonic with one to one orthogonal projection onto some plane, then unless* $X(R)$ *is the graph of a harmonic function over the plane, the function* g *in the representation of* X *is either constant or else has a zero or pole of odd order.*

It remains to be seen whether the behavior of the functions f, g and h will be of general value in studying the properties of the harmonic maps they represent.

4. *Using* Φ *in place of the Gauss map*

In attempting to study the distribution of normals to a harmonically immersed surface in E^n, I obtained the following quite disappointing result. (See [1] for a definition of quasiconformality.)

THEOREM 2. (See [10].) *If* Ω *has no isolated zeros for a complete harmonic immersion* $X : R \to E^n$, *then* $X(R)$ *must be a 2-plane if the normals omit a neighborhood of some (unoriented) direction, and if* $id : R \to R_1$ *is quasiconformal, where* R_1 *is the conformal structure determined by* I. *In particular, the normals to a complete harmonically immersed surface in* E^n *cannot omit a neighborhood of an (unoriented) direction if* $\mathcal{H} \neq 0$ *and* $id : R \to R_1$ *is quasiconformal.*

However, the proof of Theorem 2 relies on the next more interesting result. If R is any Riemann surface, and z is an arbitrary conformal parameter on R, we can associate to any immersion $X : R \to E^n$, the vector field $\phi = 2\partial X/\partial z$. Since ϕdz is invariant, it determines a mapping

$$\Phi : R \to CP^{n-1}$$

which coincides with the Gauss map if and only if X is minimal (see [9] or [15]) and which is holomorphic if and only if X is harmonic (see §3 above). The following statement looks far more like the sort of result which is known for minimal immersions in E^n than Theorem 2 does.

THEOREM 3. (See Lemma 3 in [10].) *If* $X : R \to E^n$ *is a complete, harmonic immersion then either* $X(R)$ *is a 2-plane or else* $\Phi(R)$ *comes arbitrarily close to every hyperplane* $\Sigma a_k z_k = 0$ *in* CP^{n-1}.

Theorem 3 suggests that Φ rather than the Gauss map is the appropriate map to study in connection with harmonic immersions in E^n. It may be, for example, that stability results for harmonic immersions in E^n can be tied to the behavior of the map Φ. Moreover, the map Φ may have interesting properties for certain non-harmonic immersions $X : R \to E^n$.

5. *Harmonic immersions into spheres*

In this section we discuss two theorems which characterize harmonic immersions of R into a sphere through the properties of an immersion into the containing Euclidean space. The symbol $S^n(r)$ will denote the sphere of radius r centered at the origin in E^{n+1}. We also write S^n for $S^n(1)$.

To understand the first result, note that harmonic maps (see [6]) are generally studied from one Riemannian manifold to another. Only when the first manifold has dimension two is it appropriate to replace that manifold by the Riemann surface determined by the specified Riemannian metric.

Suppose therefore that S is the underlying C^∞ 2-manifold for R. Given an immersion $X : R \to S^n$ we will consider the associated cone immersion $Y : S \times R^+ \to E^{n+1}$ defined by $Y(p, r) = rX(p)$ where p lies on

S and $r > 0$. It is known (see [7] and [9]) that X is minimal if and only if Y is minimal. To generalize that fact, we have the following, where Γ, as always, denotes the energy 1 metric for X. Note that the metric $r^2\Gamma + dr^2$ used on $S \times R^+$ becomes the induced metric for Y when X is minimal, because $\Gamma = I$ then.

THEOREM 4. (See [13].) *The immersion* $X : R \to S^n \subset E^{n+1}$ *is harmonic if and only if the cone immersion* $Y : S \times R^+ \to E^{n+1}$ *is harmonic when* $S \times R^+$ *is provided with the metric* $r^2\Gamma + dr^2$.

REMARK 5. Theorem 4 fails if one uses in place of Γ a different conformal metric g on R.

The next result was proved by Takahashi and Simon (see [16] and [18]) in the minimal case, and includes known facts (see [6] and [17]) about harmonic immersions $X : R \to S^n(r)$. By Δ_Γ we denote the Laplace Beltrami operator associated with the energy 1 metric Γ. The notation $\mathrm{Cod}(\Gamma, I)$ was explained in §2.

THEOREM 5. (See [13].) *The immersion* $X : R \to E^{n+1}$ *satisfies* $\mathrm{Cod}(\Gamma, I)$ *and* $\Delta_\Gamma X = \lambda X$ *for a function* $\lambda \neq 0$ *if and only if* $X : R \to S^n(r) \subset E^{n+1}$ *is harmonic with* $r^2 = -2/\lambda > 0$, *so that* λ *is a negative constant.*

6. *Classifying harmonic immersions*

In this section we describe results suggested by Calabi's work in [2], [3] and [4]. The proofs in [14] follow closely the exposition of Calabi's work in Lawson's book [9].

If $X : R \to E^n$ is harmonic and R is simply-connected, it makes sense to speak of the conjugate harmonic map $\hat{X} : R \to E^n$. (In Remark 1 of §3, just go from ϕ to $-i\phi$.) Thus one can associate with X the holomorphic curve $X^* : R \to C^n = E^n \times iE^n$ (with branch points wherever $\phi = 0$) defined by

$$\sqrt{2}\, X^* = (X, \hat{X}) . \tag{7}$$

Because X^* is holomorphic, it is minimal, and its induced metric I^* is its energy 1 metric. Moreover, since $I^* = |dX^*/dz|^2 dz d\bar{z}$, (4) and (7) show that I^* coincides with the energy 1 metric of X. Using a result due to Calabi (see [2]), one has the following.

THEOREM 6. *Among all noncongruent nonconstant harmonic maps of* R *into Euclidean space of any dimension which share the same energy* 1 *metric, there exists at most one holomorphic curve, and exactly one if* R *is simply connected.*

In the next result we start with any fixed holomorphic curve $\mathfrak{X}: R \to C^m$ and describe the space of noncongruent maps $X: R \to E^n$ for any n whose energy 1 metric coincides with the induced metric of $\mathfrak{X}$. Theorem 7 was proved for minimal $X: R \to E^n$ isometric to $\mathfrak{X}$ by Calabi in [4].

THEOREM 7. (See [14].) *Suppose that* $\mathfrak{X}: R \to C^m$ *is a holomorphic curve (with or without branch points) which fits into no lower dimensional affine subspace of* C^m. *If* $\mathfrak{X}$ *has induced metric* ds^2, *then the space* $\mathcal{P}(\mathfrak{X})$ *of all noncongruent harmonic maps* $X: R \to E^n$ *for any* n *which have energy* 1 *metric* $\Gamma = ds^2$ *and which fit into no lower dimensional subspace of* E^n *is naturally described as the set of all complex, symmetric* $m \times m$ *matrices* P *with* $1_m - P\bar{P} \geq 0$. *Moreover*

$$n - m = \operatorname{rank}(1_m - P\bar{P})$$

so that $m \leq n \leq 2m$. *Finally, an* X *in* $\mathcal{P}(\mathfrak{X})$ *is minimal if and only if the corresponding matrix* P *satisfies*

$$(d\mathfrak{X}/dz)^t P (d\mathfrak{X}/dz) \equiv 0 .$$

REMARK 6. Let $\mathfrak{D}$ be the Siegel domain of complex symmetric matrices P with $1_m - P\bar{P} > 0$. Theorem 7 shows that $\mathcal{P}(\mathfrak{X})$ can be naturally imbedded as the closure of $\mathfrak{D}$, with the subspace $\mathcal{M}(\mathfrak{X})$ of minimal immersions being taken to a linear variety through $P = 0$, which is the image of $\mathfrak{X}$ itself.

REMARK 7. In [3] Calabi gives necessary and sufficient conditions that a real analytic conformal metric g on a simply-connected Riemann surface R be achieved as the induced metric I for a minimal immersion of R into some Euclidean space. The considerations above show that the same necessary and sufficient conditions apply if g is achieved as the energy 1 metric Γ for an harmonic immersion of R into some Euclidean space.

TILLA KLOTZ MILNOR
DEPARTMENT OF MATHEMATICS
RUTGERS UNIVERSITY
NEW BRUNSWICK, NEW JERSEY 08903

REFERENCES

[1] L. Bers, "Quasiconformal mappings and Teichmüller's theorem," *Conference on Analytic Functions*, Princeton U. Press, Princeton (1960), 89-119.

[2] E. Calabi, "Isometric imbeddings of complex manifolds," *Annals of Math.* 58(1953), 1-23.

[3] ———, "Metric Riemann surfaces," *Contributions to the Theory of Riemann Surfaces*, Princeton U. Press, Princeton (1953), 77-85.

[4] ———, "Quelques applications l'analyse complexe aux surfaces d'aire minima," with *Topics in Complex Manifolds* (Hugo Rossi), Les Presses de l'Universite de Montreal (1968).

[5] S. S. Chern and S. I. Goldberg, "On the volume-decreasing property of a class of real harmonic maps," *Amer. J. Math.* 97(1975), 133-147.

[6] J. Eells and L. Lemaire, "A report on harmonic maps," *Bull. London Math. Soc.* 10(1978), 1-68.

[7] D. Fischer-Colbrie, "Some rigidity theorems for minimal submanifolds of the sphere," preprint.

[8] S. Kobayashi and K. Nomizu, *Foundations of Differential Geometry* II, Interscience Publishers, New York (1969).

[9] H. B. Lawson, Jr., *Lectures on Minimal Submanifolds*, Notas de Math IMPA, Rio de Janeiro (1974).

[10] T. K. Milnor, "Mapping surfaces harmonically into E^n," *AMS Proc.* 78(1980), 269-275.

[11] ———, "Harmonically immersed surfaces," *J. Diff. Geom.* 14(1979), 205-214.

[12] ———, "Abstract Weingarten Surfaces," *J. Diff. Geom.* 15(1980), 365-380.

[13] ———, "The energy 1 metric on harmonically immersed surfaces," *Michigan Math. U.* 28(1981), 341-346.

[14] ———, "Classifying harmonic maps of a surface in E^n," American J. Math. 103 (supplement), 211-218.

[15] R. Osserman, *A Survey of Minimal Surfaces*, Van Nostrand Math Studies 25, Princeton (1969).

[16] U. Simon, "Probleme der lokalen und globalen mehrdimensionalen Differentialgeometrie," *Manu. Math* 2(1970), 241-284.

[17] R. T. Smith, "Harmonic mapping of spheres," *Amer. J. Math.* 97 (1975), 229-236.

[18] T. Takahashi, "Minimal immersions of Riemannian manifolds," J. Math. Soc. Japan 18(1966), 380-385.

ESTIMATES FOR STABLE MINIMAL SURFACES IN THREE DIMENSIONAL MANIFOLDS

Richard Schoen

In this paper we study immersed surfaces M in a three dimensional manifold N which are minimal and stable, i.e. have nonnegative second variation of area for deformations of compact support. For a conformal, stable immersion $f: D_r \to N$ on the disc of radius r in the complex plane, we give in Theorem 1 a lower estimate of $|df|^2(0)$ in terms of the ratio of the geodesic distance (in the induced metric) from 0 to ∂D_r to the Euclidean radius r. An estimate of this form was first obtained by R. Osserman [11] for surfaces in $\mathbf{R}^3$ whose normals omit a fixed neighborhood on the sphere. A weaker estimate had been previously given for graphs by E. Heinz [5].

In Theorem 3 we derive a local bound on the second fundamental form of M at a point P_0 in terms of the local geometry of N near P_0 and the geodesic distance from P_0 to the boundary of M. This has the consequence that if M has no boundary in a 3-ball of N, then the intersection of M with a smaller 3-ball (of estimable radius) is a union of smooth embedded discs. The curvature estimate was done previously in special cases by Heinz [5], Osserman [11], and by the author, Simon, Yau [9].

We observe (Corollaries 2, 3) that our local estimates imply the global theorems obtained by D. Fischer-Colbrie and the author [4], and by doCarmo-Peng [3].

Seminar on Minimal Submanifolds
0-691-08324-X/83/111-16 $1.30/0 (cloth)
0-691-08319-3/83/111-16 $1.30/0 (paperback)

1. *Notation and basic formulae*

Let N be a three-dimensional oriented Riemannian manifold, and M an oriented minimal surface immersed in N. Let e_3 be the unit normal vector of M, and let A denote the second fundamental form of M, i.e. given a local basis e_1, e_2 for M, A is represented by the matrix $(h_{ij})_{1\leq i,j\leq 2}$ where

$$h_{ij} = \langle D_{e_i} e_3, e_j \rangle ,$$

D denoting covariant differentiation in N. Since M is minimal we have $h_{11} + h_{22} = 0$, and we will suppose also that M has nonnegative second variation of area for compactly supported deformations. We refer to such a surface M as a *stable* surface. The second variation inequality is (see e.g. [2]).

$$\int_M (\mathrm{Ric}(e_3) + |A|^2)\zeta^2 dv \leq \int_M |\nabla\zeta|^2 dv \tag{1}$$

for any Lipschitz function ζ with compact support on M. Here $\mathrm{Ric}(e_3)$ denotes the Ricci curvature of N in the e_3 direction, $|A|^2 = \sum_{i,j=1}^{2} h_{ij}^2$ is the square length of A, dv is the volume form for the induced metric on M, and ∇ denotes the induced connection on M. Equation (1) expresses the condition that the second variation of area for the normal variation given by ζe_3 be nonnegative.

Since the estimates we will prove come from the choice of test functions ζ in (1), we first make a few observations relevant to such choices. Let L be the operator associated to (1), i.e.

$$L\zeta = \Delta\zeta + (|A|^2 + \mathrm{Ric}(e_3))\zeta \tag{2}$$

where Δ is the Laplace operator for the induced metric on M. We note that the Gauss equation may be used to express L as (see [10])

$$L\zeta = \Delta\zeta + \left(\frac{1}{2}|A|^2 - K + S\right)\zeta \tag{3}$$

where K denotes the intrinsic (Gauss) curvature of M and $S = \frac{1}{2}\sum_{i=1}^{3} \mathrm{Ric}(e_i)$ is the scalar curvature of N. We now let ϕ be a smooth function on M, not necessarily of compact support, and we replace ζ in (1) by the function $\phi\zeta$. We note that

$$\int_M |\nabla\phi\zeta|^2\,dv = \int_M \phi^2|\nabla\zeta|^2\,dv + \frac{1}{2}\int_M (\nabla\phi^2, \nabla\zeta^2)\,dv + \int_M \zeta^2|\nabla\phi|^2\,dv$$

$$= \int_M \phi^2|\nabla\zeta|^2\,dv - \int_M \zeta^2\,\phi\Delta\phi\,dv$$

by expanding and integrating by parts. Thus (1) implies

$$\int_M \phi\cdot L\phi\,\zeta^2\,dv \le \int_M \phi^2|\nabla\zeta|^2\,dv \tag{4}$$

where ϕ, ζ are any test functions and ζ has compact support on M. Our philosophy will be to choose functions ϕ in (4) such that $\phi\cdot L\phi$ contains a positive term which we want to bound.

One choice of ϕ we will make involves suitable powers of $|A|$, so we recall the equation which $|A|$ satisfies for a minimal immersion. If N is flat, then it is well known that if ds^2 is the induced metric on M, then the metric $|A|ds^2$ is flat where it is nonvanishing. This says that $|A|$ satisfies the equation

$$\Delta \log|A| = -|A|^2 .$$

Of course, this equation is not true if N has nonzero curvature, but it is not difficult to check from well-known equations (see e.g. [9]) that

$$\Delta \log(1+|A|^2) \geq -2|A|^2 - c_1 \tag{5}$$

where c_1 is a constant depending on the curvature of N and its first covariant derivatives (but not depending on M).

2. *Main results*

Our first result concerns immersions of a disk in N. Let D_r be the open disc of radius r centered at 0 in the complex plane. Let $f: D_r \to N$ be a conformal, stable immersion. Thus the induced metric is given by

$$ds^2 = \lambda(z)\,|dz|^2\,, \qquad \lambda(z) = \frac{1}{2}\,|df|^2$$

where $z \in D_r$ is a complex number, and

$$|df|^2 = |f_x|^2 + |f_y|^2$$

is the Dirichlet integrand of f taken with respect to the metric on N. The intrinsic curvature of D_r is given by

$$K = -\frac{1}{2}\,\Delta \log \lambda$$

where $\Delta = \lambda^{-1}\left(\frac{\partial^2}{\partial x^2} + \frac{\partial^2}{\partial y^2}\right)$. We use $\rho(z)$ to denote the geodesic distance to 0, and $B_R = B_R(0)$ the geodesic ball of radius $R > 0$,

$$B_R = \{z \in D_r : \rho(z) < R\}\,.$$

For $P \in N$, let $B^3_R(P)$ be the open geodesic ball in N of radius R centered at P. We define a couple of constants $S_{P,R}$, $K_{P,R}$ determined by the curvature of N as follows

$$S_{P,R} = \max\{0, \; -\inf_{B^3_R(P)} S\}$$

$$K_{P,R} = \sup_{B^3_R(P)} \{|\text{curv.}| + |D\,\text{curv.}|\}$$

where we use "curv." as an abbreviation for the full curvature tensor of N. Our first theorem gives a lower bound on $|df|^2$ in terms of the ratio of geodesic distance to Euclidean distance in D_r.

THEOREM 1. *Let* $f: D_r \to N$ *be a stable, conformal immersion, and suppose* $B_R = B_R(0)$ *has compact closure in* D_r. *Let* $R_0 = \min\{1, R/2\}$, *and let* $a \in (0,1)$ *be a chosen number. Then we have the inequality*

$$\inf_{B_{R_0/2}} |df|^2 \geq c_2 r^{-2} R_0^{2(1-a)} R^{2a} [1 + (S_{P_0,R}) R^2]^{-a}$$

where $P_0 = f(0)$, *and* c_2 *is a positive constant depending only on* a *and* K_{P_0,R_0}.

If N is the flat $\mathbf{R}^3$, then we have the stronger result which follows immediately from Theorem 1 by scaling.

COROLLARY 1. *Let* $f: D_r \to \mathbf{R}^3$ *be a stable, conformal immersion, and suppose* B_R *has compact closure in* D_r. *Then there is an absolute constant* $c_3 > 0$ *so that*

$$|df|^2(0) \geq c_3 R^2 r^{-2}.$$

If the scalar curvature of N is nonnegative, then Theorem 1 has the following global consequence which has been proven in [4].

COROLLARY 2. *Suppose* N *has nonnegative scalar curvature, and* M *is a complete noncompact stable surface immersed in* N. *Then the universal cover of* M *is conformally equivalent to the complex plane.*

The proof of Corollary 2 follows from the method of [4] which shows that the immersion of the universal cover $\tilde{M}$ is again stable, and from uniformization which says that $\tilde{M}$ is conformally the plane or the disc. Letting $R \to \infty$ in the estimate of Theorem 1 shows that the disc is not possible.

Proof of Theorem 1. It clearly suffices, by reparametrization, to assume $r = 1$. We let $D = D_1$ be the unit disc. Let $h = \lambda^{-1/2}$, and observe that by the definition of K we have $\Delta \log h = K$, i.e.

$$\Delta h = Kh + h^{-1} |\nabla h|^2 . \tag{6}$$

We take $\phi = h$ in (4) and use (3) to get

$$\int_D (Sh^2 + \frac{1}{2} |A|^2 h^2 + |\nabla h|^2)\zeta^2 dv \leq \int_D |\nabla \zeta|^2 dx\,dy \tag{7}$$

where we have used $h^2 dv = dx\,dy$. This clearly implies for any ζ with support in B_R

$$\int_D |dh|^2 \zeta^2 dx\,dy \leq \int_D |\nabla \zeta|^2 dx\,dy + S_{P_0,R} \int_D \zeta^2 dx\,dy$$

where $|dh|^2 = (h_x)^2 + (h_y)^2$ is the standard gradient on D. This implies

$$\int_D |d(\zeta h)|^2 dx\,dy \leq 3 \int_D |\nabla \zeta|^2 dx\,dy + 2S_{P_0,R} \int_D \zeta^2 dx\,dy . \tag{8}$$

We now recall standard Sobolev inequalities on D (see e.g. [8, Theorem 3.5.3]). For any $q \in [1,2)$, ϕ with compact support in D we have

$$\left(\int_D |\phi|^{\frac{2q}{2-q}} dx\,dy\right)^{\frac{2-q}{2}} \leq c_4 \int_D |d\phi|^q dx\,dy$$

where c_4 depends on q. Setting $p = q/(2-q)$, and applying Hölder's inequality we have for any $p \in [1,\infty)$

$$(9) \qquad \left(\int_D \phi^{2p}\,dx\,dy\right)^{1/p} \le c_5 \int_D |d\phi|^2\,dx\,dy$$

for any ϕ with compact support in D, where c_5 depends on p. Applying (8), (9) with $\phi = \zeta h$, and ζ chosen to be a cutoff function of ρ which is one in $B_{R/2}$ and vanishes outside B_R, we get

$$(10) \qquad \left(\int_{B_{R/2}} h^{2p}\,dx\,dy\right)^{1/p} \le c_6 R^{-2}(1 + R^2 S_{P_0,R})$$

for any $p \in [1,\infty)$, c_6 depending only on p.

We will use (10) to give the pointwise estimate on h. In order to do this, we need to use the more delicate geometry of N inside the ball of radius R_0. For simplicity, we assume $R_0 = 1$, i.e. we assume $R \ge 2$. Easy modifications of our arguments handle the general case. We first prove integral estimates on powers of $|K|$. Note that by choosing a simple cutoff function ζ of ρ with support in B_1 in (7) we have

$$(11) \qquad \int_{B_{7/8}} |A|^2\,dx\,dy \le c_7$$

where c_7 depends on $K_{P_0,1}$. To estimate higher powers of $|A|$, we choose $\phi = h(1+|A|^2)^{1/8}$ in (4) and observe that by (5) we have

$$\Delta \log h(1+|A|^2)^{1/8} \ge K - \frac{1}{4}|A|^2 - c_8\,,$$

so by (3) we have

$$[h(1+|A|^2)^{1/8}][L(h(1+|A|^2)^{1/8})] \ge 1/4|A|^2(1+|A|^2)^{1/4}h^2 - c_8 h^2(1+|A|^2)^{1/4}$$

in B_1, where c_8 depends on $K_{P_0,1}$. Thus we have

$$\int_{B_1} |A|^{5/2}\zeta^2\,dx\,dy \le 4 \int_{B_1} |\nabla\zeta|^2(1+|A|^2)^{1/4}\,dx\,dy + 4c_8 \int \zeta^2(1+|A|^2)^{1/4}\,dx\,dy$$

for any ζ with support in B_1. Choosing a standard function ζ of ρ, we get

$$\int_{B_{3/4}} |A|^{5/2}\,dx\,dy \le c_9 \int_{B_{7/8}} (1+|A|^2)^{1/4}\,dx\,dy\ .$$

By the Hölder inequality and (11) this gives

$$\int_{B_{3/4}} |A|^{5/2}\,dx\,dy \le c_{10}\ .$$

Applying the Gauss equation then gives

$$\int_{B_{3/4}} |K|^{5/4}\,dx\,dy \le c_{11}\ . \tag{12}$$

We now adapt a standard iteration method to bound h. Observe that by (6)

$$\Delta h \ge Kh\ .$$

Multiplying by $\zeta^2 h^{2q-1}$ for $q > 1$ and ζ a compactly supported function, and integrating by parts and rearranging terms in a standard way we get

$$\int_{B_1} |\nabla\zeta h^q|^2\,dv \le c_{12}q \int_{B_1} h^{2q}|\nabla\zeta|^2\,dv + c_{12}q \int_{B_1} |K|\,h^{2q}\zeta^2\,dv\ .$$

This is the same as

$$\int_{B_1} |d(\zeta h^q)|^2 \, dx\, dy \le c_{12} q \int_{B_1} h^{2q-2} |\nabla\zeta|^2 \, dx\, dy + c_{12} q \int_{B_1} |K|\, h^{2q-2} \zeta^2 \, dx\, dy \, .$$

We now use (9) with $p = 10$, and choose ζ to be a function of ρ which is one in B_a and zero outside $B_{a+\tau}$ where $a \in [1/2, 3/4)$, $\tau \in (0, 3/4-a)$

$$\left(\int_{B_a} h^{20q} \, dx\, dy \right)^{1/10} \le c_{13} q\, \tau^{-2} \int_{B_{a+\tau}} h^{2q-2} \, dx\, dy + c_{13} q \int_{B_{a+\tau}} |K| h^{2q-2} \, dx\, dy \, .$$

Applying Holder on the right together with (12) then gives

$$\text{(13)} \qquad \left(\int_{B_a} h^{20q} \, dx\, dy \right)^{1/10} \le c_{14} q\, \tau^{-2} \left(\int_{B_{a+\tau}} h^{10q} \, dx\, dy \right)^{\frac{q-1}{5q}} .$$

We then choose sequences $q_i = 5^{-1} p\, 2^i$, $\tau_i = 2^{-i-3}$ for $i = 0, 1, 2, \cdots$ where $p > 20$ is a chosen number. We also choose $a_0 = 3/4$ and $a_i = a_{i-1} - \tau_{i-1}$ for $i = 1, 2, \cdots$. Applying (13) with $a = a_{i+1}$, $\tau = \tau_i$, $q = q_i$ then gives

$$I_{i+1} \le (c_{14} q_i \tau_i^{-2})^{1/q_i} (I_i)^{1-1/q_i}$$

where I_i is defined by

$$I_i = \left(\int_{B_{a_i}} h^{2p2^i} \, dx\, dy \right)^{1/p_2 i} .$$

Iteration then gives

$$\text{(14)} \qquad \sup_{B_{1/2}} h^2 \le c_{15} I_0^{\alpha_p} , \quad \alpha_p = \prod_{i=0}^{\infty} (1 - 1/q_i) \, .$$

It is easy to estimate

$$|1-\alpha_p| \le c_{16}\, p^{-1} \,,$$

so by (10) and (14) we see that given $\alpha \in (0,1)$, by choosing p large enough we have

$$\sup_{B_{1/2}} h^2 \le c_{17}\, [R^{-2}(1+S_{P_0,R}\, R^2)]^{\alpha}$$

which is the conclusion of Theorem 1.

We now prove a result for stable surfaces of large geodesic radius in manifolds of nonnegative Ricci curvature. In fact, we are able to estimate the maximum possible radius in terms of a (local) positive lower bound of the Ricci curvature.

THEOREM 2. *Suppose* N *has nonnegative Ricci curvature, and* M *is an immersed stable surface. Suppose* $R \ge 2$, *and* $P_0 \in M$ *is a point such that* $B_R(P_0)$ *has compact closure in* M. *There is a constant* $\varepsilon_0 \in (0,1/2]$ *depending only on* $K_{P_0,1}$ *so that*

$$\int_{B_{\varepsilon_0}(P_0)} |A|^2\, dv \le 8\pi (\log R)^{-1} \,.$$

Moreover, if in addition the Ricci curvature of N *is bounded below by a positive constant* K_0 *on* $B^3_\sigma(P_0)$ *for some* $\sigma \in (0,1/2]$, *then we have*

$$R \le \exp(c_{18}/K_0\sigma^2)$$

where c_{18} *depends only on* $K_{P_0,1}$ *and the injectivity radius of* N *at* P_0.

This theorem immediately implies the global result of [4].

COROLLARY 3. *A complete stable minimal surface in a three manifold of nonnegative Ricci curvature is totally geodesic. If the Ricci curvature is everywhere positive, then no such surface exists.*

Proof of Theorem 2. We first note that by the methods of [4], the resulting immersion of the simply-connected covering $\tilde{M}$ of M is stable, and satisfies the hypothesis of Theorem 2 for a point $\tilde{P}_0 \in \tilde{M}$ lying above P_0. Moreover, the conclusions of Theorem 2 for $\tilde{M}$ imply the same for M. Thus we may assume that M is represented by a conformal immersion $f: D \to N$ with $P_0 = f(0)$.

Given $\delta \in (0,1)$, we choose a function $\zeta = \zeta_\delta$ to be a radial function on D defined by

$$\zeta(z) = \begin{cases} 1 & \text{if } |z| < \delta \\ \dfrac{\log \delta^{1/2}/|z|}{\log \delta^{-1/2}} & \text{for } \delta \le |z| \le \delta^{1/2} . \\ 0 & \text{for } |z| > \delta^{1/2} \end{cases}$$

Direct calculation then gives

$$\int_D |d\zeta|^2 \, dx\, dy = 4\pi (\log 1/\delta)^{-1} .$$

Using this choice of ζ in (1) then shows

$$\int_{D_\delta} (\mathrm{Ric}(e_3) + |A|^2)\, dv \le 4\pi (\log 1/\delta)^{-1} \tag{15}$$

for any $\delta \in (0,1)$. We now note that by Theorem 1 with $\alpha = 1/2$, we have

$$\inf_{B_{1/2}} \lambda^{1/2} \ge \varepsilon_0 R^{1/2}$$

for a positive constant ε_0 depending only on $K_{P_0,1}$. This implies that for any curve $\gamma \subseteq B_{1/2}$ we have

$$\int_\gamma |dz| \le \varepsilon_0^{-1} R^{-1/2} \int_\gamma \lambda^{1/2} |dz|$$

which is an estimate of Euclidean length by geodesic length. Applying this to radial geodesics from 0 in B_{ε_0} implies $B_{\varepsilon_0} \subseteq D_{R^{-1/2}}$. Thus inequality (15) with $\delta = R^{-1/2}$ gives

$$\int_{B_{\varepsilon_0}} (\mathrm{Ric}(e_3) + |A|^2 dv \le 8\pi (\log R)^{-1} . \tag{16}$$

This implies the first assertion of Theorem 2. The second assertion follows also from (16) together with the following area estimate (see [7, Lemma 1]) whose proof is a modification of a standard density argument

$$\int_{B_\sigma} dv \ge c_{19}\, \sigma^2$$

for $\sigma \epsilon (0, \varepsilon_0]$ and a constant $c_{19} > 0$ depending on $K_{P_0,1}$ and the injectivity radius of N at P_0. This completes the proof of Theorem 2.

Finally we prove a result which describes the local behavior of stably immersed surfaces in three manifolds. Essentially we show that the intrinsic geometry of the surface is bounded by that of the ambient manifold even though the area of the surface inside a three-dimensional ball may be unboundedly large.

THEOREM 3. *Let* M *be an immersed stable surface in* N^3. *Given* $r_0 \epsilon (0,1]$, *and a point* $P_0 \epsilon M$ *such that* $B_{r_0}(P_0)$ *has compact closure in*

M, *then there is a constant* c_{20} *depending only on* K_{P_0,r_0} *so that*

$$|A|^2(P_0) \le c_{20} r_0^{-2} .$$

Moreover, if $M \cap B^3_{r_0}(P_0)$ *has compact closure in* M, *then there is a constant* $\varepsilon_0 > 0$ *depending only on* K_{P_0,r_0} *and the injectivity radius of* N *at* P_0 *so that* $M \cap B^3_{\varepsilon_0 r_0}(P_0)$ *is a union of embedded discs having second fundamental form of square length bounded by* $c_{21} r_0^{-2}$ *for a constant* c_{21} *depending only on* K_{P_0,r_0}.

When N is the flat $\mathbf{R}^3$, the following corollary comes from Theorem 3 together with scaling in $\mathbf{R}^3$.

COROLLARY 4. *Let* M *be a stable surface in* $\mathbf{R}^3$ *which compactly contains* $B_{r_0}(P_0)$ *for some* $P_0 \in M$, $r_0 > 0$. *There is an absolute constant* c_{22} *such that*

$$|A|^2(P_0) \le c_{22} r_0^{-2} .$$

Proof of Theorem 3. First note that by the Gauss equation we have on $B_{r_0}(P_0)$, $K \le c_{23}$ with c_{23} depending only on K_{P_0,r_0}, so by standard comparison theorems (see e.g. [1]) M has no conjugate points for a distance $\pi/\sqrt{c_{23}}$. We define a radius r_1 by

$$r_1 = \min \{r_0, \pi/4\sqrt{c_{23}}\} .$$

By composing our immersion with the exponential map at P_0 we get a stable immersion of the ball $B_{r_1}(0)$ in the tangent space to M at P_0. Let $z \in D$ be a complex coordinate centered at 0 in $B_{r_1}(0)$ compatible with the induced metric. Thus we have reduced to the case of a conformal stable immersion $f: D \to N$ such that $D = B_{r_1}$ and the function ρ^2 is smooth on D and satisfies (see [1, Theorem 5.1.4])

$$\Delta\rho^2 \ge 2 + 2r_1\sqrt{c_{23}}\cot(\sqrt{c_{23}}\, r_1) \ge \pi .$$

For a constant Λ to be chosen, we use $\phi = e^{\Lambda\rho^2}$ in (4), and observe that

$$\phi \cdot L\phi \geq (\pi\Lambda - c_{24} + |A|^2)e^{2\Lambda\rho^2} \geq 1 + |A|^2$$

where $c_{24} = \inf_{B^3_{r_1}(P_0)} \{\mathrm{Ric}\}$ and we have chosen Λ so that $\pi\Lambda \geq 1 + c_{24}$.

Thus we have by (4)

$$\int_D (1 + |A|^2)\zeta \, dv \leq e^{2\Lambda} \int_D |d\zeta|^2 \, dx \, dy$$

since $r_1 \leq 1$. We choose ζ to be a function of $|z|$ which is one on $D_{1/2}$ and vanishes on ∂D_1

$$\int_{D_{1/2}} (1 + |A|^2) dv \leq c_{25} . \tag{17}$$

To convert this to information on geodesic balls, we use Theorem 1 as we did in the proof of Theorem 2. By Theorem 1 we have $\lambda^{1/2} \geq 2\varepsilon_1 r_1$ on $B_{4^{-1}r_1}$ for a positive constant ε_1. This implies that $B_{\varepsilon_1 r_1} \subseteq D_{1/2}$, so that by (17) we have

$$\int_{B_{\varepsilon_1 r_1}} (1 + |A|^2) dv \leq c_{25} . \tag{18}$$

Thus we have an area bound and an integral curvature estimate on $B_{\varepsilon_1 r_1}$. The methods of [9] now give an estimate on $|A|^2(P_0)$. We sketch the proof for completeness. One can take $\phi = (1 + |A|^2)^{1/4}$ in (4) and use (5) to prove

$$\int_{B_{2^{-1}\varepsilon_1 r_1}} (1 + |A|^2)^{3/2} dv \leq c_{26} . \tag{19}$$

Since (5) implies that the function $g = (1+|A|^2)^{1/2}$ satisfies

$$\Delta g + c_{27}(1+|A|^2)g \geq 0 ,$$

we can use (19) together with a standard iteration procedure (see e.g. [8, Theorem 5.3.1]) to prove $|A|^2(P_0) \leq c_{28} r_1^{-2} \leq c_{29} r_0^{-2}$ provided we have a Sobolev inequality in the induced metric on B_{r_1}. The required Sobolev inequality is well known, and follows from [6, Theorem 2.1]. (One can verify the hypotheses of [6, Theorem 2.1] for their case $M = \overline{M}$, where we may decrease r_1 if necessary to verify the area bound.)

The final result of Theorem 3 follows by applying the curvature estimate of the first part at points in $B_{r_{0/2}}(P_0)$ to assert that the second fundamental form has bounded length at all such points, and noting that this implies that M is locally the graph of a function on its tangent space in N with bounded C^2 norm. This completes the proof of Theorem 3.

RICHARD SCHOEN
DEPARTMENT OF MATHEMATICS
UNIVERSITY OF CALIFORNIA
BERKELEY, CALIFORNIA 94720

REFERENCES

[1] J. Cheeger, D. Ebin, *Comparison Theorems in Riemannian Geometry*, North-Holland, 1975.

[2] S. S. Chern, Minimal submanifolds in a Riemannian manifold, Lecture notes, University of Kansas, 1968.

[3] M. doCarmo, C. K. Peng, Stable complete minimal surfaces in $\mathbb{R}^3$ are planes, Bull. AMS *1* (1979), 903-905.

[4] D. Fischer-Colbrie, R. Schoen, The structure of complete stable minimal surfaces in 3-manifolds of nonnegative scalar curvature, CPAM *33* (1980), 199-211.

[5] E. Heinz, Über die Lösungen der Minimal flächengleichung, Nachr. Akad. Wiss. Göttingen Math. Phys. Kl II (1952), 51-56.

[6] D. Hoffman, J. Spruck, Sobolev and isoperimetric inequalities for Riemannian submanifolds, CPAM *27* (1974), 715-727.

[7] W. Meeks, S. T. Yau, Topology of three dimensional manifolds and the embedding problems in minimal surface theory, to appear in Ann. of Math.

[8] C. B. Morrey, *Multiple Integrals in the Calculus of Variations*, Springer-Verlag, New York, 1966.

[9] R. Schoen, L. Simon, S. T. Yau, Curvature estimates for minimal hypersurfaces, Acta Math. *134*(1975), 275-288.

[10] R. Schoen, S. T. Yau, Existence of incompressible minimal surfaces and the topology of three dimensional manifolds with nonnegative scalar curvature, Ann. of Math. *110*(1979), 127-142.

[11] R. Osserman, An extension of certain results in function theory to a class of surfaces, Proc. NAS *45*(1959), 1031-1035.

REGULARITY OF SIMPLY CONNECTED SURFACES WITH QUASICONFORMAL GAUSS MAP

R. Schoen* and L. Simon*

We here develop a regularity theory for simply connected surfaces M embedded in $\mathbf{R}^3$ in case the Gauss map of M is (γ_1, γ_2)-quasiconformal, with γ_1, γ_2 non-negative constants. That is, in case

$$|A|^2(x) \leq -\gamma_1 K(x) + \gamma_2 , \quad x \in M , \tag{0.1}$$

where $|A|$ denotes the length of the second fundamental form of M and K is the Gauss curvature of M. (See [SL] for a discussion of the condition (0.1).)

Our main result is a Hölder estimate for the unit normal of such surfaces. In stating this result precisely, we assume (without loss of generality) that $0 \in M$, and we use the notation:

*Research of both authors was supported by an N.S.F. grant at the Institute for Advanced Study, Princeton. The second author was also partially supported by N.S.F. funds at Stanford University.

Seminar on Minimal Submanifolds
0-691-08324-X/83/127-19 $1.45/0 (cloth)
0-691-08319-3/83/127-19 $1.45/0 (paperback)

$\overline{M}$ = closure of M taken in R^3

$B_\rho = \{x \in R^3 : |x| < \rho\}$

$M_\rho = M \cap B_\rho$

M^*_ρ = component of M_ρ containing 0

ν = any continuous choice of unit normal for M

R = any real number $\in \left(0, \frac{1}{4}\gamma_2^{-1/2}\right)$ such that $(\overline{M} \sim M) \cap B_R = \emptyset$.

We also assume

(0.2) $$M \subset B_{/4\gamma_2^{-1/2}}$$

in case $\gamma_2 \neq 0$.

Our main result is then the following:

THEOREM 1. *Suppose* M *is simply connected, and suppose*

$$R^{-2}|M^*_R| \leq \mu,$$

where μ *is any positive constant and* $|M^*_R|$ *denotes the area of* M^*_R. *Suppose also that (0.1), (0.2) hold.*

Then

$$|\nu(x) - \nu(\overline{x})| \leq c(|x - \overline{x}|/R)^\alpha, \quad x, \overline{x} \in M^*_{R/2}$$

where $c > 0$ *and* $\alpha \in (0,1)$ *depend only on* γ_1, $\gamma_2 R^2$ *and* μ.

In case $\gamma_2 = 0$ and M is complete, we can let $R \to \infty$ in Theorem 1, thus obtaining the following Bernstein-type theorem:

THEOREM 2. *Suppose* M *is a simply connected complete embedded surface in* R^3 *with area growth satisfying* $|M^*_\rho| = 0(\rho^2)$ *as* $\rho \to \infty$, *and suppose that the Gauss map of* M *is* $(\gamma_1, 0)$*-quasiconformal for some constant* $\gamma_1 > 0$. *Then* M *is a plane.*

Since minimal surfaces automatically have a conformal Gauss map (i.e. a (2,0)-quasiconformal Gauss map), Theorems 1 and 2 apply in particular to simply connected minimal surfaces. Also, if M_0 is a simply connected minimal surface embedded in a Riemannian 3-manifold N, then

M_0 can be locally represented (by means of local coordinate charts of N) as a surface in R^3 with (γ_1,γ_2)-quasiconformal Gauss map, where γ_1,γ_2 depend on N (and not on M_0). Thus Theorem 1 also gives a local regularity theory for simply connected minimal surfaces embedded in a Riemannian manifold.

We should also mention that analogous results (without a-priori restrictions on area) have been obtained for *graphs* with (γ_1,γ_2)-quasiconformal Gauss map in [SL]. In [SL], γ_1 was also allowed to be negative. (One easily sees however that Theorem 1 is in general false if we allow $\gamma_1 < 0$ in the present parametric setting.)

Similar results for *stable minimal* surfaces M have been obtained, without any simple connectedness assumptions on M, in [SS], [SSY]. As far as we know, the results obtained here are the first such estimates obtained for the case of simply connected minimal surfaces without stability assumptions or restrictive assumptions on the Gauss map. For example, in [OR], Osserman obtained a curvature estimate in case the Gauss map omits a neighborhood of S^2. Similar assumptions on the Gauss map were needed to obtain the regularity results in [JE], [JS], and [SL].

§1. *Preliminaries*

Throughout this paper we shall adopt the terminology introduced above, so that M is always a simply connected surface properly embedded in R^3 with $0 \in M$ and with M satisfying the inequality (0.1). Also B_ρ, M_ρ, M^*_ρ, ν, R are as introduced above, and we continue to assume

$$M \subset B_{\gamma_2^{-1/2}/4} \tag{1.1}$$

in case $\gamma_2 \neq 0$.

Notice that if $\gamma_2 = 0$, then (0.1) guarantees that $K \leq 0$, and hence M has the convex hull property. In any case (whether γ_2 is zero or not), M at least satisfies the following weaker convex hull property:

If Γ is any Jordan curve in M, then there is a diffeomorph $\Delta \subset M$ of a closed disc, with $\partial\Delta = \Gamma$ and with

$$\Delta \subset C \tag{1.2}$$

whenever C is a closed circular cylinder containing Γ and having circular cross-sectional radius $< \frac{1}{2}\gamma_2^{-1/2}$. In particular (since a ball of radius r can be expressed as the intersection of circular cylinders having circular cross-sectional radius r), we deduce that

$$\Delta \subset B \tag{1.3}$$

whenever B is a closed ball of radius $< \frac{1}{2}\gamma_2^{-1/2}$ with $\Gamma \subset B$.

To prove (1.2) we take C as in (1.2) with $\Gamma \subset C$. If $\Delta \not\subset C$ we can find a cylinder $\tilde{C}$ with circular cross-sections concentric with those of C and with $\Delta \subset \tilde{C}$ and $(\Delta \sim \Gamma) \cap \partial\tilde{C} \neq \emptyset$. Evidently $\tilde{C}$ must then have circular cross-sections of radius $< \gamma_2^{-1/2}$ (by virtue of the facts that $M \subset B_{/4\gamma_2^{-1/2}}$ and the fact that C has circular cross-sections of radius $< \frac{1}{2}\gamma_2^{-1/2}$). But then at a point $x_0 \in (\Delta \sim \Gamma) \cap \partial\tilde{C}$ we evidently have $|A|^2 > \gamma_2$ and $K \geq 0$. But then (0.1) is impossible at x_0.

Notice that it follows from (1.3) and Sard's Theorem that, for almost all $\rho \in (0, R)$, M_ρ^* is simply connected with ∂M_ρ^* a smooth Jordan curve along which M and ∂B_ρ intersect transversally.

We also here need to recall a couple of general facts concerning surfaces $M \subset \mathbf{R}^3$. Firstly, we have the isoperimetric inequality

$$|F|^{1/2} \leq c_1 \left(|\partial F| + \int_F |\mathbf{H}| \right) \tag{1.4}$$

where F is any open subset of M with $\overline{F} \cap (M \sim \overline{M}) = \emptyset$, where ∂F is the boundary of F (assumed smooth), and where $\mathbf{H}$ is the mean curvature of M.

We here adopt the convention that $|F|$ denotes the two-dimensional Hausdorff measure of F and $|\partial F|$ denotes the one-dimensional Hausdorff measure of F, whenever F is an open subset of M.

Finally we recall the density bounds

$$\sup_{\rho\in(0,R)} \rho^{-2}|M^*_\rho| \le c_2\left(R^{-2}|M^*_R| + \int_{M^*_R}|H|^2\right). \tag{1.5}$$

(See e.g. [SL] or [GT, Chapter 15] for a proof.)

§2. *General lemma on oscillation of* ν

In this and subsequent sections, $c_3, c_4, \cdots$ will denote constants depending at most on γ_1 and $\gamma_2 R^2$, and $\rho\in(0,R)$ is arbitrary.

LEMMA 1. *Let* M *satisfy the assumptions of* §1. *Then*

$$\sup_{M^*_{\rho/2}} |\nu-\nu(0)| \le c_3(1+\rho^{-2}|M^*_\rho|)^{1/2}\left(\int_{M^*_\rho}|A|^2\right)^{1/2}. \tag{2.1}$$

REMARK. By considering examples, the reader will see that it is essential that $M^*_{\rho/2}$ rather than $M_{\rho/2}$ appears on the left of (2.1).

Proof. For convenience of notation we write

$$(1+\rho^{-2}|M^*_\rho|)\int_{M^*_\rho}|A|^2 = \varepsilon,$$

and we suppose (without loss of generality) that $\nu(\xi) = e_3 = (0,0,1)$.

Let $g(x) = \sqrt{1-\nu(x)\cdot e_3}$, $x\in M$, and note that $|\nabla g| \le |A|$ by a standard computation (see e.g. [SS]) involving the second fundamental form. By the co-area formula [FH, 3.2.22] we then have (for any $\tau\in(0,\frac{1}{4})$ and $\theta\in(0,\tau)$) that

$$(2.2)\qquad \int_{\tau-\theta}^{\tau} |\Gamma_s|\,ds \le \int_{M^*_{\rho,\tau,\theta}} |A| \le |M^*_{\rho,\tau,\theta}|^{1/2}\left(\int_{M^*_{\rho,\tau,\theta}} |A|^2\right)^{1/2},$$

where $\Gamma_s = \{x \in M^*_\rho : g(x) = s\}$, $|\Gamma_s|$ denotes one-dimensional Hausdorff measure of Γ_s, and $M^*_{\rho,\tau,\theta} = M^*_\rho \cap \{x : \tau - \theta < g(x) < \tau\}$. If we use this together with Sard's Theorem, then we can evidently find $s(\tau,\theta) \in (\tau-\theta,\tau)$ such that $\Gamma_{s(\tau,\theta)}$ either empty or smooth, with

$$(2.3)\quad |\Gamma_{s(\tau,\theta)}| \le c_4\theta^{-1}\int_{M^*_{\rho,\tau,\theta}} |A| \le c_4\theta^{-1}|M^*_{\rho,\tau,\theta}|^{1/2}\left(\int_{M^*_{\rho,\tau,\theta}} |A|^2\right)^{1/2}.$$

In particular, taking $\theta = \tau/2$, we have $s_0 \in (\tau/2,\tau)$ so that Γ_{s_0} is smooth and

$$(2.4)\qquad \begin{aligned}|\Gamma_{s_0}| &\le c_5\,\tau^{-1}|M^*_\rho|^{1/2}\left(\int_{M^*_\rho}|A|^2\right)\\ &\le c_5\,\varepsilon\tau^{-1}\rho\,.\end{aligned}$$

(Notice that for the moment τ is an arbitrary parameter; we emphasize that c_5 does not depend on τ.)

For the moment we assume

$$(2.5)\qquad c_5\,\varepsilon\tau^{-1} \le 1/16\,.$$

(Notice that this requires ε to be small in case τ is small.) Under this assumption, (2.4) implies

$$|\Gamma_{s_0}| \le \rho/16\,,$$

and it follows that we can select $\rho_0 \in (\rho/2,\rho)$ such that

$$(2.6)\qquad \Gamma_{s_0} \cap \partial M^*_{\rho_0} = \emptyset\,,$$

and such that (by Sard's Theorem), $\partial M^*_{\rho_0}$ is a smooth Jordan curve. We thus have either $g > s_0$ everywhere on $\partial M^*_{\rho_0}$ or $g < s_0$ everywhere on $\partial M^*_{\rho_0}$.

With such a choice of s_0 and ρ_0 we now define a new function $\tilde{g}$ by

$$\tilde{g} = \begin{cases} +(g|_{\bar{M}^*_{\rho_0}} - s_0) & \text{if} \quad g < s_0 \quad \text{on} \quad \partial M^*_{\rho_0} \\ -(g|_{\bar{M}^*_{\rho_0}} - s_0) & \text{if} \quad g > s_0 \quad \text{on} \quad \partial M^*_{\rho_0} . \end{cases}$$

Then $\tilde{g} < 0$ on $\partial M^*_{\rho_0}$, and we claim

$$\tilde{g}(x) \leq \tau/4 \qquad \forall x \in M^*_{\rho_0} . \tag{2.7}$$

To see this, let $\tau_0 = 0$, $\tau_k = \sum_{\nu=1}^{k} \tau/2^{\nu+2}$ (for $k \geq 1$), and for each k let $s_k \in [\tau_k, \tau_{k+1}]$ be chosen such that (2.3) holds with $s_k = s(r, \theta)$ and with τ_k, $\tau/2^{k+3}$ in place of r, θ respectively. Also, we define

$$U_k = \{x \in M^*_{\rho_0} : \tilde{g}(x) > s_k\}$$

(so that $\bar{U}_k \subset M^*_{\rho_0}$ because $\tilde{g} < 0$ on $\partial M^*_{\rho_0}$), and we note that the Gauss map $G : x \to \nu(x)$ maps ∂U_k to $\{y = (y_1, y_2, y_3) \in S^2 : \pm(1 - y_3 - s_0) = s_k\}$ and U_k to $\Delta_k \equiv \{y = (y_1, y_2, y_3) \in S^2 : \pm(1 - y_3 - s_0) > s_k\}$, where we take $+$ or $-$ according as $\tilde{g} = \pm(g|_{\bar{M}^*_{\rho_0}} - s_0)$. Since $-K$ is the *signed* area magnification factor of the Gauss map G, we thus have

$$\int_{U_k} (-K) = n|\Delta_k| \quad \text{and} \quad \int_{U_k} |K| \geq |n|\,|\Delta_k| , \tag{2.8}$$

where n is an integer. (n is in fact the constant value of the topological degree $d(G, U_k, y)$ at points $y \in \Delta_k$.) Now we have $|\Delta_k| \geq c_6 \tau$ (by virtue of the facts that $|s_k| \leq \tau/4$ and $s_0 \in (\tau/2, \tau)$). Hence the second inequality in (2.8) gives

$$c_6 |n| \tau \le \int_{U_k} |K| \le \left(\int_{U_k} |A|^2\right)^{1/2} |U_k|^{1/2} \le \varepsilon$$

so that $n = 0$ provided

$$\varepsilon < \min\{c_6^{-1}, c_5^{-1}/16\}\tau \,. \tag{2.9}$$

(Notice that this incorporates the condition (2.5).) Thus, subject to (2.9), we have

$$\int_{U_k} (-K) = 0$$

and the condition (0.1) implies

$$\int_{U_k} |A|^2 \le \gamma_2 |U_k| \,. \tag{2.10}$$

On the other hand the inequality (1.4) implies

$$|U_k|^{1/2} \le c_1 \left(|\partial U_k| + \int_{U_k} |H|\right)$$

$$\le c_1 \left(|\partial U_k| + 2\int_{U_k} |A|\right),$$

and hence, using (2.3),

$$|U_k|^{1/2} \le c_2 2^k \tau^{-1} \int_{U_{k-1}} |A| \,.$$

By Schwartz, this gives

$$|U_k| \le c_7^2 2^{2k} \tau^{-2} |U_{k-1}| \int_{U_{k-1}} |A|^2 \,. \tag{2.11}$$

Combining this with (2.10), we then have

$$|U_k| \le c_7^2 2^{2k} \tau^{-2} \gamma_2 |U_{k-1}|^2 . \tag{2.12}$$

We can iterate the relation (2.12) with $k = 2, 3, \cdots$ to deduce

$$|U_\infty|^{\frac{1}{2}^k} \le |U_k|^{\frac{1}{2}^k} \le (c_7 \tau^{-1} \sqrt{\gamma_2})^{\sum_{\nu=2}^{k} \frac{1}{2}^\nu} 2^{\sum_{\nu=2}^{k} \frac{1}{2}^\nu} |U_1| ,$$

where $U_\infty = \{x : g(x) > \tau/4\}$. Thus, letting $k \to \infty$, we deduce

$$c_8 \tau^{-2} \gamma_2 |U_1| \ge 1 \quad \text{if} \quad U_\infty \ne \emptyset . \tag{2.13}$$

On the other hand we have from (2.11)

$$|U_1| < c_9 \tau^{-2} \varepsilon^2 \rho^2 . \tag{2.14}$$

The inequalities (2.13), (2.14) are contradictory in case $U_\infty \ne \emptyset$ and

$$c_8 c_9 \gamma_2 \rho^2 \varepsilon^2 \le \tau^4 .$$

Let us select $c_{10} \ge 1$ so that

$$c_{10} \ge \max\{16c_5, c_6, \sqrt{c_8 \cdot c_9 \cdot (\gamma_2 R^2)}\}$$

and assume

$$c_{10}\, \varepsilon \le \tau^2 . \tag{2.15}$$

(Notice this incorporates the previous condition (2.5).) Then (2.13), (2.14) are contradictory unless $U_\infty = \emptyset$. Thus (2.7) is proved under the assumption (2.15). Now if $\tilde{g} = -(g - s_0)$ then (2.7) implies $g > s - \tau/4 > \tau/4$ on $M^*_{\rho_0}$, which contradicts the fact that $g(0) = 0$. Hence, subject to (2.15), we must have $\tilde{g} = g - s_0$, in which case (2.7) implies

$$g \le s_0 + \tau/4 \le 5\tau/4 \quad \text{on} \quad M^*_{\rho_0} .$$

Since $g(x) = \frac{1}{2}|\nu - \nu(0)|^2$, we have thus proved

$$|\nu - \nu(0)| \leq 2\tau \tag{2.16}$$

provided (2.15) holds. So far $\tau \in (0,1)$ is arbitrary. If $c_{10}\sqrt{\varepsilon} < 1/4$, we may take $\tau = c_{10}\sqrt{\varepsilon}$ (thus satisfying (2.15)). Then we obtain from (2.16) the inequality

$$|\nu - \nu(0)| \leq 2c_{10}\sqrt{\varepsilon} \quad \text{on} \quad M^*_{\rho_0} . \tag{2.17}$$

Thus (2.1) is proved in case $c_{10}\sqrt{\varepsilon} < 1/4$. In case $c_{10}\sqrt{\varepsilon} \geq 1/4$, the inequality (2.1) is trivial with $c_3 = 8c_{10}$, by virtue of the fact that $|\nu| \leq 1$.

§3. *Main computation for* $\int |A|^2$

In this section we let r be the non-negative function on R^3 defined by $r(x) = |x|$, and ∇ denotes the gradient operator on M. Using the usual identification of tangent space at a point $x \in R^3$ with R^3 itself, we can write

$$(\nabla f_{|M})(x) = (\nabla_1 f(x), \nabla_2 f(x), \nabla_3 f(x)) , \qquad x \in M ,$$

for any non-negative C^1 function f defined in a neighborhood of M. Here ∇_i is the operator defined by

$$\nabla_i f = (\delta_{ij} - \nu_i \nu_j) D_j f .$$

Notice that in particular we have

$$\nabla r = r^{-1}x - (r^{-1}x \cdot \nu)\nu \tag{3.1}$$

and

$$\begin{aligned} |\nabla r|^2 &= (\delta_{ij} - \nu_i\nu_j)(x_i/r)(x_j/r) = 1 - (\nu \cdot x/r)^2 \\ &= r^{-1} x \cdot \nabla r . \end{aligned} \tag{3.2}$$

Also, the matrix $(\nabla_i \nu_j)_{i,j=1,2,3}$ is related to the second fundamental form A of M by

$$A(\xi,\eta) = \Big(\sum_{i,j=1}^{3} \xi_i \eta_j \nabla_i \nu_j(x)\Big)\, \nu(x) \tag{3.3}$$

whenever $\xi = (\xi_1, \xi_2, \xi_3)$, $\eta = (\eta_1, \eta_2, \eta_3)$ are tangent vectors to M at x. Since $\nabla_i \nu_j = \nabla_j \nu_i$ and since $\nabla \nu_j$ is tangent to M for each $j = 1, 2, 3$, it is then evident that

$$|A|(x) = \Big(\sum_{i,j=1}^{3} (\nabla_i \nu_j(x))^2\Big)^{1/2} \tag{3.4}$$

In order to derive our main estimate for $\int |A|^2$, we first want to compute the quantity $\frac{d^2}{d\rho^2} \int_{M_\rho^*} |\nabla r|^2$; we shall assume that ρ is one of the (almost all, by Sard's Theorem) values in $(0, R)$ such that M and ∂B_ρ intersect transversally.

By the co-area formula ([FH, 3.2.22]) we have

$$\frac{d^2}{d\rho^2} \int_{M_\rho^*} |\nabla r|^2 = \frac{d}{d\rho} \int_{L_\rho} |\nabla r| , \tag{3.5}$$

$L_\rho = \partial M_\rho^*$. By a standard computation (essentially the computation for the first variation of length of the family L_ρ as one-dimensional submanifolds of M) we have

$$\frac{d}{d\rho} \int_{L_\rho} |\nabla r| = -\int_{L_\rho} \kappa_g + \int_{L_\rho} D_v |\nabla r| .$$

Here $v = |\nabla r|^{-2} \nabla r$ is the velocity vector associated with the flow on M which generates the curves L_ρ, and κ_g denotes the geodesic curvature of L_ρ in M; that is $\kappa_g = \langle v/|v|, \boldsymbol{\kappa}\rangle$, where $\boldsymbol{\kappa}$ is the curvature vector of L_ρ, considered as a space curve in $\mathbf{R}^3$. (For the above computation it is best to think of $\kappa_g = \langle v/|v|, \mathbf{h}\rangle$, where $\mathbf{h}$ is the mean curvature vector of L_ρ considered as a submanifold of M.)

Thus we have (again using the co-area formula)

$$\frac{d^2}{d\rho^2}\int_{M_\rho^*} |\nabla r|^2 = -\int_{L_\rho} \kappa_g + \frac{d}{d\rho}\int_{M_\rho^*} |\nabla r|\, D_v |\nabla r|$$

$$= -\int_{L_\rho} \kappa_g + \frac{1}{2}\frac{d}{d\rho}\int_{M_\rho^*} D_v |\nabla r|^2 .$$

By the Gauss-Bonnet formula, this can be expressed

$$\frac{d^2}{d\rho^2}\int_{M_\rho^*} |\nabla r|^2 = \frac{1}{2}\int_{M_\rho^*} (-K) + 2\pi + \frac{1}{2}\frac{d}{d\rho}\int_{M_\rho^*} D_v |\nabla r|^2 ,$$

where we have used the fact that M_ρ^* is simply connected. Re-arranging we thus have

$$\tag{3.6} \frac{1}{2}\int_{M_\rho^*} (-K) = \frac{d^2}{d\rho^2}\int_{M_\rho^*} |\nabla r|^2 - 2\pi - \frac{1}{2}\frac{d}{d\rho}\int_{M_\rho^*} D_v |\nabla r|^2 .$$

On the other hand, by the first variation formula for M, we have

$$\tag{3.7} \int_M \operatorname{div}_M X = \int_M \langle X, H\rangle ,$$

whenever X is a C^1 vector field defined in a neighborhood of M, with $X = 0$ on $\overline{M} \sim M$. Here H is the mean curvature vector of M (so that

$|\mathbf{H}| \le 2|A|$) and $\operatorname{div}_M X(x)$, for any $x \in M$, is defined to be $\sum_{i=1}^{2} \langle \tau_i, D_{\tau_i} X(x) \rangle$, where τ_1, τ_2 is any orthonormal basis for the tangent space of M at x.

Substituting $X(x) = \phi_\rho(r)\, x/r \; (r = |x|)$ in (3.7), where ϕ_ρ is a smooth approximation of the characteristic function of B_ρ and $\operatorname{spt} \phi_\rho \subset B_\rho$, one now easily sees (using (3.7) with M_ρ^* in place of M) that

$$\frac{d}{d\rho} \int_{M_\rho^*} |\nabla r|^2 = \int_{M_\rho^*} (r^{-1} + \mathbf{H} \cdot x/r)$$

and hence

$$\frac{d^2}{d\rho^2} \int_{M_\rho^*} |\nabla r|^2 = \rho^{-1} \frac{d}{d\rho} \int_{M_\rho^*} (1 + x \cdot \mathbf{H}) .$$

Substituting this relation in (3.6), we then finally get the identity

$$\text{(3.8)} \qquad \frac{1}{2} \rho \int_{M_\rho^*} (-K) = \frac{d}{d\rho} |M_\rho^*| - 2\pi\rho + \frac{d}{d\rho} \int_{M_\rho^*} (x \cdot \mathbf{H} - \frac{1}{2} r D_v |\nabla r|^2) .$$

(It is perhaps worth emphasizing that this identity is valid for any simply connected M as in §1, without any assumption of the form (0.1).)

Next we want to bound $D_v|\nabla r|^2$ in terms of $|A|$. To do this we can directly compute (using (3.2)) that

$$\begin{aligned} D_v|\nabla r|^2 &= - |\nabla r|^{-2} \nabla_i r \, \nabla_i (\nu \cdot x/r)^2 \\ &= - 2|\nabla r|^{-2} (\nabla_i r)(\nabla_i \nu_j)(\nu \cdot x/r)(x_j/r) \\ &\quad - 2|\nabla r|^{-2} (\nabla_i r)\, \nu_j (\nabla_i (x_j/r))(\nu \cdot x/r) . \end{aligned}$$

Since $D_k(x_j/r) = r^{-1}(\delta_{jk} - (x_j/r)(x_k/r))$ and since $(x_j/r)\nabla_i \nu_j = (x_j/r)(\nabla_j \nu_i) = (\nabla_j r)(\nabla_j \nu_i)$, we then have

$$\begin{aligned} D_v|\nabla r|^2 &= -2|\nabla r|^{-2}(\nabla_i r)(\nabla_j r)(\nabla_i \nu_j)(\nu \cdot x/r) \\ &\quad + 2|\nabla r|^{-2} \sum_{i=1}^{3} (\nabla_i r)^2 (\nu \cdot x/r)^2 \\ &= -2(\nabla_i r/|\nabla r|)(\nabla_j r/|\nabla r|)(\nu \cdot x/r)(\nabla_i \nu_j) \\ &\quad + 2(\nu \cdot x/r)^2 . \end{aligned}$$

By (3.4) we then have

$$|D_v|\nabla r|^2| \le 2|\nu \cdot x/r|\,|A| + 2(\nu \cdot x/r)^2 ,$$

and it follows from (3.8) that

$$\text{(3.9)} \quad \tfrac{1}{2}\rho \int_{M_\rho^*} (-K) \le \frac{d}{d\rho}|M_\rho^*| - 2\pi\rho + \frac{d}{d\rho}\int_{M_\rho^*} r(4|\nu \cdot x/r|\,|A| + (\nu \cdot x/r)^2) .$$

(3.9) has been derived for almost all $\rho \in (0, R)$. Since $\int_{M_\rho^*} f$ is an increasing function of ρ whenever f is non-negative, we can integrate in (3.9) to obtain

$$\frac{1}{2}\int_a^b \rho \int_{M_\rho^*} (-K)\,d\rho \le |M_b^* \sim M_a^*| - \pi(b^2 - a^2) + \int_{M_b^* \sim M_a^*} (4|\nu \cdot x/r|\,|A| + (\nu \cdot x/r)^2) .$$

Using (0.1) together with (1.5), we thus obtain

$$(3.10)\quad \frac{1}{2}(b-a)\,a\int_{M_a^*}|A|^2 \le \gamma_1(|M_b^*\sim M_a^*| - \pi(b^2-a^2)) + \gamma_1\int_{M_b^*\sim M_a^*}(4r|\nu\cdot x/r|\,|A| + (\nu\cdot x/r)^2) + \frac{1}{2}\gamma_2\,|M_b^*|\,b(b-a)\,.$$

By Schwartz's inequality, this implies

$$(1-a/b)(a/b)\int_{M_a^*}|A|^2 \le c_{12}(R^{-2}|M_R^*|+1) + c_{12}\left(\int_{M_b^*}|A|^2\right)^{1/2}(R^{-2}|M_R^*|+1)^{1/2},\quad b>a\ge R/2\,.$$

Taking $a=\rho_k=\sum_{\nu=1}^{k}2^{-\nu}R$, $b=\rho_{k+1}\equiv\sum_{\nu=1}^{k+1}2^{-\nu}R$, we have then that *either* $\int_{M_{R/2}^*}|A|^2\le(R^{-2}|M_R^*|+1)$ *or*, for every $k=1,2,\cdots$, we have

$$\int_{M_{\rho_k}^*}|A|^2 \le 2^{k+1}c_{12}(R^{-2}|M_R^*|+1)^{1/2}\left(\int_{M_{\rho_{k+1}}^*}|A|^2\right)^{1/2}.$$

The latter alternative, upon iteration, implies that

$$(3.11)\qquad \int_{M_{R/2}^*}|A|^2 \le c_{13}(R^{-2}|M_R^*|+1)\,.$$

Thus we finally obtain (3.11) in any case, regardless of which of the two alternatives above holds. Notice that then (3.10), together with the density bound (1.5), implies

$$(3.12)\quad \frac{1}{2}(b-a)a\int_{M_a^*}|A|^2 \le \gamma_1(|M_b^* \sim M_a^*| - \pi(b^2-a^2)) + c_{14}\left\{(1+R^{-2}|M_R^*|)^{1/2} b\left(\int_{M_b^*}(\nu\cdot x/r)^2\right)^{1/2} + \int_{M_b^*}(\nu\cdot x/r)^2\right\} + c_{15}b^2(b/R)^2 .$$

§4. *Proof of main theorem*

We note first that it suffices, subject to the assumption that

$$R^{-2}|M_R^*| \le \mu$$

(μ constant), to prove that there is a $\theta \in (0,\frac{1}{2})$ depending only on μ, γ_1, $\gamma_2 R^2$ such that

$$(4.1)\qquad \sup_{M_{\theta R}^*}|\nu - \nu(0)| \le \frac{1}{4} .$$

Indeed, once this is established we can use standard estimates ([NL] or [SL]) for quasiconformal maps to deduce

$$(4.2)\qquad |\nu(x)-\nu(\overline{x})| \le c(|x-\overline{x}|/R)^{\alpha}$$

for each $x, \overline{x} \in M_{\theta R}^*$, with $c > 0$ and $\alpha \in (0,1)$ depending only on γ_1, $\gamma_2 R^2$, μ. By translating the origin, it then follows that (4.2) holds with $\theta = \frac{1}{2}$, as required.

We now proceed to prove (4.1). We first show that, for any given $\varepsilon_0 > 0$ and $\beta \in (0,1)$, there is a positive integer N (depending only on ε_0 and μ) such that

(4.3) $$\int_{M^*_{R/2}\cap\{x:\beta R/2^{k+1}\le|x|\le\beta R/2^k\}} |A|^2 < \varepsilon_0 \quad \text{for some} \quad k \in \{1,\cdots,N\}\,.$$

Indeed if (4.3) is false, then we have, writing $\mu = c_{13}(1+\mu)$ and using (3.11),

$$\tilde{\mu} \ge \int_{M^*_{R/2}} |A|^2 \ge \sum_{k=1}^{N} \int_{M^*\cap\{x:\beta R/2^{k+1}\le|x|\le\beta R/2^k\}} |A|^2$$

$$\ge N\varepsilon_0\,,$$

so that $N \le \tilde{\mu}\varepsilon_0^{-1}$. Thus (4.3) holds for $N > \tilde{\mu}\varepsilon_0^{-1}$. Now (with k such that the inequality in (4.3) holds) we let W be the component of $M^*_{R/2} \cap \{x : \frac{7}{8}\theta_0 R < |x| < \theta_0 R\}$ which contains $\partial M^*_{\theta_0 R}$, where $\theta_0 \in (\frac{5}{8}2^{-k}\beta, \frac{7}{8}2^{-k}\beta)$ is chosen to ensure that $\partial M^*_{\theta_0 R}$ is smooth. (This can be done by Sard's Theorem.) Since W can be covered by balls of radius $\frac{1}{64}\theta_0 R$ which have centers in W and which are contained in $\{x : 2^{-(k+1)}\beta R < |x| < 2^{-k}\beta R\}$, and since the number of balls in such a covering can be taken to be bounded (by an absolute constant, independent of k, ε_0, γ_1, γ_2 and R), it then follows from (4.3) and Lemma 1 that

(4.4) $$\sup_W |\nu - \nu(\xi)| \le c_{15}\sqrt{\varepsilon_0}$$

where ξ is any fixed point of W. Using this together with the convex hull property (1.3) and the fact that M is embedded, we deduce immediately that

(4.5) $$\sup_{x\in W} \left|\frac{x}{|x|}\cdot\nu(\xi)\right| \le c_{16}(\sqrt{\varepsilon_0} + \sqrt{\gamma_2}\,\theta_0 R)\,,$$
$$|W| - \pi((\theta_0 R)^2 - (\tfrac{5}{8}\theta_0 R)^2) \le c_{17}(\varepsilon_0 + \gamma_2(\theta_0 R)^2)(\theta_0 R)^2\,.$$

Then the estimate (1.5), together with the estimate (3.12) (used with $a = \frac{5}{8}\theta_0 R$, $b = \theta_0 R$), implies

$$\int_{\frac{5}{8}\theta_0 R}^{\theta_0 R} \rho \int_{M_\rho^*} |A|^2 d\rho \le c_{18}(\sqrt{\varepsilon_0} + \beta^2) 2^{-2k}$$

with c_{18} depending on μ as well as γ_1 and $\gamma_2 R^2$. This gives

$$\int_{M^*_{\theta_0 R/2}} |A|^2 \le c_{19} \sqrt{\varepsilon_0}$$

if we take $\beta^2 \le \sqrt{\varepsilon_0}$. We then have

$$\sup_{M^*_{\theta_0 R/4}} |\nu - \nu(0)| \le c_{20}\, \varepsilon_0^{1/4}$$

by Lemma 1. Hence choosing ε_0 small enough to ensure $c_{20}\, \varepsilon_0^{1/4} < 1/4$, we then have (4.1) with $\theta = \theta_0/4$. (With θ_0 depending on γ_1, $\gamma_2 R^2$ and μ.)

§5. *Concluding remarks*

Many of the computations here extend directly to the case of simply connected M immersed (rather than embedded) in $\mathbf{R}^3$. (In fact the only place where we made really essential use of the embeddedness of M was in deducing (4.5) from (4.4).) In particular, there is a version of the result of §2 which is valid for immersed surfaces. Nevertheless, neither the main result (Theorem 1) nor the Bernstein result (Theorem 2) are valid for immersed surfaces. (Enneper's surface, for example, demonstrates this.)

R. SCHOEN
MATHEMATICS DEPARTMENT
UNIVERSITY OF CALIFORNIA
BERKELEY, CALIFORNIA 94720
U.S.A.

L. SIMON
MATHEMATICS DEPARTMENT
RESEARCH SCHOOL OF PHYSICAL SCIENCES
AUSTRALIAN NATIONAL UNIVERSITY
CANBERRA A.C.T. 2600
AUSTRALIA

REFERENCES

[FH] H. Federer, Geometric Measure Theory. Springer-Verlag (1969).

[GT] D. Gilbarg, N. Trudinger, Elliptic partial differential equations of Second Order. Springer-Verlag (1977).

[JE] H. Jenkins, On 2-dimensional variation problems in parametric form. Arch. Rat. Mech. Anal. *8*(1961), 181-206.

[JS] H. Jenkins and J. Serrin, Variational problems of minimal surface type I. Arch. Rat. Mech. Anal. *12*(1963), 185-212.

[NL] L. Nirenberg, On nonlinear elliptic partial differential equations and Hölder continuity. Comm. Pure Appl. Math. *6*

[OR] R. Osserman, On the Gauss curvature of minimal surfaces. Trans. A.M.S. *96*(1960), 115-128.

[SS] R. Schoen and L. Simon, Regularity of Stable Minimal Hypersurfaces. Preprint, Institute for Advanced Study (1980).

[SSY] R. Schoen, L. Simon and S.-T. Yau, Curvature Estimates for Minimal Hypersurfaces. Acta. Math. *134*(1975), 276-288.

[SL] L. Simon, A Hölder Estimate for Quasiconformal Maps between surfaces in Euclidean space. Acta. Math. 139(1977), 19-51.

CLOSED MINIMAL SURFACES IN HYPERBOLIC 3-MANIFOLDS

Karen K. Uhlenbeck

Introduction

A hyperbolic manifold is a complete Riemannian manifold M without boundary of constant curvature -1. The universal cover of a hyperbolic manifold is n-dimensional hyperbolic space H^n and the deck transformations induce a representation of the fundamental group $\rho : \pi_1 M \to \text{Iso}\, H^n$. We are interested in hyperbolic 2 and 3-dimensional manifolds. In these cases (the orientation preserving) isometry group of H^2 is $PSL(2,R)$, or the real 2×2 matrices with determinant 1 under the identification $A = -A$. The isometry group $\text{Iso}\, H^2$ lies naturally in $\text{Iso}\, H^3 = PSL(2,C)$, the group of complex 2×2 matrices with determinant 1 under the identification $A = -A$. This group may also be identified with the conformal group of $S^2 = \partial H^3$, the group of Möbius or fractional linear transformations, and the connected component of the identity in the Lorentz group $SO(3,1)$. The study of hyperbolic manifolds is to a great extent the study of these group representations.

This article contains some of the background material for the study of closed minimal surfaces in hyperbolic 3-manifolds. We present some Kleinian group theory and some initial results which show that the space of quasi-Fuchsian groups can be parameterized by the closed minimal surfaces with their second fundamental forms lying in the associated hyperbolic 3-manifolds. All proofs are quite elementary.

Seminar on Minimal Submanifolds
0-691-08324-X/83/147-22 \$1.60/0 (cloth)
0-691-08319-3/83/147-22 \$1.60/0 (paperback)

My interest in this subject grew out of questions put to me by W. Abikoff. I am indebted to W. Abikoff and H. Masur for introducing me to the subject of Kleinian groups. Many mathematicians including E. Calabi, J. Kazdan, B. Maskit, J. Sacks, and W. Thurston made valuable contributions to the ideas in this paper.

§1. *Closed geodesics in Riemann surfaces*

In this section we motivate our investigation of minimal surfaces in hyperbolic 3-manifolds by describing an important role of geodesics in the study of Riemann surfaces. This construction is common in the literature. See, for example, the papers of L. Keen [9], [10] and W. Thurston's notes [16].

Throughout this article, Σ is a fixed closed Riemann surface of genus $g \geq 2$. The *Teichmüller space* $\mathcal{T}(\Sigma)$ is defined to be the space of tensors representing conformal structures under the equivalence induced by $\mathrm{Diff}_0\Sigma$, the group of diffeomorphisms *isotopic to the identity*. This space $\mathcal{T}(\Sigma)$ is also called the space of marked conformal structures in the sense that a presentation of the fundamental group may be consistently chosen modulo inner automorphisms. The Teichmüller space $\mathcal{T}(\Sigma)$ should be carefully distinguished from the *Riemann space* $\mathcal{R}(\Sigma)$, the space of tensors for conformal structures on Σ under the equivalence induced by $\mathrm{Diff}\,\Sigma$, the group of all diffeomorphisms. The *mapping class group* $\mathrm{Map}\,\Sigma = \mathrm{Diff}\,\Sigma/\mathrm{Diff}_0\Sigma$ acts on $\mathcal{T}(\Sigma)$ and its orbit space is the Riemann space $\mathcal{R}(\Sigma) = \mathcal{T}(\Sigma)/\mathrm{Map}\,\Sigma$. The Teichmüller space is a complex cell of dimension $3g-3$ while the Riemann space is not a manifold but has the analytic cell decomposition of an orbit space.

We shall consistently use Σ to denote the topological surface and Σ_μ, $\mu \in \mathcal{T}(\Sigma)$ to denote the surface with a marked conformal structure. If we forget the marking, the surface Σ_μ is equivalently a surface with conformal structure, a Riemann surface, a complex (1-dim)manifold, or a hyperbolic surface. These concepts are equivalent and we use them interchangeably. The fact that every Riemann surface Σ_μ is obtained as $\Sigma_\mu = H^2/G$ is a "uniformization theorem" due to Poincaré.

Several sets of coordinates for $\mathcal{T}(\Sigma)$ can be obtained using geodesics. Consider $\Sigma_\mu \in \mathcal{T}(\Sigma)$ to have a distinguished set of 3g–3 simple, closed non-intersecting, non-homotopically equivalent curves $\{\gamma_1, \cdots, \gamma_{3g-3}\}$ which divide the surface up into 2g–2 "pairs of pants" $\Sigma = P_1 \cup \cdots \cup P_{2g-2}$.

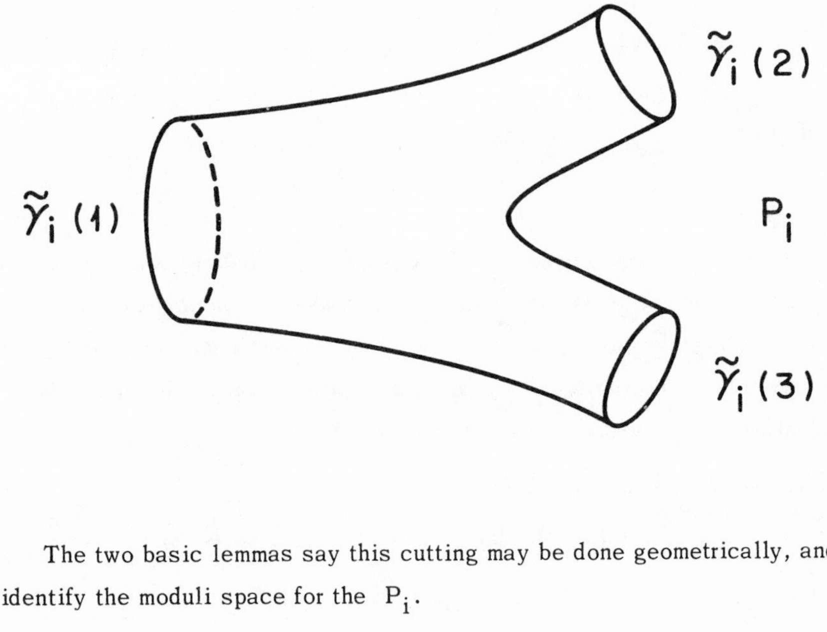

The two basic lemmas say this cutting may be done geometrically, and identify the moduli space for the P_i.

LEMMA 1.1. *Consider* Σ_μ *as a hyperbolic manifold. The set of closed geodesics* $\{\tilde{\gamma}_1, \cdots, \tilde{\gamma}_{3g-3}\}$ *which are homotopic to* $\{\gamma_1, \cdots, \gamma_{3g-3}\}$ *are also simple and non-intersecting.*

LEMMA 1.2. *The three ordered lengths* $(|\tilde{\gamma}_{i(1)}|, |\tilde{\gamma}_{i(2)}|, |\tilde{\gamma}_{i(3)}|)$ *are moduli for the space of marked conformal structures on* P_i.

THEOREM 1.3. *A set of real coordinates for the cell* $\mathcal{T}(\Sigma)$ *is the* 3g–3 *numbers* $(\mu_1, \cdots, \mu_{3g-3}) = (\ln|\tilde{\gamma}_1|, \cdots, \ln|\tilde{\gamma}_{3g-3}|)$ *with the corresponding* (3g–3) *twist numbers* $(\theta_1, \cdots, \theta_{3g-3})$ *describing the angle at which the geodesic boundaries are attached.*

This of course depends on the fact that 2-manifolds may be glued across geodesic boundaries of the same length since they are "flat." Also, note the change in parameter $\theta_i \to \theta_i + 2n\pi$ describes a Dehn twist about $\tilde{\gamma}_i$, and is an element of map Σ. This is an easy explanation of why the angles must range over all real values.

Naturally, one might wonder if a similar construction has any importance in higher dimensions. From experience, one expects a great deal of difference between dimension two and dimension three. Both the hyperbolic manifolds and the codimension one minimal submanifolds behave very differently.

§2. *Quasi-Fuchsian manifolds*

The differential geometer may have a preconceived notion that compact manifolds are easier to deal with. The famous theorem of Mostow indicates this is a mistake in this case. There is no deformation theory for compact 3-manifolds, whereas there is a great deal of deformation theory for their closed minimal submanifolds.

THEOREM 2.1 (Mostow). *If* $f: M^3 \to \tilde{M}^3$ *is a diffeomorphism between hyperbolic* 3*-manifolds with finite volume, then* f *is isotopic to an isometry.*

We abandon for this article the hope of identifying the minimal surfaces in compact hyperbolic manifolds to proceed in a more profitable direction. If M is a complete hyperbolic 3-manifold, then M may be identified with the representation $\rho: \pi_1 M \to PSL(2, C)$ given by deck transformations on $H^3 \to M$. Such a group $G = \rho(\pi_1 M)$ is a discrete, discontinuous group acting on H^3 and every such group corresponds to an $M^3 = H^3/G$. Conformally H^3 is the Poincaré ball in R^3, and $PSL(2, C)$ acts on $\partial H^3 = S^2$ in its disguise as the Möbius transformations. Discrete groups $G \subset PSL(2, C)$ have traditionally been classified by the behavior of their action as Möbius transformations on S^2. For more information on what follows see [3], [7]. Let $\Omega = \Omega(G) \subset S^2$ be the maximal open set on which G acts discontinuously. The set $L = L(G) = S^2 - \Omega$ is called the *limit set of* G. If $\Omega = \emptyset$, G is called a *Kleinian group of the first kind.* If $\Omega = \emptyset$, G is a *Kleinian group of the second kind.* Most authors find it convenient and traditional to refer to only the second kind as simply *Kleinian.*

If the limit set L of G is contained in a round circle $S^1 \subset S^2$, then G is conjugate to a subgroup of PSL(2, R) and is called *Fuchsian.* We are interested only in cases where $L = S^1 \subset S^2$, which are called *Fuchsian of the first kind.* When G is Fuchsian of the first kind, $\Omega/G = S^2 - S^1/G = H^2_+/G \cup H^2_-/G = \Sigma_\mu \cup \Sigma_{\bar{\mu}}$, where Σ_μ and $\Sigma_{\bar{\mu}}$ are conjugate (anti-holomorphic) surfaces with a certain number of branch points and punctures. Naturally any theory of minimal surfaces will be complicated, but probably not impossibly, by branch points and punctures.

A *quasi-Fuchsian group* G is a group whose limit set lies in a Jordan curve. (If the curve has finite one-dimensional Hausdorff measure, the group is Fuchsian.) If G is the fundamental group of a compact surface Σ, then $\Omega = \Omega_1 \cup \Omega_2$ and $\Omega_1/G = \Sigma_1$, $\Omega_2/G = \Sigma_2$ where $(\Sigma_1, \Sigma_2) \in \mathcal{T}(\Sigma) \times \mathcal{T}(\Sigma)$ is a pair of Riemann surfaces. The following uniformization theorem is due to Bers [1], [2].

THEOREM 2.2 (Bers). *Every pair of Riemann surfaces* $(\Sigma_1, \Sigma_2) \in \mathcal{T}(\Sigma) \times \mathcal{T}(\Sigma)$ *can be represented as* $S^2 - L/G = \Sigma_1 \cup \Sigma_2$ *for* G *quasi-Fuchsian. This representation is unique up to conjugation in* PSL(2, C).

A fundamental tool in the original proof was an existence and regularity theorem for elliptic systems of first order in two variables due to C. B. Morrey [13]. These two deformation theorems are just the tip of a large, nearly complete theory of the classification and deformation of complete hyperbolic manifolds. The classical function theoretical results of Ahlfors, Bers, Marden, Maskit and many others have recently been added to by the more topological approach of Thurston and Sullivan.

There is one additional concept which is important to us. If L is the limit set of a Kleinian group, let $H_0 \subset H^3$ be the convex hull of L in $H^3 \cup S^2$. Then $M_0 = H_0/G \subset M = H^3/G$. If M_0 has finite volume, M (or G) is said to be *geometrically finite*. In any case, M_0 has convex boundary. When M_0 is compact, the existence theory for minimal surfaces applies. If M is quasi-Fuchsian and Σ is a compact surface (without punctures), then $M_0 \subset M$ is compact. We shall be primarily concerned with quasi-Fuchsian groups, for which we hope to find a parameter space of minimal surfaces of $6g-6$ complex dimensions corresponding to the dimension of $\mathcal{T}(\Sigma) \times \mathcal{T}(\Sigma)$. The correspondence between the Kleinian group theory and 3-manifolds can be read about in papers of Marden [12] and Thurston [16].

§3. *Existence and uniqueness theorems*

The existence of immersed minimal surfaces is straightforward. The basic results of Schoen-Yau [15] and Sacks-Uhlenbeck [14] apply to a large class of hyperbolic manifolds, including compact manifolds and quasi-Fuchsian manifolds. It is a result of Gulliver [6] that branch points do not occur for area minimizing surfaces in 3-manifolds.

THEOREM 3.1. *Let M_0 be a compact hyperbolic 3-manifold with convex boundary such that $\pi_1\Sigma \simeq G \subset \pi_1 M_0$. Then there exists an immersed area minimizing surface $\Sigma \subset M$ with $\pi_1\Sigma = G$.*

COROLLARY 3.2. *If M is quasi-Fuchsian, $M \simeq \Sigma \times R$, then M contains an area minimizing immersed surface $\Sigma \subset M_0 \subset M$ with $\pi_1\Sigma = \pi_1 M$.*

One predicts this theorem is true also in the case that Σ is not compact, but contains punctures. This has not been investigated. The branched minimal immersions of the type discussed correspond to the critical points of a proper function on $\mathcal{T}(\Sigma)$. If the solutions are not unique, there exist unstable (not area minimizing) mean curvature zero branched immersions [14].

There are partial uniqueness results. Denote the normal bundle to $\Sigma \subset M$ by $T^{\perp}\Sigma = \Sigma \times R$. Then $\exp : T^{\perp}\Sigma \to M$ is onto if M is complete. If $\Sigma \subset M$ has mean curvature zero, then the second fundamental form $h \in C^{\infty}(T^*\Sigma \otimes T^*\Sigma)$ has eigenvalues $\pm\lambda$ with respect to the metric induced on Σ by the immersion. Here $\lambda^2 \in C^{\infty}(\Sigma)$ and $\lambda \geq 0$. The following is proved by writing down the pull-back metric $\exp^* M$ on $T^{\perp}\Sigma$.

THEOREM 3.3. *If* M *is complete, hyperbolic,* $\Sigma \subset M$ *minimal and* $|\lambda(x)| \leq 1$ *for all* $x \in \Sigma$, *then*

(a) $\exp T^{\perp}\Sigma \simeq \tilde{M} \to M$, *where* $\tilde{M}$ *is the cover of* M *corresponding to* $\pi_1\Sigma \subset \pi_1 M$.

(b) $\tilde{M}$ *is quasi-Fuchsian.*

(c) $\Sigma \subset M$ *is area minimizing;* $\Sigma \subset \tilde{M}$ *is the only closed minimal surface of any type in* $\tilde{M}$.

(d) $\Sigma \subset \tilde{M}$ *is embedded.*

(e) $\Sigma \subset M$ *is totally geodesic if and only if* $\tilde{M}$ *is Fuchsian.*

One does not know to what extent this theorem is true for $\lambda(x) > 1$, since it is the method and not the result which breaks down. However, we do know that the uniqueness theorem is not always valid. We certainly expect quasi-Fuchsian manifolds to contain only finitely many minimal surfaces. Our example is not quasi-Fuchsian and contains an infinite number of immersed minimal surfaces.

We use the remarkable construction of Jørgenson and Thurston for compact hyperbolic manifolds which fiber by Σ over S^1. Briefly, if $\sigma : \Sigma \to \Sigma$ is a diffeomorphism, then its mapping torus is the compact manifold $\Sigma \times R/\tilde{\sigma} = M_{\sigma}$, where the equivalence $\tilde{\sigma}$ is given by $(x, \lambda)\,\tilde{\sigma}(\sigma(x), \lambda+1)$. Their result is that M_{σ} has a hyperbolic structure if no diffeomorphism isotopic to σ leaves a finite set of non-trivial, non-intersecting loops invariant. However, from our basic Theorem 3.1, there exists an area minimizing immersion $\Sigma \subset M_{\sigma}$ with $0 \to \pi_1\Sigma \to \pi_1 M_{\sigma} \to Z \to 0$.

Now consider $\tilde{M}$ $\Sigma \times R \to M_{\sigma}$ to be the cylinder, the regular cover of M_{σ} corresponding to the normal subgroup $\pi_1\Sigma \subset \pi_1 M_{\sigma}$.

THEOREM 3.4. *The complete hyperbolic manifold* $\tilde{M}$ *contains an infinite number of area minimizing surfaces* Σ_i, *the* Z *lifts of the area minimizing immersion of* $\Sigma \subset M_\sigma$.

The manifold $\tilde{M}$ is not quasi-Fuchsian, but in the Thurston model it lies on the boundary of the quasi-Fuchsian manifolds. If the area minimizing surfaces in $\tilde{M}$ are isolated, we expect a large number of minimal surfaces in quasi-Fuchsian manifolds near $\tilde{M}$. Conversely, the uniqueness of minimal surfaces in all quasi-Fuchsian manifolds would probably imply that $\tilde{M}$ is fibered by minimal surfaces. This is unlikely, but unknown as yet.

§4. *The Gauss-Codazzi equations*

The Riemann curvature tensor $R_{ijk\ell}$ on a 3-manifold has 6 components. Due to the Bianchi identities, these components are not completely independent. In a collar neighborhood of an immersed surface $\Sigma \subset M$, there exist normal coordinates induced by $\exp : T^{\perp}\Sigma \to M$ in a neighborhood on which $(-\nu, \nu) \times \Sigma \subset T^{\perp}\Sigma \to M$ is (locally) a diffeomorphism. If coordinates $\{(x^1, x^2)\}$ are introduced on Σ, then $\exp(x^3, (x^1, x^2)) = (x^1, x^2, x^3)$ induces a coordinate patch in M.

Much of what follows is true if M has constant curvature [11]. However, we assume M is hyperbolic. There are three curvature equations of the form

$$R_{i3j3} = -g_{ij} \qquad i = (1,2);\ j = (1,2)\,.$$

These are the equations used to prove Theorem 3.3. Once the metric and second fundamental form have been given, these 3 equations determine uniquely a metric on $T^{\perp}\Sigma$ which may be singular due to conjugate points of $\exp : T^{\perp}\Sigma \to M$. However the metric is unique and non-singular in some collar coordinate patch $(-\nu, \nu) \times \Sigma \subset M$.

Three curvature equations remain, which act as constraint equations for the metric and second fundamental form on Σ. When these remaining 3 equations are satisfied a metric of constant curvature -1 is determined from this data in the collar $(-\nu, \nu) \times \Sigma$ by the three $R_{i3j3} = -g_{ij}$ equations. It is due to the Bianchi identities that the three constraint curvature equations are valid away from Σ when they hold on Σ. A more global form of this unique construction occurs in Section 5.

Given a minimal immersion $\Sigma \subset M$, assign to $\Sigma = \Sigma_\mu$ the conformal structure induced by the immersion. Two of the remaining curvature equations have the form $R_{ijk3} = 0$ and are called the Codazzi equations. These can be written rather neatly in a way which depends on the complex structure of Σ_μ but not the metric length. To see this, choose isothermal (complex) coordinates $(x^1, x^2) = x^1 + ix^2 = z$ on Σ_μ. Let $h = h_{11}(dx^1)^2 + h_{22}(dx^2)^2 + 2h_{12}dx^1dx^2$ be the second fundamental form of $\Sigma \subset M$. Since Σ is minimal, $h_{11} = -h_{22}$ and

$$h = \mathrm{Re}(h_{11} - ih_{12})(dz)^2 = \mathrm{Re}\ \alpha\ .$$

The form $\alpha = (h_{11} - ih_{12})dz^2$ is invariantly defined as a quadratic differential, or a section of $T^*\Sigma \otimes_C T^*\Sigma$. It was known to Bianchi and observed by many others that the Codazzi equations for a minimal surface are equivalent to the Cauchy-Riemann equations for α [11].

THEOREM 4.1. *Let $\Sigma_\mu \subset M$ be a minimal immersion. Then the second fundamental form of the immersion is the real part of a holomorphic quadratic differential on Σ_μ.*

The space of holomorphic quadratic differentials Q_μ over Σ_μ is a complex vector space of complex dimension $3g-3$. The total bundle

$$Q(\Sigma) = \bigcup_{\mu \in \mathfrak{T}(\Sigma)} Q_\mu$$

is a complex vector bundle over $\mathfrak{T}(\Sigma)$ which has a natural identification with the cotangent bundle of Teichmuller space $T^*\mathfrak{T}(\Sigma)$. The $12g-12$ real dimension of the total space of this bundle agrees with the dimension of the space of quasi-Fuchsian manifolds (groups).

We finally come to the last, least tractible of the six curvature equations, the Gauss equation for intrinsic curvature on Σ_μ, or $R_{1212} = -g_{11}g_{22} + g_{12}^2$. The conformal surface Σ_μ carries its natural hyperbolic metric g^μ. The induced metric from $\Sigma_\mu \subset M$ can be written $g_{ij} = e^{2u} g_{ij}^\mu$. Let Δ be the Laplace operator and $|\alpha|^2$ the norm of α, both in the hyperbolic metric on Σ_μ.

THEOREM 4.2. *The Gauss equation may be written*

$$\Delta u + 1 - e^{2u} + |\alpha|^2 e^{-2u} = 0 .$$

This equation is an elliptic quasi-linear equation similar to those studied by Kazdan [8].

The linearized operator for the Gauss equation at u is

$$L\phi = \Delta\phi - 2e^{2u}\phi + 2|\alpha|^2 e^{-2u}\phi .$$

This operator is identical to the operator for second variation of area.

$$d^2A(\phi,\phi) = \int_{\Sigma_\mu} (|d\phi|^2 + 2e^{2u}\phi^2 - 2|\alpha|^2 e^{-2u}\phi^2)\, dx_\mu$$

$$= -\int_{\Sigma_\mu} \phi L\phi \, dx_\mu .$$

This gives special results for stable minimal surfaces. Consider the one-parameter family of Gauss equations for metrics on a minimal surface Σ_μ of fixed conformal type and second fundamental form $\mathrm{Re}\,\tau\alpha = \tau\,\mathrm{Re}\,\alpha$ for α fixed, $\tau \in R^+$. $F(u,\tau) = \Delta u + 1 - e^{2u} - \tau^2|\alpha|^2 e^{-2u} = 0$. Using the implicit function theorem one finds a smooth curve of solutions $\sigma : [0, q_0) \to R^+ \times C^\infty(\Sigma)$ solving $F(\sigma) = 0$ with $\sigma(0,0) = (0,0)$. Let $\sigma(q) = (\tau(q), u(q)$.

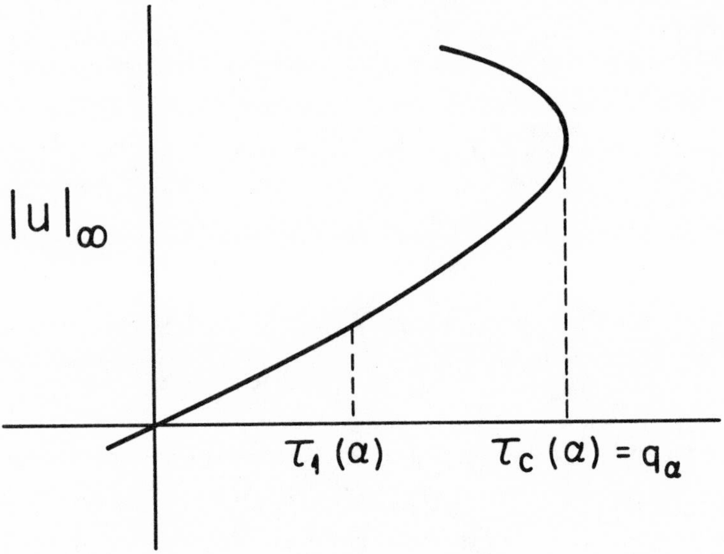

Diagram 4.3. Bifurcation diagram for Gauss equation.

THEOREM 4.4. *There exists a smooth curve* $\sigma: [0, q_0) \to R^+ \times C^\infty(\Sigma)$ *solving* $F(\sigma(q)) = F(u(q), \tau(q)) = 0$ *with* $\sigma(0) = (0, 0)$. *The parameterization may be chosen with* $\tau^2(q) = q(2\tau_c(\alpha) - q)$ *for* $\tau_c(\alpha) < q_0 < 2\tau_c(\alpha)$ *and* $0 \le q < q_0$. *Furthermore*

(a) $\frac{\partial u}{\partial q}(q, x) = \dot{u}(q, x) < 0, \ 0 \le q < q_0$.

(b) *A minimal surface* Σ_μ *in any* M *with second fundamental form* $\tau\alpha$ *is stable if and only if its metric is* $e^{2u(q)} g^\mu$ *for* $\tau^2 = q(2\tau_c(\alpha) - q)$ *and* $0 \le q \le \tau_c(\alpha)$.

In other words, the metric of every stable minimal surface can be found by looking at these branches of solutions of the Gauss equation. The inequality (a) implies that the eigenvalues $\pm\lambda = \pm\tau|\alpha|e^{-2u}$ of the second fundamental form increase in size along this solution curve $\sigma(q)$ for $0 < q < \tau_c$. The set of solutions for which the pointwise inequality $+\lambda \le 1$

is true occurs in a smaller interval $0 \le \tau \le \tau_1(\alpha) < \tau_c(\alpha)$ of the parameter τ in which the stable surfaces occur.

COROLLARY 4.5. *If* Σ_μ *is a minimal surface with second fundamental form* $\mathrm{Re}(\tau\alpha)$, *then the hypotheses of Theorem 3.3* $(\lambda(x) \le 1)$ *are true if and only if the induced metric* $e^{2u}g^\mu$ *is given by* $u = u(q)$, $\tau^2 = q(2\tau_c(\alpha) - q)$ *in the interval* $0 \le \tau \le \tau_1(\alpha)$ *where* $0 \le q \le \tau_c(\alpha) - \sqrt{\tau_c^2(\alpha) - \tau_1^2(\alpha)}$.

This gives us two disk bundles which lie in $Q(\Sigma)$. $Q_1(\Sigma) \subset Q_c(\Sigma) \subset Q(\Sigma)$.

$$Q_c(\Sigma) = \{\alpha \in Q_\mu(\Sigma) : \tau_c(\alpha) \le 1,\ \mu \in \mathcal{T}(\Sigma)$$
$$Q_1(\Sigma) = \{\alpha \in Q_\mu(\Sigma) : \tau_1(\alpha) \le 1,\ \mu \in \mathcal{T}(\Sigma)\ .$$

COROLLARY 4.6. *Every element in* $Q_c(\Sigma)$ *corresponds uniquely to a formal solution* $(e^{2u}g^\mu, \alpha)$ *of the Gauss-Codazzi equations for a stable minimal surface* Σ_μ *in a hyperbolic manifold. Elements* (μ, α) *of* $Q_1(\Sigma)$ *correspond in one-to-one fashion to quasi-Fuchsian manifolds with unique minimal surfaces* Σ_μ *with second fundamental form* α *and* $\lambda(x) \le 1$.

These maps are analytic, although we have not carried out this computation here. Also, the shape of the boundaries of $Q_c(\Sigma)$ and $Q_1(\Sigma)$ has not been investigated.

§5. *Induced group representations*

It is a fundamental theorem in differential geometry that every formal solution of the Gauss-Codazzi equations for immersion of H^2 in H^3 is realized uniquely (up to isomorphism) by an immersion. Given $\{(\mu, \alpha, u) : (\mu, \alpha) \in Q(\Sigma)$ and u a solution of the Gauss equation$\}$ we can lift the data $(e^{2u}g^\mu, \alpha)$ on Σ to $(e^{2\tilde{u}}1, \tilde{\alpha})$ on H^2.

THEOREM 5.1. *There exists a unique minimal immersion* $i = i(\mu, \alpha, u) : H^2 \to H^3$ *such that* $i(0) = 0$, $di_0 : T_0H^2 = R^2 \to R^2 \subset R^3 = T_0H^3$. *The immersion* i *satisfies*:

(a) $\mathrm{Re}\,\tilde{a}$ *is the second fundamental form.*

(b) *The induced metric is* $e^{2\tilde{u}}I$ *on* H^2.

The immersion depends analytically upon the data $(\mu, a, u) \in \{(\mu, a, u) : (\mu, a) \in Q(\Sigma)$ *and* $\Delta u + 1 - e^{2u} - |a|^2 e^{-2u} = 0$ *on* $\Sigma_\mu\}$.

The uniqueness tells us that every other minimal immersion with the same data differs from the given one by an isometry of H^3. Such isometries induce a representation of the deck transformation $\pi_1(\Sigma)$ in $PSL(2, C)$.

COROLLARY 5.2. *Given* $(\mu, a) \in Q(\Sigma)$ *and* u *a solution of the Gauss equation, there exists a representation* $\rho : \pi_1\Sigma \to PSL(2, C)$ *leaving the minimal immersion* $i(\mu, a, u)$ *invariant. The representation* ρ *depends analytically on the data*

$$\begin{array}{ccc} \pi_1\Sigma \circlearrowright H^2 & \xrightarrow{\ i\ } & H^3 \supset \rho(\pi_1\Sigma) \\ \downarrow & & \downarrow \\ \Sigma & \dashrightarrow & H^3/\rho(\pi_1\Sigma) \simeq M . \end{array}$$

Diagram 5.3. Induced representations of $\pi_1\Sigma$.

We made definitions at the end of Section 4 which allow us to consider only stable minimal immersions. In this case the metric is determined by the conformal structure and the second fundamental form.

THEOREM 5.4. *There exists a continuous map* $\mathcal{P} : Q_c(\Sigma) \to \mathrm{Rep}(\pi_1\Sigma, PSL(2, C))$ *such that* $\mathcal{P}(\mu, a) = G$ *leaves invariant a stable minimally immersed* $H^2 \subset H^3$, $H^2/G = \Sigma_\mu$ *and the second fundamental form of* H^2 *in* H^3 *is lift of* $\mathrm{Re}\, a$ *from* Σ_μ *to* H^2.

The question remains of deciding when the representation $\mathcal{P} = \mathcal{P}(\mu, a)$ is discrete and the three manifold M exists in Diagram 5.3. From the methods in Theorem 3.3, we know that formal solutions of the Gauss-Codazzi equations can always be realized locally as the data for some minimal embedding in an incomplete hyperbolic manifold. This manifold may not have a completion. However, we have the following partial information.

THEOREM 5.5. $\mathcal{P} : Q_c(\Sigma) \to \mathrm{Rep}(\pi_1 \Sigma, \mathrm{PSL}(2, C))$ *has in its image the quasi-Fuchsian groups. Under* $\mathcal{P}$, *the sub-disk bundle* $Q_1(\Sigma) \subset Q_c(\Sigma)$ *is homeomorphic to a neighborhood of* $\mathrm{Rep}(\pi_1 \Sigma, \mathrm{PSL}(2, R))$. *If* $\mathcal{P}(\mu, \alpha)$ *is quasi-Fuchsian, then* $M = H^3/\mathcal{P}(\mu, \alpha)$ *contains* Σ_μ *as a stable minimal surface with second fundamental form* α.

There are many questions left unanswered: this last theorem is anticlimactic.

1. Which elements of $Q_c(\Sigma)$ correspond to discrete groups? To quasi-Fuchsian groups? Is this set connected? (This has relevance for the embedding problems.)
2. If $\mathcal{P}(\mu, \alpha)$ is discrete, is $\mathcal{P}(\mu, e^{i\theta}\alpha)$ also discrete? Is $\mathcal{P}$ a complex analytic mapping? (Probably no.)
3. Which minimal area immersions lie in compact manifolds? (Hard to impossible.)
4. Which minimal area immersions are unique, embedded, injective on the fundamental group?
5. What is the relation between the Hausdorff dimension of the limit set of $\mathcal{P}(\mu, \alpha)$ on S^2 and the data (μ, α) ?

Appendix: Proofs

None of the proofs of the theorems in §§3-5 are difficult. They were omitted from the text in order to emphasize the structure of the theory. For convenience, each theorem is listed here in order, although in some cases we refer to another author for a proof.

Theorem 3.1. This theorem follows from the published work of Schoen and Yau, who include ∂M_0 with mean curvature ≥ 0 and more general hypotheses on M_0 [15].

Corollary 3.2. If M is quasi-Fuchsian and $\pi_1 M = \pi_1 \Sigma$ for Σ a compact surface, the maximal convex submanifold $M_0 \subset M$ is compact. Theorem 3.1 applies to M_0.

Theorem 3.3. The coordinate system used in this proof is the one described in Section 4. Use $\exp : T^{\perp}\Sigma \to M$ to pull back the metric from M to $T^{\perp}\Sigma \simeq \Sigma \times R$. For $(x,t) \in \Sigma \times R$, the pullback metric has the form

$$G(x,t) = \begin{pmatrix} g(x,t) & 0 \\ 0 & 1 \end{pmatrix} = \begin{pmatrix} g_{11}(x,t) & g_{12}(x,t) & 0 \\ g_{12}(x,t) & g_{22}(x,t) & 0 \\ 0 & 0 & 1 \end{pmatrix} .$$

If the coordinate system on $(\Sigma, 0)$ is isothermal, $g(x,0) = \{g_{ij}(x,0)\} = e^{2v(x)} I$. The matrix

$$h(x,t) = \{h_{ij}(x,t)\} = \frac{1}{2}\{\partial/\partial t g_{ij}(x,0)\}$$

is the second fundamental form of $(\Sigma, 0) \subset M$. Since Σ is minimal, $h_{11} + h_{22} = 0$. Then $\pm\lambda(x) = \pm\sqrt{-\det h(x)\, e^{-2v(x)}}$ are the eigenvalues of the second fundamental form with respect to $e^{2v(x)} I$.

The three curvature equations $R_{i3j3} = -g_{ij}$ can be written explicitly in $T^{\perp}\Sigma$ as

$$\frac{1}{2}(\partial/\partial t)^2 g_{ij}(x,t) - \frac{1}{4}\,\partial/\partial t g_{i\ell}(x,t)\, g^{\ell k}(x,t)\, \partial/\partial t g_{kj}(x,t) = -g_{ij}(x,t) .$$

For fixed x, this is a second order system of differential equations with the explicit matrix solution

$$g(x,t) = e^{2v(x)}(\cosh t\, I + \sinh t\, e^{-2v(x)} h(x))^2 .$$

This metric is non-singular in a collar neighborhood of Σ in any case. If $\lambda(x) = \sqrt{-\det h(x)\, e^{-2v}} \le 1$, it is non-singular for all t. This directly proves (a), (d), and (e).

To show $\tilde{M} \simeq T^{\perp}\Sigma$ is quasi-Fuchsian for $\lambda(x) < 1$, note the direct, bounded metric comparison:

$$(1 - \|\lambda\|_{\infty}) \cosh^2 t\, g(x,0) \leq g(x,t) \leq (1 + \|\lambda\|_{\infty}) \cosh^2 t\, g(x,0)\,.$$

The base metric g compares uniformly with the constant curvature metric g^{μ} on $\tilde{M}$. A Fuchsian metric on $\Sigma \times R$ has the form

$$\begin{pmatrix} \cosh^2 t\, g^{\mu} & 0 \\ 0 & 1 \end{pmatrix}.$$

Therefore $\tilde{M}$ is quasi-isometrically equivalent to a Fuchsian manifold and is therefore quasi-Fuchsian.

To prove (c), compute the mean curvature of $(\Sigma, t) \subset T^{\perp}\Sigma$. In the explicit coordinates used already, mean curvature is

$$H(x,t) = \frac{2 \cosh t \sinh t (1 - \lambda^2(x))}{\cosh^2 t - \sinh^2 t\, \lambda^2(x)}\,.$$

Suppose $\Sigma_0 \subset \tilde{M}$ is any compact normal surface and let $t_{\infty} = \max_{(t,x)\epsilon\Sigma_0} t$ and $t_{-\infty} = \min_{(t,x)\epsilon\Sigma_0} t$. If $\lambda^2 \leq 1$, by maximum and minimum principles, $t_{\infty} \leq 0$ and $t_{-\infty} \geq 0$. This gives $t_{-\infty} = t_{\infty} = 0$ and $\Sigma_0 \subset \Sigma$.

Theorem 3.4. The map $\Sigma \to \Sigma \times R \to \Sigma \times R/\tilde{\sigma} = M_{\sigma}$ induces an injection on the fundamental group of Σ. By references [14] and [15], there exists a minimal immersion of Σ in M_{σ} which is homotopic to the fiber of M_{σ}. The lifts of this area minimizing surface to the cover $\tilde{M} \simeq \Sigma \times R$ are also area minimizing (in the lifted hyperbolic metric). This is discussed in [5] and in Section 5.

Theorem 4.1. This theorem is true for all local mean curvature zero immersions of surfaces in constant curvature 3-manifolds. See in particular reference [11].

Theorem 4.2. Let $K \in C^\infty(\Sigma)$ be the intrinsic curvature in the metric induced by the immersion $\Sigma \subset M$. Then $K(x) = e^{-2u(x)}(-\Delta u(x) - 1)$ describes the relation between K and the change in metric form hyperbolic g^μ_{ij} to $g_{ij} = e^{2u} g^\mu_{ij}$. On the other hand $K(x) = -1 - \lambda^2(x)$ describes the curvature of a surface in a 3-manifold of constant curvature -1 and extrinsic curvatures $\pm\lambda(x)$. Since $\lambda(x) = |\alpha(x)|\, e^{-2u(x)}$ where $|\alpha(x)|$ is the metric length in the metric of constant curvature -1, we get

$$K = e^{-2u}(-\Delta u - 1) = -1 - |\alpha|^2 e^{-4u}\,.$$

Theorem 4.4. Let $L^2_k = L^2_k(\Sigma)$ be the space of functions on Σ with k derivatives in $L^2(\Sigma)$. Then

$$F(u,\tau) = \Delta u + 1 - e^{2u} - \tau^2\, |\alpha|^2 e^{-2u}$$

defines a smooth map

$$F : L^2_{k+2}(\Sigma) \times R \to L^2_k(\Sigma)\,.$$

The differential

$$dF(u,\tau)(\dot u,\dot\tau) = \Delta \dot u - 2(e^{2u} - \tau^2|\alpha|^2 e^{-2u})\,\dot u + 2\tau\dot\tau\, |\alpha|^2 e^{-2u}$$

is onto when the linear operator

$$L(u,\tau) = -\Delta + 2(e^{2u} - \tau^2|\alpha|^2 e^{-2u})$$

has positive eigenvalues. When L is non-negative, its kernel (and co-kernel) are one-dimensional and generated by a positive function. In this case $dF(u,\tau)(0,\dot\tau) = 2\tau\dot\tau e^{-2u}$ spans the deficient direction of L and dF is still onto.

By the implicit function theorem, every solution of $F(u,\tau) = 0$ with $L = L(u,\tau)$ non-negative lies in a one-parameter family of solutions in $L^2_k(\Sigma) \times R$. Moreover, if $\dot u = \partial u/\partial q$, $\dot\tau = \partial\tau/\partial q$

$$dF(u,\tau)(\dot u,\dot\tau) = -L\dot u + 2\tau\dot\tau\, |\alpha|^2 e^{-2u}\,.$$

By the maximum principle $\dot{u} < 0$ if L has positive eigenvalues and $\dot{\tau} > 0$. This shows the existence of the solution curve and (a).

The curve $\sigma(q)$ of solutions for which $L = L(u,\tau)$ is positive can end in only two ways, since one can check the existence of uniform bounds on solutions with this property. Either $\tau = 0$ and we are on the left side of the branch described, or $L = L(u,\tau)$ has a one-dimensional kernel and $\dot{\tau} = 0$ at the end.

$$dF(u,\tau)(\dot{u},\dot{\tau}) = -L\dot{u} = 0 .$$

Differentiate this equation in the parameter q again.

$$\begin{aligned} 0 &= dF(u,\tau)(\ddot{u},\ddot{\tau}) + d^2F(u,\tau)(\dot{u},\dot{\tau})(\dot{u},\dot{\tau}) \\ &= -L\ddot{u} - 4(e^{2u}+\tau^2|\alpha|^2 e^{-2u})\dot{u}^2 - 2\ddot{\tau}\tau|\alpha|^2 e^{-2u} . \end{aligned}$$

Take the inner product with $\dot{u}$. For the first term

$$\int_\Sigma \dot{u}(L\ddot{u}) = \int_\Sigma (L\dot{u})\ddot{u} = 0 .$$

Then

$$\int_\Sigma 4(e^{2u}+\tau^2|\alpha|^2 e^{-2u})u^3 + 2\ddot{\tau}\tau \int_\Sigma |\alpha|^2 e^{-2u}\dot{u} = 0 .$$

Since $\dot{u}$ is the first eigenfunction of $L(u,\tau)$, it has a single sign, and $\ddot{\tau} < 0$. If we choose $\dot{\tau} \geq 0$ along the curve, then $\dot{u} < 0$, $u < 0$ by the maximum principle, and $\|\dot{u}\|_\infty > 0$. Therefore we can follow every solution with $L(u,\tau) \geq 0$ down to $u = 0$, $\tau = 0$. Only the one branch of solutions exists for $L(u,\tau) \geq 0$. Since $\ddot{\tau} < 0$ where $\dot{\tau} = 0$ and $\tau = \tau_\alpha$, we can reparameterize so $\tau^2(q) = q(2\tau_c(\alpha) - q)$.

The connection with stable minimal surfaces follows from the fact that if $\Sigma_\mu \subset M$ is a stable minimal surface with second fundamental form real $\tau\alpha$ and metric $e^{2u}g^\mu_{ij}$, then

$$d^2A(\phi,\phi) = \int_\Sigma \phi L(u,\tau)\phi \geq 0 .$$

The stable minimal surfaces are exactly those with $L(u,\tau) \geq 0$.

Corollary 4.5. Since $u(q,x)$ is strictly decreasing and $\tau(q)$ is increasing, $\lambda(x) = \tau(q)|\alpha|e^{-2u(x,q)}$ increases along q. Define $\tau_1(\alpha) = \min\{\tau : \tau(q)|\alpha|e^{-2u(x,q)} = 1 \text{ for some } x, 0 < q < q_c\}$.

Corollary 4.6. $Q_c(\Sigma)$ is defined to be the set of elements in $Q(\Sigma)$ with a formally stable solution to the Gauss equation. For $(\mu,\alpha) \in Q_1(\Sigma)\,\lambda(x) \leq 1$ by definition. The metric constructed in Theorem 3.3 gives a complete hyperbolic metric on $\Sigma \times R \simeq T^{\perp}\Sigma$. The hypotheses of Theorem 3.4 apply to this manifold.

Proof of Theorem 5.1. We include a proof of this theorem. We show that the immersion in fact depends analytically on the data. This and other results of this type are essentially the same as Calabi's work on surfaces in spheres [4].

Let $H^3 = \{w \in E^{3,1} : (w\cdot w) = -1\}$. Choose an identification $H^2 \simeq D^2 = \{x \in R^2 : |x| < 1\}$. Write $\tilde{\alpha}$ and $e^{2\tilde{u}} g^{\mu} = e^{2v} I$ in these coordinates, using polar coordinates in $H^2 \simeq D^2$. We write down equations for an orthogonal frame on H^2 which is singular at the origin.

$$\vec{e}_0 = \vec{w} \qquad\qquad \vec{e}_n = \vec{e}_0 \wedge \vec{e}_r \wedge \vec{e}_\theta$$

$$\vec{e}_r = e^{-v}\vec{w}_r$$

$$\vec{e}_\theta = \frac{1}{r}e^{-v}\vec{w}_\theta$$

Given the formal second fundamental form $\tilde{\alpha} = (\tilde{\alpha}_{rr}, \tilde{\alpha}_{r\theta}, \tilde{\alpha}_{\theta\theta})$ and the metric $e^v I$, we can write down a system of ordinary differential equations.

$$\begin{aligned}
\dot{\vec{e}}_0 &= e^v \vec{w}_r \\
\dot{\vec{e}}_r &= v_\theta \vec{e}_\theta + \tilde{\alpha}_{rr} \vec{e}_n + e^v \vec{e}_0 \\
\dot{\vec{e}}_\theta &= -v_\theta \vec{e}_r + \tilde{\alpha}_{r\theta} \vec{e}_n \\
(\dot{\vec{e}}_n) &= -\tilde{\alpha}_{rr} \vec{e}_r - \tilde{\alpha}_{r\theta} \vec{e}_\theta \,.
\end{aligned}$$

Given initial conditions $\vec{e}_0(0,\theta) = (0,0,0,1)$, $\vec{e}_r(0,\theta) =$ $(\cos\theta, \sin\theta, 0, 0, 0)\, \vec{e}_\theta(0,\theta) = (\sin\theta, -\cos\theta, 0, 0)$ and $\vec{e}_n(0,\theta) =$ $(0,0,1,0)$ we can solve uniquely for the frame as a function of the parameter where $\cdot = \partial/\partial r$. From the Gauss-Codazzi equations, $\rho = \vec{e}_0 : H^2 \to H^3$ is a complete minimal immersion with the desired metric and second fundamental form. Note that v and α are analytic since they both solve non-linear elliptic equations. Therefore $\rho : H^2 \to H^3$ is also analytic in x. Also, from the form of the equations, $\vec{e}_0$ is also analytic in the functional parameters $\tilde{\alpha}$ and v.

Corollary 5.2. The definition of ρ can be made from the immersion in Theorem 5.1. Let $\gamma \in \pi_1 \Sigma$ and $w : H^2 \to \tilde{M}$ be the minimal immersion from Theorem 5.1. Then $w \circ \gamma$ is also a minimal immersion taking $x_0 \to w(\gamma(x_0))$ and $T_{x_0}(H^2)$ to $T_{w(\gamma(x_0))}$ under $dw_{\gamma(x_0)} \circ d\gamma_{x_0}$. Let $\rho(\gamma)$ be the unique element of $PSL(2,C)$ satisfying

$$\begin{aligned}
\rho(\gamma)\, w(0) &= w(\gamma(0)) \\
d\rho(\gamma)_{w(0)} dw_0 &= dw_{\gamma(0)} \circ d\gamma_0 \,.
\end{aligned}$$

The map $\rho(\gamma)$ is unique and analytic in the given data. Since w and $\rho(\gamma)^{-1} w \circ \gamma$ agree on $0 \in H^2$ and have the same differential, then $w = \rho(\gamma)^{-1} \circ w \circ \gamma$ as required.

Theorem 5.4. In Corollary 4. ,we obtained a map from $Q_c(\Sigma)$ to formal solutions of the Gauss-Codazzi equations. Compose the map with the representations constructed in Corollary 5.2. By construction, this minimal surface has the correct conformal structure and second fundamental form. To show it is stable, consider the first eigenfunction of the stability operator L, so $L\phi + \lambda\phi = 0$ for $\lambda \leq 0$, $\phi > 0$. Lift to $\tilde{L}$ and $\tilde{\phi}$ on the covering minimal surface $H^2 \subset H^3$. Then $\tilde{L}\tilde{\phi} + \lambda\tilde{\phi} = 0$ for $\lambda < 0$, $\phi > 0$. Since $\tilde{L}$ is the second variation operator, this implies stability of $H^2 \subset H^3$ [5].

Theorem 5.5. By Corollary 3.2, every quasi-Fuchsian manifold M contains a stable minimal immersion. Let (μ, α) be the conformal type and second fundamental form of this stable minimal surface. Then $H^3/\mathcal{P}(\mu, \alpha) = M$. By Theorem 3.4 and Corollary 4.6, $\mathcal{P}|Q_1(C) \to \mathrm{Rep}(\pi_1\Sigma, PSL(2,C))$ is one-to-one, onto a neighborhood of the Fuchsian groups. Although we have not entirely shown the details, it is easy to see that the map is analytic. It is locally invertible due to the strict-stability of the unique minimal surfaces. The implicit function theorem implies their local smooth dependence on their ambient quasi-Fuchsian manifolds.

KAREN K. UHLENBECK
DEPARTMENT OF MATHEMATICS
UNIVERSITY OF ILLINOIS AT CHICAGO
CHICAGO, ILLINOIS 60680

REFERENCES

[1] L. Bers, Simultaneous uniformization, Bull. Amer. Math. Soc. *66*(1960), 94-97.

[2] ———, Uniformization, moduli and Kleinian groups, Bull. London Math. Soc. *4*(1972), 257-300.

[3] ———, What is a Kleinian group? in a crash course in Kleinian groups, Springer Lecture Notes *400*(1974), 1-13.

[4] E. Calabi, Minimal immersions of surfaces in Euclidean spheres, J. Diff. Geo. *1*(1967), 111-127.

[5] D. Fischer-Colbrie and R. Schoen, The structure of complete stable minimal surfaces of non-negative scalar curvature (to appear).

[6] R. Gulliver III, Regularity of minimizing surfaces of prescribed mean curvature, Ann. of Math. *97*(1973), 275-305.

[7] W. Harvey (editor), Discrete groups and automorphic functions, Proc. NATO Conference, Cambridge, England (1975).

[8] J. Kazdan and F. Warner, Existence and conformal deformation of metrics with prescribed Gaussian and scalar curvatures, Ann. of Math. *101* (1975), 317-331.

[9] L. Keen, Intrinsic moduli on Riemann surfaces, Ann. of Math. *84* (1966), 404-420.

[10] ———, On Fricke moduli, Annals of Math. Studies *66* (1971), 205-224.

[11] H. B. Lawson, Jr., Complete minimal surfaces in S^3, Ann. of Math. *92* (1970), 335-374.

[12] A. Marden, Kleinian groups and 3-dimensional topology, in a crash course on Kleinian groups, Springer Lecture Notes *400* (1974), 108-121.

[13] C. B. Morrey, On the solution of quasi-linear elliptic partial differential equations, Trans. Amer. Math. Soc. *43* (1938), 126-166.

[14] J. Sacks and K. Uhlenbeck, Minimal immersions of Riemann surfaces, Trans. Amer. Math. Soc. *271* (1982), 639-652.

[15] R. Schoen and S. T. Yau, Existence of incompressible minimal surfaces and the topology of 3-manifolds with non-negative curvature, Ann. of Math. *110* (1979), 127-142.

[16] W. Thurston, Notes on the topology of 3-manifolds, Princeton Math. Dept.

MINIMAL SPHERES AND OTHER CONFORMAL VARIATIONAL PROBLEMS

Karen K. Uhlenbeck

In this article, the relationship between energy integrals and area or volume integrals is described. We also state some of the results on existence of closed minimal surfaces in Riemannian manifolds due to Schoen-Yau and Sacks-Uhlenbeck.

In our general framework M is a compact Riemannian domain manifold without boundary of dimension n. The ambient Riemannian manifold N is compact, and its dimension is usually not too important. For convenience we assume $\partial N = \emptyset$, although in the most general results, ∂N may be non-empty if it has the correct curvature in N [15].

Let $s : M \to N$. Then $ds(x) : T_x M \to T_{s(x)} N$, $TN \simeq T^*N$ and $ds^*(x) : T^*_{s(x)} N \to T^*_x M$. There is a class of variational problems with Lagrangian depending on the eigenvalues of the symmetric composition map with respect to the metric tensor.

$$ds^*(x)\, ds(x) : T_x M \to T^*_x M \, .$$

The most familiar is area or volume.

$$A(s) = \int_M (\det(ds^*(x)\, ds(x)))^{1/2} * 1 \, .$$

Seminar on Minimal Submanifolds
0-691-08324-X/83/169-08 $0.90/0 (cloth)
0-691-08319-3/83/169-08 $0.90/0 (paperback)

As one knows from the minimal length problem (where $M = S^1$), the difficulty with this integral is that there is no preferred parameterization; the group of diffeomorphisms of M leaves A invariant and there is too large a critical set. In the geodesic case, one considers the *energy integral* instead [6]. For $M = S^1$, solution curves have minimal (critical) length and constant velocity

$$E(s) = \int_M \text{trace } ds^*(x) ds(x) * 1 .$$

In general dimensions the family of integrals

$$E_p(s) = \int_M \left(\frac{1}{n} \text{trace } ds^*(x)\, ds(x)\right)^{p/2} * 1$$

may help clarify the situation. All these integrals have the same critical curves for $n = 1$.

The best possible behavior for a variational problem is that it satisfy a Morse theory or a Ljusternik-Shnirelmann theory. Without going into the details which are well-documented in the literature, this means the number of critical points or the topology of the critical set depends on topological information from the function space on which the variational problem is worked [9], [10], [11]. The Palais-Smale condition is conceptually the easiest of the conditions to describe which imply the topological results. This condition is automatic for a proper function on a finite dimensional manifold. It replaces the local compactness of finite dimensions in infinite dimensions.

Let $F : X^\infty \to R$ be a differentiable function on a complete Banach (Finsler) manifold X^∞. Then F satisfies the *Palais-Smale condition* if a sequence $f_\alpha \in X^\infty$ with $|F(f_\alpha)| \leq B$ and $\lim_{\alpha\to\infty} |dF(f_\alpha)| = 0$ has a convergent subsequence $f_{\alpha'} \to f$ in X^∞ [11].

In our applications $X^\infty = L_1^p(M, N)$, the Sobolev manifold of maps $s: M \to N$ whose first derivatives lie in L^p. This is a complete smooth Banach manifold in the range of the Sobolev embedding $L_1^p(M, N) \subset C^0(M, N)$, or for $p > \dim M = n$.

THEOREM [8]. *If* $p > \dim M = n$, *then* $E_p : L_1^p(M, N) \to R$ *satisfies the Palais-Smale condition.*

The dimension of M is important. Weaker versions of the following result, related to Simons' theorem for Yang-Mills fields in dimensions larger than 4, have been known for quite a long time [17].

THEOREM. *Let* $M = S^n$ *and* $p < \dim M$. *Then* $E_p : C^1(M, N) \to R$ *can take on no non-trivial local minimum.*

We see the borderline is $p = n$, which is exactly the case where the integral E_n depends on the conformal structure of M and not the Riemannian distance in M. This is the integral which has some relationship with area.

THEOREM. $A(s) \le E_n(s)$. *Equality holds if and only if* s *is weakly conformal* ($d^*s(x) \cdot ds(x)$ *is a multiple of the metric tensor*). *If* s *is weakly conformal,* s *is a critical point of* A *if and only if it is a critical point of* E_n.

There is nothing mysterious about the proof. Let $(\lambda_1, \cdots, \lambda_n)$ be the eigenvalues of $d^*s(x) ds(x)$. The inequality follows from integrating the inequality

$$\prod_{i=1}^{n} |\lambda_i| \le \left(\frac{1}{n} \sum_{i=1}^{n} \lambda_i^2\right)^{n/2} .$$

Equality holds in the integrals exactly when all the eigenvalues λ_i are equal almost everywhere. This is our definition of weakly conformal. The last statement is a calculus fact extended to infinite dimensions. Restated in one variable; assume $f(x) \le g(x)$ and $f(x_0) = g(x_0)$. Then $f'(x_0) = 0 \iff g'(x_0) = 0$.

As we remarked before, E_n depends on the conformal structure of M. The integral A does not. Define the *Teichmüller space* $\mathcal{T}(M)$ of conformal structures as the space of conformal tensors on M under the identification induced by the action of the identity component of the diffeomorphism group $\mathrm{Diff}_0 M$. Now E_n is regarded as a function depending on both the map $s : M \to N$ and the conformal structure $\mu \in \mathcal{T}(M)$.

THEOREM. *If* E_n *has a critical point* (s, μ) *at* $s : M \to N$ *and* $\mu \in \mathcal{T}(M)$, *the map* $s : M_\mu \to N$ *is a weakly conformal minimal immersion.*

The proof for $n > 2$ is the same as for $n = 2$ [13]. We have chosen a preferred parameterization for the minimal area or volume maps as a conformal one. However, very little is known about $\mathcal{T}(M)$ for $n = \dim M > 2$. For $\dim M = 2$, $M = \Sigma$ a surface, then $E = E_2$ is the usual energy integral whose critical points are harmonic maps. The function theorists have studied $\mathcal{T}(\Sigma)$ for Σ a Riemann surface and given us enough information to use this last theorem for $n = 2$ [1], [4]. In fact, in the case $\Sigma = S^2$ (also in very symmetric equivariant problems), $\mathcal{T}(\Sigma)$ is a point and the harmonic maps actually agree with the conformal branched immersions.

From now on we confine our attention to $M = \Sigma$ a surface and $n = 2$. It is true that the same results are true for $n > 2$. We cannot as yet identify these results with the minimal volume problem; they have not been useful geometrically in studying the geometry of N. The parameterization of the map $s : M \to N$ enters into most variational problems in an essential way. This makes the geometry of M important and obscures the geometry of the ambient manifold N.

We give now a brief outline of the perturbation technique which was used to prove the existence of minimal 2-spheres [13]. This method is similar to one used in studying the Yamabe problem and some variation of it probably works on all conformally invariant problems [18].

Recall that $E_{2\alpha}$ satisfies the Palais-Smale condition on $L_1^{2\alpha}(\Sigma, N)$ for $\alpha > 1$. For technical reasons

$$\tilde{E}_{2\alpha}(s) = \int_{\Sigma} (1 + \text{trace } d^*s\, ds)^{\alpha} * 1$$

is used because its Euler-Lagrange equations are uniformly elliptic. So $\tilde{E}_{2\alpha}$ has lots of smooth critical points s_α. We keep track of them as $\alpha \to 1$. By $B(x_0, \rho) \subset \Sigma$ we mean a small disk of radius ρ about x_0 in the surface Σ. The main a priori estimate on solutions s_α of the Euler-Lagrange equations of $\tilde{E}_{2\alpha}$ uses the exact Sobolev embedding $L_2^p(B(x_0, \rho), N) \subset L_1^q(B(x_0, \rho), N)$ for $1/p - 1/2 + 1/q = 0$. $E = E_2$ is here the usual energy.

A Priori Estimate. There exists $\varepsilon > 0$ such that if $E(s_\alpha | B(x_0, \rho)) < \varepsilon$ and s_α is a critical map for $\tilde{E}_{2\alpha}$, then

$$\max_{x \in B(x_0, \rho/2)} |ds_\alpha(x)|^2 \le \rho^{-2} \mathcal{K} E(s_\alpha | B(x_0, \rho)) .$$

The main convergence theorem uses this a priori estimate on disks in the complement of a finite number of points $(x_1, \cdots, x_\ell) \in \Sigma$. By weak compactness and passing to subsequences, if $E(s_\alpha) \le B$, we show $s_\alpha \to s$ in $C^{1,\beta}(\Sigma - \{x_1, \cdots, x_\ell\}; N)$. Here $s \in C^\infty(\Sigma, N)$ is a smooth but possibly trivial harmonic map. Around the points x_i where convergence fails, we reparameterize by a (conformal !) dilation. Let

$$\tilde{s}_\alpha(y) = s_\alpha(x_\alpha + y/m_\alpha)$$

for $x_\alpha \to x_i \in \Sigma$, $m_\alpha \to \infty$. Then by choosing things properly

$$s_\alpha \to s_i : R^2 = S^2 - \{\infty\} \to N$$

converges. The map s_i is a smooth, *non-trivial*, harmonic map which extends smoothly over the point at ∞ to give $s_i : S^2 \to N$ harmonic.

THEOREM (4.4 and 4.6 of [13]). *Let* $E(s_\alpha) \le B$ *and* s_α *be a critical point for* $\tilde{E}_{2\alpha}$. *Then there exists a subsequence* $s_\alpha \to s$ *in*

$C^{1,\beta}(\Sigma - \{x_1, \cdots, x_\ell\}; N)$ *for* $s : \Sigma \to N$ *harmonic. Moreover, there is a non-trivial harmonic map* $s_i : S^2 \to N$ *associated to each* $x_i \in \Sigma$ *at which the convergence* $s_\alpha \to s$ *is not uniform.*

This theorem illustrates a general philosophy for conformally invariant variational problems. There is an obstruction to the convergence of the critical points of the nearly variational problems in the Palais-Smale range (here $\alpha > 1$ or $p > n$) to solutions of the borderline, conformally invariant problem. It consists of solutions lying over the conformally flat (usual) spheres $S^n = R^n \cup \{\infty\}$. These solutions occur in the dilation of small neighborhoods of points blown up to cover $R^n = S^n - \{\infty\}$.

These methods and others yield a number of results on the existence of closed minimal surfaces in Riemannian manifolds. There are very important results due to Meeks and Yau [5] and Gulliver [3] on the embeddedness of these solutions for $\dim N = 3$. In general dimensions these theorems are for minimal branched immersions. As such, they have very little in common with the theorems obtained via geometric measure theory [12].

THEOREM [13]. *If* $\pi_2 N \neq 0$, *then there exist free homotopy classes* $\Lambda_i \in [S^2, N]$ *such that* $s_i \in \Lambda_i$ *and* $s_i : S^2 \to N$ *is an area minimizing branched immersion. Representatives* $\lambda_i \in \Lambda_i$, $\lambda_i \in \pi_2 N$ *form a generating set for* $\pi_2 N$ *as a* $Z[\pi_1(N)]$ *module.*

THEOREM [15], [13]. *If genus* $\Sigma \geq 1$, $u : \Sigma \to N$ *and* $u_* : \pi_1 \Sigma \to \pi_1 N$ *is an injection, then there exists an area minimizing branched immersion* $s : \Sigma \to N$ *with* u_* *and* s_* *equivalent on* $\pi_1 \Sigma$.

There are two results on the existence of possibly unstable minimal surfaces.

THEOREM [13]. *If* $\pi_i(N) \neq 0$, $i \geq 2$, *then there exists a minimal branched immersion* $s : S^2 \to N$.

THEOREM [14]. *Assume* N *has negative sectional curvature,* $u: \Sigma \to N$, $u_*: \pi_1 \Sigma \to \pi_1 N$ *an injection. Let* $K = \{\tau \in \mathrm{Map}\, \Sigma : u \circ \tau \sim u\}$. *If* $\mathcal{T}(\Sigma)/K$ *is a manifold and* u *is not homotopic to a finite covering of another surface in* N, *then the geometrically distinct minimal immersions homotopic to* u *correspond to the critical points of a proper map on* $\mathcal{T}(\Sigma)/K$.

There are hyperbolic manifolds N for which $\mathcal{T}(\Sigma)/K$ is a manifold with the topology of S^1. Negative sectional curvature does not imply the stability of closed minimal surfaces.

KAREN K. UHLENBECK
DEPARTMENT OF MATHEMATICS
UNIVERSITY OF ILLINOIS AT CHICAGO
CHICAGO, ILLINOIS 60680

REFERENCES

[1] W. Abikoff, Degenerating families of Riemann surfaces, Annals of Math. *105*(1977), 29-44.

[2] J. P. Bourguignon, H. B. Lawson, and J. Simons, Stability and gap phenomena for Yang-Mills fields, Proc. Nat. Acad. Sci. U.S.A. *76* (1979), 1550-1553.

[3] R. Gulliver, Regularity of minimizing surfaces of prescribed mean curvature, Annals of Math. *97*(1973), 275-305.

[4] L. Keen, Collars on Riemann surfaces, Annals of Math Studies *79* (1974), 263-268.

[5] W. H. Meeks and S. T. Yau, Topology of three dimensional manifolds and embedding problems in minimal surface theory, Annals of Math. *112*(1980), 441-484.

[6] J. Milnor, Morse theory, Annals of Math. Studies *51* (1963).

[7] C. B. Morrey, Jr., Multiple integrals in the calculus of variations, Springer, New York (1966).

[8] R. S. Palais, Foundations of global non-linear analysis, Benjamen, New York (1968).

[9] ________, Ljusternik-Schnirelmann theory on Banach manifolds, Topology *5*(1966), 115-132.

[10] ________, Critical point theory and the manimax principle, AMS Proc. of Sym. in Pure Math. XV(1970), 185-212.

[11] R.S. Palais and S. Smale, A generalized Morse theory, Bull. Amer. Math. Soc. *70*(1964), 165-171.

[12] J. Pitts, Existence and regularity of minimal manifolds, Bull. Amer. Math. Soc. *82*(1976), 503-504.

[13] J. Sacks and K. Uhlenbeck, The existence of minimal 2-spheres, Ann. of Math. *113*(1981), 1-24.

[14] ———, Minimal immersions of closed Riemann surfaces, Trans. Amer. Math. Soc. *271*(1982), 639-652.

[15] R. Schoen and S.T. Yau, Existence of incompressible minimal surfaces and the topology of three dimensional manifolds with non-negative scalar curvature, Annals of Math. *110*(1979), 127-142.

[16] ———, Incompressible minimal surfaces, three dimensional manifolds with non-negative scalar curvature and the positive mass conjecture in general relativity, Proc. Nat. Acad. Sci. *75* (6)(1978), 2567.

[17] Y.L. Xin, Some results on stable harmonic maps (preprint).

[18] H. Yamabe, On a deformation of Riemannian structures on compact manifolds, Osaka Math. J. *12*(1960), 21-37.

MINIMAL HYPERSURFACES OF SPHERES WITH CONSTANT SCALAR CURVATURE

Chia-Kuei Peng* and Chuu-Lian Terng*

Introduction

Let M be an n-dimensional closed minimally immersed hypersurface in the unit sphere S^{n+1}, and let h denote the second fundamental form of M, a symmetric covariant two tensor on M. We denote the square of the length of h by S. Then S is intrinsic and given by

$$S = n(n-1) - R ,$$

where R is the scalar curvature of M. In particular, S is constant if and only if M has constant scalar curvature. It is well known that if $0 \leq S \leq n$, then $S \equiv 0$ or $S \equiv n$ (J. Simons [11]). This result is based on an identity for the Laplacian of h, using that M is minimal. The minimal hypersurfaces with $S = 0$ are the equatorial n-spheres in S^{n+1}, S.S. Chern, DoCarmo and S. Kobayashi ([5]) and B. Lawson ([8]) proved independently that the Clifford tori are the only minimal hypersurfaces with $S \equiv n$. It would seem interesting to study the closed minimal hypersurfaces in the sphere with $S =$ constant, or equivalently those with constant scalar curvature. Naturally, the first question is the following: Is there a next

*Research supported in part by NSF grant MCS-77-18723(02).

Seminar on Minimal Submanifolds
0-691-08324-X/83/177-22 \$1.60/0 (cloth)
0-691-08319-3/83/177-22 \$1.60/0 (paperback)

larger value for S, and if so what is it? (This question was raised in [5].) In this paper we give a partial answer to this problem.

There is one minimal hypersurface among each of E. Cartan's isoparametric families of hypersurfaces in S^{n+1}, which has constant principal curvatures and hence the length of its second fundamental form is constant. In Section 1, we use results of E. Cartan ([1]) and H. Munzner ([9]) to compute the value S for these examples. It turns out that S can only be 0, n, $2n$, $3n$ or $5n$.

In Section 2, we compute the Laplacian of the covariant derivative of the second fundamental form of M. We also introduce the higher order trace functions of h and compute their Laplacians. We will use these computations to prove our main theorem:

THEOREM. *Let* $M^n (n \geq 3)$ *be a closed minimally immersed hypersurface in* S^{n+1} *with* $S = \text{constant}$. *If* $S > n$, *then* $S > n + \frac{1}{12n}$.

Moreover, for the case of $n = 3$, we have the following sharp result which will be proved in Section 3.

THEOREM. *Let* M^3 *be a closed minimally immersed hypersurface in* S^4 *with* $S = \text{constant}$. *If* $S > 3$, *then* $S \geq 6$. *Moreover,* $S = 6$ *is assumed in the example of Cartan* [2] *and Hsiang* [6].

We would like to express our gratitude to Professor S.S. Chern, W.Y. Hsiang and S.T. Yau for their encouragement in this work.

§1. *Isoparametric families of hypersurfaces*

In this section, we will outline some results concerning isoparametric families of hypersurfaces.

Geometrically an isoparametric family is a "parallel" family of hypersurfaces.

Let $f: M^n \to S^{n+1}$ be an immersed hypersurface in S^{n+1}, and e_{n+1} the unit normal field to M^n in S^{n+1}. For each $t > 0$, let $f_t(x)$, $x \in M$

be the point of S^{n+1} on the geodesic from $f(x)$ starting in the direction $e_{n+1}(x)$, which has the geodesic distance t from $f(x)$. Explicitly, we have

$$f_t(x) = (\cos t) f(x) + (\sin t) e_{n+1}(x) .$$

If we denote by $\lambda_k = \cot \theta_k$ $(k=1, \cdots, n)$ the principal curvatures of f at a point x, then it is easy to derive that the principal curvatures for the immersion f_t at the point $f_t(x)$ are given by $\frac{\lambda_k}{1-t\lambda_k} = \cot(\theta_k - t)$. Thus we have the following proposition:

PROPOSITION 1 ([10]). f_t *has constant mean curvature for each* t *if and only if* f *has constant principal curvatures.*

From this, we can give a geometric definition of an isoparametric family of hypersurfaces as follows.

DEFINITION 1 ([10]). In S^{n+1}, an isoparametric family of hypersurfaces is a family of parallel hypersurfaces $f_t : M^n \to S^{n+1}$ obtained from a hypersurface $f : M \to S^{n+1}$ with constant principal curvatures.

In what follows f_t will be as in Definition 1. $\lambda_i = \lambda_i(t)$, $i = 1, \cdots, p$, will denote the distinct principle curvatures of f_t in increasing order, and m_i the multiplicity of λ_i. Then E. Cartan [1] obtained the following identities: For each i, $1 \leq i \leq p$,

$$\sum_{j \neq i} m_j \frac{1 + \lambda_i \lambda_j}{\lambda_i - \lambda_j} = 0 . \tag{1.1}$$

These identities give some restrictions on the number of distinct principle curvature and their multiplicities. In fact, H. F. Münzner has obtained the following theorem:

THEOREM 1 ([9]). *Let* $\lambda_1(t) < \lambda_2(t) < \cdots < \lambda_p(t)$ *be the distinct principle curvatures of an isoparametric family, and* $m_1, \cdots, m_p$ *their multiplicities. Then*

(1) p *can be only* $1, 2, 3, 4,$ *or* 6.

(2) *if* $p = 3$, *then* $m_1 = m_2 = m_3$

if $p = 4$ *or* 6, *then* $m_1 = m_3 = m_5$, $m_2 = m_4 = m_6$.

(3) $\lambda_k(t) = \cot [\frac{1}{p} \{(k-1)\pi + \arccos t\}]$.

It is easy to see then there is one minimal hypersurface among each isoparametric family of hypersurfaces in S^{n+1}; namely choose suitable t so that

$$\sum_{k=1}^{p} m_k \lambda_k(t) = \sum_{k=1}^{p} m_k \cot\left[\frac{1}{p}\{(k-1)\pi + \arccos t\}\right] = 0 . \tag{1.2}$$

By using (1.1), (1.2) and Theorem 1, and a direct computation, we have:

COROLLARY 1. *Let* M *be an* n*-dimensional minimally immersed hypersurfaces in* S^{n+1} *with constant principle curvatures, then the square* S *of the length of the second fundamental form of* M *can only be* 0, n, $2n$, $3n$, *or* $5n$. *More precisely, if* M *has* p *distinct principle curvatures then* $S = (p-1)n$ *and scalar curvature* $R \geq 0$.

REMARK. An equivalent analytic definition of isoparametric family of hypersurfaces is the following:

DEFINITION 2 ([1]). An isoparametric family of hypersurfaces in S^{n+1} is a family of level hypersurfaces defined by a smooth function $f: S^{n+1} \to \mathbb{R}$ such that $\|df\|^2$ and Δf are functions of f.

§2. *Formulas involving the second fundamental form of a minimal hypersurface in the sphere*

In this section, we derive some formulas by using moving frame, essentially following Chern's computation ([4]). Then we derive formulas for higher order covariant derivatives of the second fundamental form.

Let M be an n-dimensional manifold immersed in an $(n+1)$-dimensional Riemannian manifold N. We choose a local orthonormal frame field $e_1, \cdots, e_{n+1}$ in N such that, restricted to M, the $e_1, \cdots, e_n$ are tangent to M, and we let $\omega_1, \cdots, \omega_{n+1}$ be the dual coframe. Then the structure equations of N are given by

$$(2.1)\qquad \begin{aligned} d\omega_A &= -\sum_B \omega_{AB} \wedge \omega_B, \qquad \omega_{AB} + \omega_{BA} = 0 \\ d\omega_{AB} &= -\sum_C \omega_{AC} \wedge \omega_{CB} + \Omega_{AB}, \end{aligned}$$

where

$$\Omega_{AB} = \frac{1}{2} \sum_{C,D} K_{ABCD} \omega_C \wedge \omega_D$$

$$K_{ABCD} + K_{ABDC} = 0 .$$

The Ricci tensor and the scalar curvature are defined respectively by

$$(2.2)\qquad K_{AB} = K_{BA} = \sum_C K_{ACBC}$$

$$(2.3)\qquad K = \sum_A K_{AA} = \sum_{A,C} K_{ACAC} .$$

For S^{n+1}, we have

$$(2.4)\qquad K_{ABCD} = \delta_{AC}\delta_{BD} - \delta_{AD}\delta_{BC}$$

$$(2.5)\qquad K_{AB} = n\delta_{AB}$$

$$(2.6)\qquad K = n(n+1) .$$

If we restrict these forms to M, then $\omega_{n+1} = 0$. Since $0 = d\omega_{n+1} = -\sum_{i=1}^{n} \omega_{n+1,i} \wedge \omega_i$, by Cartan's lemma we can write

$$(2.7)\qquad \omega_{n+1,i} = \sum_j h_{ij}\omega_j, \qquad h_{ij} = h_{ji} .$$

Here and from now on, the range of summation is from 1 to n. From these formulas, we obtain

$$d\omega_i = -\sum_j \omega_{ij} \wedge \omega_j, \qquad \omega_{ij} + \omega_{ji} = 0. \tag{2.8}$$

$$d\omega_{ij} = -\sum_k \omega_{ik} \wedge \omega_{kj} + \frac{1}{2}\sum_{k,\ell} R_{ijk\ell}\,\omega_k \wedge \omega_\ell, \tag{2.9}$$

where

$$R_{ijk\ell} = K_{ijk\ell} + h_{ik}h_{j\ell} - h_{i\ell}h_{jk}. \tag{2.10}$$

The symmetric 2-form

$$h = \prod = \sum_{ij} h_{ij}\,\omega_i\,\omega_j \tag{2.11}$$

and the scalar

$$H = \frac{1}{n}\sum_i h_{ii} \tag{2.12}$$

are called the second fundamental form and the mean curvature of the immersed manifold M respectively; and if H is identically zero, then M is said to be minimal.

In what follows, $N = S^{n+1}$.

Define the covariant derivative Dh of h (with component h_{ijk}) by

$$\sum_k h_{ijk}\,\omega_k = dh_{ij} - \sum_m h_{im}\,\omega_{mj} - \sum_m h_{mj}\,\omega_{mi}. \tag{2.13}$$

Exterior differentiate (2.7) and use the structure equations, we obtain

$$\sum_{k,j} h_{ijk}\,\omega_k \wedge \omega_j = 0. \tag{2.14}$$

Thus

$$h_{ijk} = h_{ikj}. \tag{2.15}$$

Next we exterior differentiate (2.13) and define $h_{ijk\ell}$ by

$$\sum_{\ell} h_{ijk\ell}\omega_{\ell} = dh_{ijk} - \sum_m h_{mjk}\omega_{mi} - \sum_m h_{imk}\omega_{mj} - \sum_m h_{ijm}\omega_{mk} . \tag{2.16}$$

Then

$$\sum_{k,\ell}\left(h_{ijk\ell} - \frac{1}{2}\sum_m h_{im}R_{mjk\ell} - \frac{1}{2}\sum_m h_{mj}R_{mik\ell}\right)\omega_k \wedge \omega_{\ell} = 0 . \tag{2.17}$$

$$h_{ijk\ell} - h_{ij\ell k} = \sum_m h_{im}R_{mjk\ell} + \sum_m h_{mj}R_{mik\ell} . \tag{2.18}$$

In this paper, we shall also need the third covariant derivative of the second fundamental form. So similarly, exterior differentiate (2.16) and define $h_{ijk\ell m}$ by

$$\sum_m h_{ijk\ell m}\omega_m = dh_{ijk\ell} - \sum_m h_{mjk\ell}\omega_{mi} - \sum_m h_{imk\ell}\omega_{mj} - \sum_m h_{ijm\ell}\omega_{mk} - \sum_m h_{ijkm}\omega_{m\ell} . \tag{2.19}$$

and

$$h_{ijk\ell m} - h_{ijkm\ell} = \sum_r h_{rjk}R_{ri\ell m} + \sum_r h_{irk}R_{rj\ell m} + \sum_r h_{ijr}R_{rk\ell m} . \tag{2.20}$$

As above, denote by S the square of the length of h, i.e.

$$S = \sum_{ij} h_{ij}^2 . \tag{2.21}$$

It is easy to see that for a minimal hypersurface M in S^{n+1}

$$S = n(n-1) - R , \tag{2.22}$$

where R is the scalar curvature of M, which shows that S is intrinsic.

Next we compute Δh (following [4], [11]) and also $\Delta(Dh)$. By definition

$$(\Delta h)_{ij} = \Delta h_{ij} = \sum_k h_{ijkk}$$

From (2.15) we have

$$\Delta h_{ij} = \sum_k h_{kijk},$$

and from (2.18), we have

$$\Delta h_{ij} = \sum_k h_{kikj} + \sum_k \left(\sum_m h_{mi} R_{mkjk} + \sum_m h_{km} R_{mijk} \right).$$

Since M is minimal, so $\sum_k h_{kk} = 0$, together with (2.10), we have

$$\Delta h_{ij} = (n-s) h_{ij}. \tag{2.23}$$

So

$$\frac{1}{2} \Delta S = \sum_{ij} h_{ij} \Delta h_{ij} + \sum_{ijk} h_{ijk}^2 = (n-s) s + \sum_{ijk} h_{ijk}^2. \tag{2.24}$$

By integrating (2.24) on M, it is evident that if $0 \le S \le n$, then either $S \equiv 0$ or $S \equiv n$.

Next we compute the Laplacian of the covariant derivative of the second fundamental form of M. Using (2.18), (2.20) a long but straightforward computation gives

$$\begin{aligned} \Delta h_{ijk} &= \sum_\ell h_{ijk\ell\ell} \\ &= (2n+3-s) h_{ijk} + 2 \sum_{m,\ell} (h_{im\ell} h_{mk} h_{j\ell} + h_{jm\ell} h_{mi} h_{\ell k} + h_{km\ell} h_{mi} h_{\ell j}) \\ &\quad - \sum_{m,\ell} (h_{ijm} h_{m\ell} h_{\ell k} + h_{ikm} h_{m\ell} h_{\ell j} + h_{jk\ell} h_{m\ell} h_{mi}). \end{aligned} \tag{2.25}$$

Hence

$$(2.26)\quad \frac{1}{2}\Delta\Big(\sum_{ijk} h_{ijk}^2\Big) = \sum_{ijk} h_{ijk}\Delta h_{ijk} + \sum_{i,j,k,\ell} h_{ijk\ell}^2$$

$$= (2n+3-s)s(s-n) - 3\sum_{\substack{i,j,k\\ \ell,m}} h_{ijk}h_{ij\ell}h_{km}h_{m\ell}$$

$$+ 6\sum_{\substack{i,j,k\\ \ell,m}} h_{ijk}h_{i\ell m}h_{j\ell}h_{km} + \sum_{i,j,k,\ell} h_{ijk\ell}^2 .$$

From now on, we assume that $S \equiv$ constant. Then (2.24) implies that

$$(2.27)\quad \sum_{ijk} h_{ijk}^2 = S(S-n)$$

is also a constant.

For any point $p \in M$, we can choose a frame field $e_1, \cdots, e_n$ so that (h_{ij}) is diagonalized at that point, say

$$(2.28)\quad h_{ij} = \lambda_i \delta_{ij} .$$

Under such frame field, at p, (2.26) becomes

$$(2.29)\quad \sum h_{ijk\ell}^2 = s(s-n)(s-2n-3) + 3(A-2B) ,$$

where

$$(2.30)\quad A = \sum_{i,j,k} h_{ijk}^2\lambda_i^2 , \qquad B = \sum_{i,j,k} h_{ijk}^2\lambda_i\lambda_j .$$

The crucial point now is to give a lower bound for $\sum_{ijk\ell} h_{ijk\ell}^2$ in terms of S. First, we see that

$$(2.31)\quad \sum_{i,j,k,\ell} h_{ijk\ell}^2 \geq 3\sum_{i\neq j} h_{ijij}^2 + \sum_i h_{iiii}^2 .$$

Define

$$t_{ij} = h_{ijij} - h_{jiji} . \tag{2.32}$$

Using (2.15) and (2.18),

$$t_{ij} = (\lambda_i - \lambda_j)(1+\lambda_i\lambda_j) . \tag{2.33}$$

We obtain

$$\sum_{i\neq j} h_{ijij}^2 = \sum_{i<j} h_{ijij}^2 + (h_{ijij} - t_{ij})^2 \tag{2.34}$$

$$= \sum_{i\neq j}\left(h_{ijij} - \frac{t_{ij}}{2}\right)^2 + \frac{1}{4}\sum_{i\neq j} t_{ij}^2 .$$

But from (2.33),

$$\sum_{i\neq j} t_{ij}^2 = \sum_{i\neq j} (\lambda_i - \lambda_j)^2(1+\lambda_i\lambda_j)^2 \tag{2.35}$$

$$= 4\left[nS - 2S^2 + S\sum_i \lambda_i^4 - \left(\sum_i \lambda_i^3\right)^2\right]$$

$$= \frac{2}{S}\sum_k \left[S\lambda_k^2 - \lambda_k\left(\sum_i \lambda_i^3\right) - S\right]^2 .$$

Let f_m denote the mth symmetric function of the principal curvatures, i.e.

$$f_m = \sum_i \lambda_i^m = \operatorname{tr}(h_{ij})^m .$$

They are smooth function on M.

By using (2.34) and (2.35) together with the inequality (2.31), it gives

$$\sum_{i,j,k,\ell} h_{ijk\ell}^2 \geq 3\sum_{i\neq j}\left(h_{ijij} - \frac{1}{2}t_{ij}\right)^2 + \sum_i h_{iiii}^2 + \frac{3}{2S}\sum_{K=1}^{n} (S + f_3\lambda_k - S\lambda_k^2)^2 . \tag{2.36}$$

The next lemma is the basic step.

LEMMA 1. *Let* $M^n \subset S^{n+1}$ *be closed minimal hypersurface with* $S = \text{Constant}$, *and* $S > n$. *Then there exists a point* p *and a principal curvature* λ_k *such that*

$$f_3\lambda_k - S\lambda_k^2 \geq -\frac{2}{3n} s^2 . \tag{2.37}$$

Proof. If the conclusion is not valid, then on the whole manifold M and for each principal curvature λ_k, $1 \leq k \leq n$, we have

$$f_3\lambda_k - S\lambda_k^2 < -\frac{2}{3n} S^2 \tag{2.38}$$

or

$$S\lambda_k^2 - f_3\lambda_k - \frac{2}{3n} S^2 > 0 . \tag{2.39}$$

From (2.39), it follows that $\lambda_k \neq 0$, for all k. Hence, the determinant of h_{ij} cannot change sign and the following function is well defined

$$F = \log \det(h_{ij}) . \tag{2.40}$$

(If $\det(h_{ij}) < 0$, we can use the function $F = \log[-\det(h_{ij})]$; the proof of lemma is the same.)

We compute the Laplacian of F as follows

$$F_k = \sum_{i,j} h^{ij} h_{ijk} \tag{2.41}$$

$$\Delta F = \sum_k F_{kk} = \sum_{i,j,k} (h_k^{ij} + h^{ij} h_{ijkk}) . \tag{2.42}$$

Notice that

$$h_{,k}^{ij} = -\sum_{m,\ell} h^{im} h^{\ell j} h_{m\ell k} \tag{2.43}$$

and

(2.44) $$\Delta h_{ij} = \sum_k h_{ijkk} = -(s-n)\,h_{ij}\,.$$

We obtain

(2.45) $$\begin{aligned}\Delta F &= -\sum_{i,j,k,m,n} h^{im}h^{jn}h_{mnk}h_{ijk} - n(s-n)\\ &= -\sum_{i,j,k}\frac{1}{\lambda_i\lambda_j}h_{ijk}^2 - n(s-n)\,.\end{aligned}$$

Next we will prove that if (2.39) is true, then ΔF would be negative everywhere. Since it is impossible for a smooth function on any compact manifold, Lemma 1 will be proved.

In fact, by using (2.27), (2.45) can be rewritten to

(2.46) $$\Delta F = -\sum_{A,B,C}\frac{1}{\lambda_A\lambda_B}h_{ABC}^2 - \frac{n}{S}\sum_{A,B,C}h_{ABC}^2\,.$$

We use roman indices and greek indices to denote the negative principal curvature and positive principal curvature respectively.

Since we assume that $S > n$, then, at least, there exists one term h_{ABC} which is not vanishing, so, it suffices to check that all of the coefficients are negative.

The only terms which must be checked are those involving $h_{i\alpha C}$. Now, let us fix the indices (i,α) and, without loss of generality, assume $C = j$. We pick out these terms from (2.46). There are two cases:

(I) $i = j$ $$-\frac{2}{\lambda_i\lambda_\alpha}h_{i\alpha i}^2 - \frac{3n}{S}h_{i\alpha i}^2$$

(II) $i \neq j$ $$-\frac{2}{\lambda_i\lambda_\alpha}h_{i\alpha j}^2 - \frac{6n}{S}h_{i\alpha j}^2$$

But, from (2.39), we have

(2.47) $$|\lambda_i\lambda_\alpha| > \frac{2}{3n}S\,.$$

It follows that

$$(2.48)\qquad -\frac{2}{\lambda_i\lambda_\alpha}-\frac{3n}{S}<0\,.$$

The proof is complete.

Due to Lemma 1, it is easy to prove the following:

THEOREM 1. *For each* n, *there exists a positive number* $C(n)$. *Such that if* M *is a compact minimally immersed hypersurface in* S^{n+1} *with the square* S *of the length of the second fundamental form a constant bigger* n, *then* $S > n + C(n)$.

Proof. From (2.36) and Lemma 1, there exists a point p on M such that at p we have

$$(2.49)\qquad \sum_{i,j,k,\ell} h_{ijk\ell}^2 \geq \frac{3S}{2}\left(1-\frac{2S}{3n}\right)^2 .$$

But in the right hand of (2.29), every term is bounded above by some function of S having a factor $(S-n)$. Hence the theorem follows.

To derive the gap size $C(n)$ explicitly, we estimate the right hand of (2.29). Notice that h_{ijk} are symmetric for all of the indices,

$$(2.50)\qquad 3(A-2B) = \sum_{i,j,k} h_{ijk}^2[\lambda_i^2+\lambda_j^2+\lambda_k^2-2(\lambda_i\lambda_j+\lambda_j\lambda_k+\lambda_k\lambda_i)]$$

$$= \sum_{i\neq j\neq k}[\quad] + 3\sum_{i=j\neq k}[\quad] + \sum_{i=j=k}[\quad].$$

By using $\Sigma\lambda_i^2 = S$, we have

$$(2.51)\qquad \sum_{i\neq j\neq k} = \sum_{i\neq j\neq k} h_{ijk}^2[2(\lambda_i^2+\lambda_j^2+\lambda_k^2)-(\lambda_i+\lambda_j+\lambda_k)^2] \leq 2S\sum_{i\neq j\neq k} h_{ijk}^2$$

$$(2.52)\qquad \sum_{i=j\neq k} = \sum_{i\neq k} h_{iik}^2[\lambda_k^2-2\lambda_i\lambda_k] \leq \sum_{i\neq k} h_{iik}^2[3(\lambda_i^2+\lambda_k^2)-2(\lambda_i+\lambda_k)^2] \leq 3S\sum_{i\neq k} h_{iik}^2$$

and

$$\sum_{i=j=k} = -3\sum_i h_{iii}^2\lambda_i^2 \leq 0 . \tag{2.53}$$

Moreover, we have

$$\sum_{i,j,k} h_{ijk}^2 = \sum_{i\neq j\neq k} h_{ijk}^2 + 3\sum_{i=j\neq k} h_{ijk}^2 + \sum h_{iii}^2 = S(s-n) . \tag{2.54}$$

These imply

$$3(A-2B) \leq 3S\sum_{i,j,k} h_{ijk}^2 = 3S^2(s-n) . \tag{2.55}$$

By substituting inequalities (2.55) and (2.49) into (2.29), it gives

$$\frac{3}{2}\,S\left(1-\frac{2}{3n}\,S\right)^2 \leq S(s-n)(s-2n-3) + 3S^2(s-n) . \tag{2.56}$$

i.e.

$$\frac{3}{2}\,S\left(1-\frac{2}{3n}\,S\right)^2 \leq S(s-n)(4s-2n-3) . \tag{2.57}$$

Hence, if $s > n$, $C(n)$ in Theorem 1 can be chosen to be $\frac{1}{12n}$. Because Corollary 1 of Section 1, one may guess that $2n$ is the next value. However, we can only prove this for $n = 3$.

§3. *Sharp estimate for the three dimensional case*

In this section, our main result is

THEOREM 2. *Let M^3 be a three dimensional compact minimally immersed hypersurface in S^4 with the square S of the length of the second fundamental form a constant. If $S > 3$, then $S \geq 6$.*

In order to prove this theorem, we will make use of the special property for three dimension. As in the last section, let us introduce the symmetric functions.

(3.1) $$f_3 = \sum_{i,j,k} h_{ij}h_{jk}h_{ki} = \sum_i \lambda_i^3$$

and

(3.2) $$f_4 = \sum_{i,j,k,\ell} h_{ij}h_{jk}h_{k\ell}h_{\ell i} = \sum_i \lambda_i^4 .$$

By using (2.23) and Ricci formula (2.18), we can compute the Laplacian of the above functions as follows:

(3.3) $$\Delta f_3 = 3\left[(n-s)f_3 + 2\sum_{i,j,k} h_{ijk}^2\lambda_i\right]$$

(3.4) $$\Delta f_4 = 4\left[(n-s)f_4 + 2\sum_{i,j,k} h_{ijk}^2\lambda_i^2 + \sum_{i,j,k} h_{ijk}^2\lambda_i\lambda_j\right] .$$

From now on we always assume $n = 3$. Notice that in this case we have

(3.5) $$f_4 = \sum_{i=1}^{3} \lambda_i^4 = \frac{1}{2} S^2 .$$

Hence, (3.4) gives

(3.6) $$2A + B = \frac{1}{2} S^2(s-3) .$$

Where, $A = \Sigma\, h_{ijk}^2\lambda_i^2$ and $B = \Sigma\, h_{ijk}^2\lambda_i\lambda_j$ (as (2.30)).

It follows that (2.29) can be rewritten as

(3.7) $$\begin{aligned}\sum h_{ijk\ell}^2 &= S(S-3)(S-9) + 3(A-2B)\\ &= S(S-3)(S-9) + 4(2A+B) - 5(A+2B)\\ &= 3S(S-3)^2 - 5(A+2B) .\end{aligned}$$

Then, from the symmetricity of h_{ijk} we have

(3.8) $$\begin{aligned}A + 2B &= \frac{1}{3}\sum_{i,j,k} h_{ijk}^2(\lambda_i^2+\lambda_j^2+\lambda_k^2+2\lambda_i\lambda_j+2\lambda_j\lambda_k+2\lambda_k\lambda_i)\\ &= \frac{1}{3}\sum_{i,j,k} h_{ijk}^2(\lambda_i+\lambda_j+\lambda_k)^2 \geq 0 .\end{aligned}$$

Thus, we obtain the following inequality

(3.9) $$\sum h_{ijk\ell}^2 \leq 3S(S-3)^2 .$$

Using (2.31), (2.34), (2.35), we see that in order to obtain the best estimate of S we should compute (3.9) at the minimum point of $(f_3)^2$. In what follows we study the function f_3 and prove that the minimum of $(f_3)^2$ is zero under the assumption $S > 3$.

LEMMA 2. (1) $f_3 = \text{const}$ *if and only if* M *has constant principal curvature.*

(2) $-\sqrt{\frac{S^3}{6}} \leq f_3 \leq \sqrt{\frac{S^3}{6}}$ *and equality is reached if and only if two of the principal curvatures are equal.*

The above lemma tells us that $f_3 = \text{const}$ implies M has constant principal curvature. In this case, S can be only $0, 3, 6$. (Corollary 1 of Section 1.) Hence, we will only consider the case of $f_3 \neq \text{const}$. The next lemma is the main ingredient in Theorem 2.

LEMMA 3. *If* $S > 3$ *and* $f_3 \neq \text{const}$, *then there exists a point* q *in* M *so that*

(3.10) $$f_3(q) = 0 .$$

Proof. Suppose not. Without loss of generality, we can assume that $f_3 > 0$ everywhere. Consider the point p_0, such that

(3.11) $$f_3(p_0) = \min_M f_3 > 0 .$$

At the first, notice that all of the principal curvature λ_1, λ_2 and λ_3 are different at the point p_0. Otherwise, from Lemma 2, we would have

(3.12) $$\text{Min } f_3 = \max f_3 .$$

This shows that $f_3 = \text{const}$ which is contrary to the assumption. Next, due to the maximal principle, we have

(3.13) $$\nabla f_3(p_0) = 0$$

and

(3.14) $$\Delta f_3(p_0) \geq 0 .$$

(3.13) implies that

(3.15) $$\sum_i h_{iik}\lambda_i^2 = 0 \qquad k=1,2,3 .$$

By differentiating $\sum_i h_{ii} = 0$ and $\sum_{i,j} h_{ij}^2 = S = \text{const}$, we obtain

(3.16) $$\sum_i h_{iik} = 0 \qquad k=1,2,3$$

and

(3.17) $$\sum_i h_{iik}\lambda_i = 0 \qquad k=1,2,3 .$$

It is easily seen that the determinant of the system of equations (3.15), (3.16) and (3.17)

(3.18) $$\begin{vmatrix} 1 & 1 & 1 \\ \lambda_1 & \lambda_2 & \lambda_3 \\ \lambda_1^2 & \lambda_2^2 & \lambda_3^2 \end{vmatrix} \neq 0 \text{ since the } \lambda_i \text{ are distinct.}$$

Thus this system has only the trivial solution, namely

(3.19) $$h_{iik} = 0 \qquad i.k=1.2.3 .$$

On the other hand, from (3.14) and (3.3), it gives

(3.20) $$2 \sum_{i,j,k} h_{ijk}^2\lambda_i \geq (s-3)f_3 > 0 .$$

However, it follows from (3.19) that

$$(3.21)\quad 3\sum_{i,j,k} h_{ijk}^2\lambda_i = \sum_{i,j,k} h_{ijk}^2(\lambda_i+\lambda_j+\lambda_k) = \sum_{i\neq j\neq k} h_{ijk}^2(\lambda_i+\lambda_j+\lambda_k) = 0\,.$$

which is contrary to (3.20). The proof is complete.

Before proving the theorem, we need some formulas at the point q satisfying $f_3(q) = 0$.

LEMMA 4. *At the point* q *as in Lemma 3 where* $f_3(q) = 0$ *we have*

$$(3.22)\quad \lambda_1 = -\sqrt{\frac{S}{2}}, \quad \lambda_2 = 0 \quad \textit{and} \quad \lambda_3 = \sqrt{\frac{S}{2}}$$

$$(3.23)\quad h_{331} = h_{111} \qquad h_{112} = -\frac{1}{2}h_{222} \qquad h_{113} = h_{333}$$

$$h_{221} = -2h_{111} \qquad h_{332} = -\frac{1}{2}h_{222} \qquad h_{223} = -2h_{333}$$

$$(3.24)\quad \sum_{i,j,k} h_{ijk}^2 = 6h_{123}^2 + 16h_{111}^2 + \frac{5}{2}h_{222}^2 + 16h_{333}^2 = S(S-3)$$

and

$$(3.25)\quad \sum_{i,j} h_{ij2}^2 \geq \frac{1}{3}S(S-3)\,.$$

Proof. (3.24) is trivial. (3.25) can be verified from (3.16), (3.17) and (3.22). By substituting (3.23) into Σh_{ijk}^2, it gives (3.24). To prove (3.25), it is enough to notice that

$$(3.26)\quad \sum_{i,j} h_{ij2}^2 = 2h_{123}^2 + 8h_{111}^2 + \frac{3}{2}h_{222}^2 + 8h_{333}^2\,.$$

Now we are in a position to prove Theorem 2.

Proof of Theorem 2. Since $f_3(q) = 0$, (2.36) gives

$$(3.27)\quad h_{ijk\ell}^2 \geq 3\sum_{i\neq j}\left(h_{ijij} - \frac{1}{2}t_{ij}\right)^2 + \frac{3}{4}S(S^2-4S+6)\,.$$

By combining it with (3.9), we obtain

$$3S(S-3)^2 \geq 3\sum_{i\neq j}\left(h_{ijij}-\frac{1}{2}t_{ij}\right)^2+\frac{3}{4}S(S^2-4S+6)\,. \tag{3.28}$$

What remains to be done will be to estimate the first term in the right hand of (3.28). In fact, at the point q, we have

$$t_{12}=t_{23}=-\sqrt{\frac{S}{2}}\,. \tag{3.29}$$

Then, notice that

$$\begin{aligned}\sum_{i\neq j}\left(h_{ijij}-\frac{1}{2}t_{ij}\right)^2 &\geq 2\left[\left(h_{1212}-\frac{1}{2}t_{12}\right)^2+\left(h_{2323}-\frac{1}{2}t_{23}\right)^2\right]\\ &=\left[h_{1212}+h_{2323}-\frac{1}{2}(t_{12}+t_{23})\right]^2+\left[h_{1212}-h_{2323}-\frac{1}{2}(t_{12}-t\right.\\ &\geq (h_{1212}-h_{2323})^2\,.\end{aligned} \tag{3.30}$$

Differentiate $\Sigma h_{ij}^2=S$, we obtain

$$\sum_{i,j}h_{ij\ell\ell}h_{ij}+\sum_{i,j}h_{ij\ell}^2=0 \qquad \ell=1.2.3\,. \tag{3.31}$$

By substituting (3.22) into (3.31), it gives

$$\sqrt{\frac{S}{2}}\,(h_{11\ell\ell}-h_{33\ell\ell})=\sum_{i,j}h_{ij\ell}^2 \qquad \ell=1,2,3\,. \tag{3.32}$$

In particular, we have

$$\sqrt{\frac{S}{2}}\,(h_{1122}-h_{3322})=\sum_{i,j}h_{ij2}^2\,. \tag{3.33}$$

It follows that

$$(3.34)\qquad h_{1212}-h_{2323} = h_{1122}-h_{2233} = h_{1122}-(h_{3322}+t_{23})$$
$$= \sqrt{\frac{2}{S}}\sum_{i,j} h_{ij2}^2 + \frac{S}{2}$$
$$\geq \sqrt{\frac{2}{S}}\left(\frac{1}{3}\, S(S-3) + \frac{S}{2}\right).$$

The inequality above is due to (3.25).

Now it follows from (3.28), (3.30) and (3.34) that

$$(3.35)\qquad 3S(S-3)^2 \geq \frac{6}{S}\left[\frac{1}{3}\, S(S-3)^2 + \frac{S}{2}\right]^2 + \frac{3}{4}\, S(S^2-4S+6).$$

Namely

$$(3.36)\qquad S(S-6)(19s-42) \geq 0.$$

It is clear that if $S > 3$, then $S \geq 6$.

§4. *Some remarks*

In this paper, the basic point is to estimate a lower bound for $\sum_{ijk\ell} h_{ijk\ell}^2$ in terms of the length of the second fundamental form. Naturally, another question is how to estimate the upper bound for $\sum h_{ijk\ell}^2$. In this regard, we have also obtained the following result:

THEOREM. *Let M^3 be a closed minimally immersed hypersurface with constant scalar curvature R in S^4. If M has three distinct principal curvature at every point (or, equivalently, if $|\sum_i \lambda_i^3| < \sqrt{\frac{S^3}{6}}$ on M), then $R = 6 - S \geq 0$.*

The proof of this theorem will be reported in a forthcoming paper, and we suspect that the condition about the $\sum_i \lambda_i^3$ might be superfluous.

Because of the theorem above and the examples of minimal hypersurfaces with constant principal curvatures, the following problems are interesting.

PROBLEM 1. Does there exist a constant $\beta(n)$ depending only on n such that $S \leq \beta(n)$ for all minimal hypersurfaces of S^{n+1} ?

PROBLEM 2. Does there exist a closed minimal hypersurface of S^{n+1} with negative constant scalar curvature?

PROBLEM 3. Does there exist a closed minimal hypersurface of S^{n+1} with non-positive (not identically zero) Scalar curvature?

It follows from Gauss-Bonnet theorem that there is no closed minimal surface of S^3 with negative scalar curvature.

CHIA-KUEI PENG
SCHOOL OF MATHEMATICS
INSTITUTE FOR ADVANCED STUDY
PRINCETON, NEW JERSEY 08540

CHUU-LIAN TERNG
SCHOOL OF MATHEMATICS
INSTITUTE FOR ADVANCED STUDY
PRINCETON, NEW JERSEY 08540

REFERENCES

[1] E. Cartan, Familles de surfaces isoparametriques dans les espaces à courbure constante, Annali di Mat. *17*(1938), 177-191.

[2] ________, Sur des familles remarquables d'hypersurfaces isoparamétriques dans les espaces sphériques, Math. Z. *45*(1939), 335-367.

[3] ________, Sur quelques familles remarquables d'hypersurfaces, C.R. Congrès Math. Liege, 1939, 30-41.

[4] S.S. Chern, Minimal submanifolds in a Riemannian manifold. Mimeographed Lecture Notes, Univ. of Kansas, 1968.

[5] S.S. Chern, M. Docarmo and S. Kobayashi, Minimal submanifolds of a sphere with second fundamental form of constant length. In functional analysis and related fields, edited by F. Browder, Springer-Verlag, Berlin, 1970.

[6] W.Y. Hsiang, Remarks on closed minimal submanifolds in the standard Riemannian m-sphere, J. Diff. Geom. *1*(1967), 257-267.

[7] W.Y. Hsiang and H.B. Lawson, Minimal submanifolds of low cohomogeneity, J. Diff. Geom. *5*(1971), 1-38.

[8] H. B. Lawson, Local rigidity theorems for minimal hypersurfaces, Ann. of Math. *89*(1969), 187-197.

[9] H. F. Münzner, Isoparametrische Hyperfläche in sphären, to appear in Math. Ann.

[10] K. Nomizu, Elie Cartan's work on isoparametric families of hypersurfaces, Proceedings of Symposia in Pure Mathematics *27*(1975), 191-200.

[11] J. Simons, Minimal varieties in Riemannian manifolds, Ann. of Math., *88*(1968), 62-105.

REGULAR MINIMAL HYPERSURFACES EXIST ON MANIFOLDS IN DIMENSIONS UP TO SIX*

Jon T. Pitts

We discuss recent developments in the theory of existence and regularity of minimal surfaces on Riemannian manifolds. This note is a general research announcement of results which will appear in [P5]. Our principal theorem is the following:

REGULARITY THEOREM. *If* $3 \leq n \leq 6$ *and if* M *is an* n*-dimensional, smooth, compact, Riemannian manifold, then* M *supports a nonempty, smooth, compact,* (n–1)*-dimensional, imbedded, minimal submanifold (without boundary).*

In these dimensions this theorem is a complete answer to a fundamental question: for what positive integers k and n does a smooth, compact, n-dimensional, Riemannian manifold support a regular, closed, minimal submanifold of dimension k ? Classically the only case in which there were satisfactory answers of great generality was when $k = 1$ and n was arbitrary (existence of closed geodesics). The first breakthrough to higher dimensions without severe restrictions on the ambient manifold came in 1974 when we established a precursor of the theorem above, valid when $k = 2$ and $n = 3$. (This was announced in [P1], and later revised

*Research supported in part by a grant from the National Science Foundation.

Seminar on Minimal Submanifolds
0-691-08324-X/83/199-07 $0.85/0 (cloth)
0-691-08319-3/83/199-07 $0.85/0 (paperback)

and distributed in [P2], [P3], [P4].) Much of the method we used then was peculiar to the case $k = 2$. We have developed new estimates, more powerful and more general, with which the theory has been extended to the dimensions of the regularity theorem. This is the first general existence theorem of this type valid for regular k-dimensional surfaces when $k > 2$. (In case $k = 1$, one may refer to [LF], [LS], [K], among others. [SU] and [SY] contain interesting developments in the case $k = 2$. The explicit constructions in [L1] are also noteworthy.)

The theory has two major parts. The first part is a *general existence theory* for minimal surfaces, applicable on arbitrary compact Riemannian manifolds in all dimensions and codimensions. The second part is the *regularity theory* in which the special dimension-dependent estimates are derived and applied to produce the regularity theorem above. Both parts are formulated in the context of geometric measure theory. In particular we shall be talking about surfaces of three kinds: *manifolds*, *integral currents* (oriented surfaces with integer densities and integer topological multiplicities), and *integral varifolds* (unoriented surfaces with integer densities). The basic reference is [F]; the standard reference for varifolds is [AW].

General existence theory

The general existence theory has its roots in [A1] and [A2]. Almgren demonstrated [A1] that on a compact manifold M, the homotopy groups of the integral cycle groups on M are isomorphic to the homology groups of M. We may now formulate the idea underlying the variational calculus. If, for example, M is an orientable submanifold of some Euclidean space, $n = \dim(M)$, and $1 \leq k \leq n$, then (in the notation of [F], particularly [F, 4.4.8])

$$\pi_{n-k}(\tilde{\mathcal{Z}}_k(M); \{0\}) \simeq H_n(M; Z) \simeq Z,$$

hence there exists a nonzero $(n-k)$-dimensional homotopy class Π of k-dimensional surfaces, and

$$0 < L = \inf\{\sup \text{image}(\mathbf{M}\circ\psi) : \psi \in \Pi\} < \infty .$$

In principle, one can choose $\psi_c \in \Pi$ such that

$$\sup \text{image}(\mathbf{M}\circ\psi_c) = L ;$$

and ψ_c having been chosen, one might reasonably suppose that at least one of the surfaces in the set

$$\{T : \mathbf{M}(T) = L\} \cap \text{image}(\psi_c)$$

is a k-dimensional minimal surface. Although examples show that this program cannot be implemented quite so simply, Almgren was able nevertheless to construct an analogous variational calculus in the large, from which he concluded [A2] that M supports a nonzero stationary integral varifold in all dimensions not exceeding dim(M). We construct a similar variational calculus and prove what turns out to be a critical extension; namely, that M supports nonzero stationary integral varifolds having an additional variational property.

The almost minimizing property

A k-dimensional varifold V on a manifold M has the *almost minimizing property* provided that for each point $p \in M$, there exists a positive number r such that if $0 < s < r$ and $U = M \cap \{x : s < \text{distance}(p,x) < r\}$, then there exist a sequence of positive real numbers $\delta_1, \delta_2, \cdots,$ decreasing to 0 and a sequence of currents $T_1, T_2, \cdots,$ in $\tilde{\mathcal{Z}}_k(M)$ such that

(1) in the open set U the sequence of associated integral varifolds $|T_1|, |T_2|, \cdots,$ converges to V in the weak topology on varifolds; and

(2) if i is a positive integer, $S_0, \cdots, S_m$, is a finite sequence of currents in $\tilde{\mathcal{Z}}_k(M)$, $S_0 = T_i$,

$$\begin{aligned} &\text{spt}(S_j - S_{j-1}) \subset U && \text{for each } j = 1,2,\cdots,m , \\ &\mathbf{M}(S_j - S_{j-1}) \le \delta_i && \text{for each } j = 1,2,\cdots,m , \\ &\mathbf{M}(S_j) \le \mathbf{M}(T_i) + \delta_i && \text{for each } j = 1,2,\cdots,m , \end{aligned}$$

then

$$M(T_i) - i^{-1} \leq M(S_m) .$$

Intuitively one considers a varifold with the almost minimizing property to be one which may be approximated arbitrarily closely by integral currents which are themselves very nearly locally area minimizing. The main existence theorem is the following:

GENERAL EXISTENCE THEOREM. *If* M *is any smooth compact Riemannian manifold and if* $1 \leq k \leq \dim(M)$, *then* M *supports a nonzero,* k*-dimensional, stationary, integral varifold which possesses the almost minimizing property.*

Regularity theory

The study of almost minimizing varifolds began in the first place because Almgren's theorem on the existence of stationary integral varifolds is insufficient to settle the question of existence of *regular* minimal surfaces on manifolds. This is because varifolds which are stationary and integral have in general essential singularities, possibly of positive measure. If, in addition, the varifold has the almost minimizing property, then it possesses strong local stability properties which yield estimates on the singular sets. In particular, these estimates imply that the singular set is empty for hypersurfaces of n-dimensional manifolds when $3 \leq n \leq 6$. We discuss briefly how these estimates are derived.

Regular stable hypersurfaces

Stable surfaces (minimal surfaces whose second variation of area is nonnegative) have been vigorously investigated in recent years (see [S], [L2], among many). For our purposes, the salient property of regular stable hypersurfaces is the curvature estimate of Schoen, Simon, and Yau [SSY]. In particular, if $3 \leq n \leq 6$ and if N is a smooth stable (n–1)-dimensional submanifold of a compact n-dimensional manifold, then locally the principal curvatures of N are uniformly bounded by a number depending essentially only on n and the mass of N. This curvature estimate of [SSY] and the compactness and monotonicity arguments of geometric measure theory are combined to yield precise descriptions of the geometry of stable surfaces. Here are two examples.

DECOMPOSITION THEOREM. *If* $3 \leq n \leq 6$, *and if* M *is a sufficiently planar, smooth,* n*-dimensional submanifold of a Euclidean space, then a stable* (n–1)*-dimensional submanifold of* M *lying sufficiently near a cone decomposes as the union of disjoint minimal graphs of functions over an open subset of a single* (n–1)*-dimensional plane.*

We also derive several precise analytic estimates of these minimal graphs. Simple examples (e.g. the unstable catenoid in $\mathbf{R}^3$) show that the decomposition theorem is false if the submanifold is merely minimal.

COMPACTNESS THEOREM. *If* $3 \leq n \leq 6$; *if* M *is a smooth, compact,* n*-dimensional, Riemannian manifold and* $\mathfrak{U}$ *is an open cover for* M ; *and if* $\mathcal{S}$ *is the space of all* (n–1)*-dimensional integral varifolds on* M *which have regular support and which are stable in each open set in* $\mathfrak{U}$; *then all closed, uniformly mass bounded subsets of* $\mathcal{S}$ *are compact in the weak topology on varifolds.*

Regularity of almost minimizing varifolds

As noted above, a varifold V having the almost minimizing property may be approximated by integral currents which are nearly locally area minimizing. We apply a novel continuation argument which depends on the careful construction of certain comparison surfaces to V whose existence is guaranteed by the almost minimizing property. As a consequence, the regularity problem reduces to consideration of varifolds which are locally the limits of sequences of currents which are themselves *actually* locally area minimizing. In the dimensions of our regularity theorem, locally area minimizing currents have regular support, thus one completes the regularity theory by invoking an appropriate modification of the compactness theorem above.

Addendum

In new work, R. Schoen and L. Simon (*Regularity of Stable Minimal Hypersurfaces* (preprint)) obtain strong curvature estimates for stable hypersurfaces of all dimensions, substantially improving the estimates in [SSY] on which we have relied. As a consequence, they extend the decomposition theorem, the compactness theorem, and particularly the regularity theorem (appropriately formulated) to apply also to n-dimensional manifolds, $7 \leq n < \infty$.

JON T. PITTS
DEPARTMENT OF MATHEMATICS
UNIVERSITY OF ROCHESTER
ROCHESTER, NEW YORK 14627

REFERENCES

[AW] W.K. Allard, *On the first variation of a varifold*, Ann. of Math. (2) 95(1972), 417-491.

[A1] F.J. Almgren, Jr., *The homotopy groups of the integral cycle groups*, Topology 1(1962), 257-299.

[A2] ———, The theory of varifolds. Mimeographed notes, Princeton (1965).

[F] H. Federer, Geometric measure theory, Springer-Verlag, New York, 1969.

[K] W. Klingenberg, Lectures on closed geodesics, Springer-Verlag, New York, 1978.

[L1] H.B. Lawson, Jr., *The global behavior of minimal surfaces in* S^n, Ann. of Math. (2) 92(1970), 224-237.

[L2] ———, *Minimal varieties in real and complex geometry*, Les Presses de L'Universite de Montreal, Montreal, 1974.

[LS] L. Lusternik and L. Schnirelmann, *Sur le problème de trois géodésiques fermées sur les surfaces de genre* 0, C.R. Acad. Sci. Paris 189(1929), 269-271.

[LF] L.A. Lyusternik and A.I. Fet, *Variational problems on closed manifolds*, Dokl. Akad. Nauk. SSSR (N.S.) 81(1951), 17-18.

[P1] J. Pitts, *Existence and regularity of minimal surfaces on Riemannian manifolds*, Bulletin of Amer. Math. Soc., Vol. 82, No. 3, May 1976.

[P2] J. Pitts, *Existence of minimal surfaces on compact manifolds I (Almost minimizing varifolds)* (preprint).

[P3] ———, *Existence of minimal surfaces on compact manifolds II (Three dimensional compact manifolds contain two-dimensional minimal submanifolds)* (preprint).

[P4] ———, *The decomposition theorem for two-dimensional stable hypersurfaces* (preprint).

[P5] ———, Existence and regularity of minimal surfaces on Riemannian manifolds (to appear, Princeton University Press).

[SU] J. Sacks and K. Uhlenbeck, *The existence of minimal immersions of 2-spheres* (preprint).

[SSY] R. Schoen, L. Simon, and S. T. Yau, *Curvature estimates for minimal hypersurfaces*, Acta Math. **134** (1975), 276-288.

[SY] R. Schoen and S. T. Yau, *Existence of incompressible minimal surfaces and the topology of three dimensional manifolds with non-negative scalar curvature*, Ann. of Math. (2) **110** (1979), 127-142.

[S] J. Simons, *Minimal varieties in Riemannian manifolds*, Ann. of Math. (2) **88** (1968), 62-105.

AFFINE MINIMAL SURFACES

Chuu-Lian Terng

This is a survey article on affine differential geometry. We study the basic properties of hypersurfaces in R^{n+1} invariant under the group of unimodular affine transformations. Details can be found in many references, in particular [1], [3], [4], [5]. These hypersurfaces have an affine invariant metric tensor, which depends on second order partial derivatives of the embedding functions. Therefore one can consider the corresponding area functional and its Euler-Lagrangian equation is a fourth order elliptic P.D.E. The critical points of this area functional are the affine minimal hypersurfaces, which have some properties analogous to those of minimal surfaces in Euclidean space. The affine minimal equation arises naturally from geometric considerations and therefore it provides a good model example for the study of non-linear fourth order elliptic P.D.E.

Let A^{n+1} be the unimodular affine space of dimension $n+1$, i.e. the space with real coordinates $x^1, \cdots, x^{n+1}$, and the volumn element $dV = dx^1 \wedge \cdots \wedge dx^{n+1}$. The group G which preserves the volumn element is the unimodular affine group, i.e.

(1) $$x^{*\alpha} = \sum_{\beta} C^{\alpha}_{\beta} x^{\beta} + d^{\alpha}, \qquad 1 \le \alpha, \beta \le n+1,$$

where

(2) $$\det(C^{\alpha}_{\beta}) = 1 .$$

Seminar on Minimal Submanifolds
0-691-08324-X/83/207-10 $1.00/0 (cloth)
0-691-08319-3/83/207-10 $1.00/0 (paperback)

In the space A^{n+1}, distance and angle have no meaning, but there are notions of volumn and parallelism.

Let X be the position vector of the hypersurface sitting in A^{n+1}. Let $X, e_1, \cdots, e_{n+1}$ be an affine frame on M such that $e_1, \cdots, e_n$ are tangent to M at X, and

(3) $$(e_1, \cdots, e_{n+1}) = \det(e_1, \cdots, e_{n+1}) = 1 \,.$$

We can write

(4) $$dX = \sum_{\alpha} \omega^{\alpha} e_{\alpha}$$

(5) $$de_{\alpha} = \sum_{\beta} \omega_{\alpha}^{\beta} e_{\beta} \,.$$

The ω^{α}, ω_{α}^{β} are the Maurer-Cartan forms of G, and therefore satisfy

(6) $$\sum_{\alpha} \omega_{\alpha}^{\alpha} = 0 \,.$$

The structure equations of A^{n+1} give

(7) $$d\omega^{\alpha} = \sum_{\beta} \omega^{\beta} \wedge \omega_{\beta}^{\alpha}$$

(8) $$d\omega_{\alpha}^{\beta} = \sum_{r} \omega_{\alpha}^{r} \wedge \omega_{r}^{\beta} \,.$$

If we restrict the forms to the hypersurface M, we have

(9) $$\omega^{n+1} = 0 \,,$$

so

(10) $$\sum_{i=1}^{n} \omega^{i} \wedge \omega_{i}^{n+1} = 0 \,.$$

By Cartan's lemma, we have

(11) $$\omega_i^{n+1} = \sum_j h_{ij}\omega^j\,, \qquad h_{ij} = h_{ji}\,.$$

Let

(12) $$H = \det(h_{ij})\,,$$

then it follows that

(13) $$II = |H|^{\frac{-1}{n+2}} \sum_{ij} h_{ij}\,\omega^i\omega^j$$

is an affine invariant quadratic form. From now on we assume that M is non-degenerate, i.e. $\operatorname{rank}(h_{ij}) = n$, then II defines a pseudo-Riemannian structure on M, called the affine metric on M, and its volumn element is

(14) $$dv = |H|^{n+2}\,\omega^1 \wedge \cdots \wedge \omega^n\,.$$

A direct computation using the non-degenerate of II shows that we can pick e_{n+1} so that

(15) $$\omega_{n+1}^{n+1} + \frac{1}{n+2}\, d\log|H| = 0\,.$$

The e_{n+1} defined by this equation is determined up to a scalar function and

(16) $$\nu = |H|^{\frac{1}{n+2}}\, e_{n+1}$$

is invariant under affine frame change; ν is called the affine normal vector. This affine normal has an interesting geometric interpretation ([1]):

Given $P \in M$, let $B(t)$ be the region of M bounded between the tangent plane TM_P and a parallel plane at distance t from TM_P, and

(17) $$\begin{aligned} C(t) &= \text{the center of mass of } B(t) \text{ w.r.t. II} \\ &= \left(\int_{B(t)} X\, dv\right) \Big/ \int_{B(t)} dv\,. \end{aligned}$$

Then $C'(0)$ is in the direction of the affine normal at P.

Next, we will derive the Fubini-Pick form for M. Exterior differentiation of (11) gives:

$$\sum_j \left(dh_{ij} + h_{ij}\,\omega_{n+1}^{n+1} - \sum_k h_{ik}\,\omega_j^k - \sum_k h_{kj}\,\omega_i^k \right) \wedge \omega^j = 0\,. \tag{18}$$

Define

$$\sum_k h_{ijk}\,\omega^k = dh_{ij} + h_{ij}\,\omega_{n+1}^{n+1} - \sum_k h_{ik}\,\omega_j^k - \sum_k h_{kj}\,\omega_i^k\,. \tag{19}$$

It follows from (18) that

$$h_{ijk} = h_{ikj}\,. \tag{20}$$

The cubic form

$$J = \sum_{ijk} h_{ijk}\,\omega^i \omega^j \omega^k \tag{21}$$

is called the Fubini-Pick form, which measures the deviation of the affine connection ω_i^j from the Levi-Civita connection $\tilde{\omega}_i^j$ of the affine invariant metric II. In fact

$$\tilde{\omega}_i^j - \omega_i^j = \frac{1}{2} \sum_{k,\ell} h^{jk} h_{ik\ell}\,\omega^\ell\,, \tag{22}$$

where

$$(h^{ij}) = (h_{ij})^{-1}\,. \tag{23}$$

Let Δ be the Laplace-Beltrami operator for II, then a standard computation gives

$$\Delta X = n\nu\,. \tag{24}$$

This gives another interpretation of the affine normal vector; namely

$$\nu = \frac{\Delta X}{n}\,. \tag{25}$$

By exterior differentiation of (15), we get

(26) $$\sum_i \omega_{n+1}^i \wedge \omega_i^{n+1} = 0 \,,$$

which gives

(27) $$\omega_{n+1}^i = \sum_j \ell^{ij} \omega_j^{n+1} \,, \qquad \ell^{ij} = \ell^{ji}$$

The quadratic form

(28) $$\mathrm{III} = |H|^{\frac{1}{n+2}} \omega_{n+1}^i \omega_i^{n+1}$$

is an affine invariant, called the third fundamental form.

With these fundamental forms one can get an analogue of the fundamental theorem of Euclidean hypersurfaces: The three fundamental forms determine M up to an affine transformation, and given three such forms satisfying certain integrability condition similar to the Gauss-Codazzi equations, there is an affine hypersurface having them as fundamental forms. (See [3] for details.)

The trace of III w.r.t. II, i.e.

(29) $$L = \frac{1}{n} |H|^{\frac{-1}{n+2}} \sum_{ij} h_{ij} \ell^{ij} \,,$$

is also an affine invariant, and is called the affine mean curvature. A hypersurface is called affine minimal, if $L = 0$. That these are the critical points of the variational problem given by the affine area functional is seen as follows.

Consider the hypersurface given by the graph of f, i.e.

(30) $$X(x^1,\cdots,x^n) = (x^1,\cdots,x^n, f(x^1,\cdots,x^n)) \,.$$

Then (3), (4), (5), (6) hold, if we set

$$
(31)\qquad
\begin{aligned}
\omega^{i} &= dx^{i}\\
\omega^{n+1} &= dx^{n+1} - \sum_{i} \frac{\partial f}{\partial x^{i}}\, dx^{i}\\
e_{i} &= \Big(0, \cdots, \underset{i^{th}}{1}, 0, \cdots, 0, \frac{\partial f}{\partial x^{i}}\Big)\\
e_{n+1} &= (0, \cdots, 0, 1)
\end{aligned}
$$

with

$$
(32)\qquad
\begin{aligned}
\omega_{i}^{j} &= 0\\
\omega_{i}^{n+1} &= \sum_{j} \frac{\partial^{2} f}{\partial x^{i}\partial x^{j}}\, dx^{j}\,.
\end{aligned}
$$

So

$$
(33)\qquad h_{ij} = \frac{\partial^{2} f}{\partial x^{i}\partial x^{j}}
$$

$$
(34)\qquad \mathrm{II} = |H|^{-\frac{1}{n+2}} \sum_{ij} \frac{\partial^{2} f}{\partial x^{i}\partial x^{j}}\, dx^{i} dx^{j}\,.
$$

To find the affine normal, we let

$$
(35)\qquad e_{i}^{*} = e_{i}
$$

$$
(36)\qquad e_{n+1}^{*} = e_{n+1} + \sum_{i} a^{i} e_{i}\,,
$$

where e_{n+1}^{*} is in the affine normal direction, then a^{i} are determined by solving (15) for these new frames. We get

$$
(37)\qquad a^{i} = -\frac{1}{n+2}\, h^{ij}\, \frac{\partial}{\partial x^{j}} (\log |H|)\,.
$$

Therefore

$$
(38)\qquad \nu = |H|^{\frac{1}{n+2}} \Big(e_{n+1} + \sum_{i} a^{i} e_{i}\Big)
$$

is the affine normal vector. Moreover, L can be computed directly,

$$(39)\quad L = \frac{1}{n}|H|^{n+2}\left\{\sum_i \frac{\partial a^i}{\partial x^i} - \sum_{ij} h_{ij} a^i a^j\right\}$$

$$= \frac{1}{n(n+2)^2}|H|^{n+2}\left\{\sum_{ij}(n+1)h^{ij}\frac{\partial \log|H|}{\partial x^i}\frac{\partial \log|H|}{\partial x^j} - (n+2)h^{ij}\frac{\partial^2 \log|H|}{\partial x^i \partial x^j}\right\}.$$

Consider the area integral

$$(40)\qquad A = \int |H|^{\frac{1}{n+2}} dx^1 \wedge \cdots \wedge dx^n .$$

Then the Euler-Lagrangian equation is

$$(41)\qquad 0 = \frac{1}{n+2}\sum_{ij}\frac{\partial^2}{\partial x^i \partial x^j}\left(|H|^{\frac{1}{n+2}} h^{ij}\right)$$

$$= \frac{n(n+1)}{n+2} L .$$

This is a fourth order elliptic P.D.E.

If f is quadratic, then h_{ij} = constant, hence $L = 0$. It is interesting to study the analogue of the Berstein problem, which is raised in [5], namely if f is a solution of (41) defined on the whole space R^n such that the Hessian matrix of f is always positive definite, is f a quadratic polynomial? This is the analogue of Berstein problem in affine geometry.

Now suppose $e_1, \cdots, e_{n+1}$ are as above with e_{n+1} in the affine normal direction. Then

$$(42)\qquad U = |H|^{\frac{-1}{n+2}} e_1 \wedge \cdots \wedge e_n$$

defines an affine invariant map ([3], [4]) from M to $\Lambda^n TM$, and is called the affine conormal. A direct computation gives

$$\Delta U = nLU \;, \tag{43}$$

and consequently there is no closed affine minimal hypersurface in A^{n+1}.

Next we consider affine surfaces in A^3, so that U can be identified with $H^{-1/4} e_1 \times e_2$. It follows from the above local formulas that the affine surfaces of elliptic type (i.e. II is positive definite) are determined by U alone. More explicitly, let $U(x_1, x_2)$ be a map from a domain D of A^2 to A^3 such that U, ΔU are everywhere linearly dependent and $\left(U, \frac{\partial U}{\partial x_1}, \frac{\partial U}{\partial x_2}\right) \neq 0$ in D. Then there is an affine surface X in A^3 having x_1, x_2 as isothermal coordinates for II such that U is the affine conormal, $\frac{U_1 \times U_2}{\|X_1 \times X_2\|}$ the affine normal vector, and $\frac{1}{2}\frac{\Delta U}{U}$ is its affine mean curvature. Moreover,

$$X(x_1, x_2) = \int_{(0,0)}^{(x_1, x_2)} U \times U_2 \, dx_2 - U \times U_1 \, dx_1 \; . \tag{44}$$

If we apply the above formulas to affine minimal surfaces, then U is harmonic and we obtain an affine analogue of the Schwartz representation of minimal surfaces in R^3. In fact, there is a correspondence between holomorphic curves in C^3 and affine minimal surfaces, i.e. given a holomorphic curve $W(z) = W(x+iy)$ with $U = \mathrm{Re} W$ and $\left(U, \frac{\partial U}{\partial x}, \frac{\partial U}{\partial y}\right) \neq 0$ then

$$X(x,y) = \frac{i}{4} \int (W + \overline{W}) \times (W_z \, dz - \overline{W}_z \, d\bar{z}) \tag{45}$$

is affine minimal, and U is the affine normal vector.

Suppose $f : R^2 \to R$ is a smooth solution of (41) with positive definite Hessian, then the graph given by f is an affine minimal surface of elliptic type. The affine metric

(46) $$\mathrm{II} = \left(\det\left(\frac{\partial^2 f}{\partial x^i \partial x^j}\right)\right)^{-\frac{1}{4}} \sum_{ij} \frac{\partial^2 f}{\partial x_i \partial x_j} dx^i dx^j$$

has non-negative curvature, which is equal to one-eighth of the square of the length of the cubic form J. Moreover, if II is complete then II is conformally equivalent to the standard metric on $\mathbb{R}^2$. However, the affine conormal is

(47) $$U = H^{-\frac{1}{4}}\left(-\frac{\partial f}{\partial x^1}, -\frac{\partial f}{\partial x^2}, 1\right),$$

where

(48) $$H = \det\left(\frac{\partial^2 f}{\partial x^i \partial x^j}\right);$$

so by (43) we have that $H^{-1/4}$ is a positive harmonic function on $\mathbb{C}$, hence a constant. Then it follows from [2], [7] that f is quadratic. Therefore whether Berstein's theorem is true in A^3 is equivalent to whether formula (46) defines a complete metric on $\mathbb{R}^2$ when f is an entire solution of (41).

Finally, we remark on an analogue of Bäcklund's theorem for affine surfaces ([6]). The classical Bäcklund theorem says the following: Let M, M^* be two surfaces in $\mathbb{R}^3$ and $\ell : M \to M^*$ a diffeomorphism. Suppose that

(1) The line joining P and $P^* = \ell(P)$ is a common tangent line for M and M^* for all $P \in M$.

(2) The distance $\|\overrightarrow{PP^*}\| = r$ is a constant.

(3) The angle between the unit normals at P and P^* is a constant θ.

Then both M and M^* have Gaussian curvature $-\frac{\sin^2\theta}{r^2}$. Moreover, if M is a surface in $\mathbb{R}^3$ with constant Gaussian curvature $-\frac{\sin^2\theta}{r^2}$, and v_0 is a unit vector in TM_{P_0}, which is not in the principal curvature direction, then there is a surface M^* and ℓ as above such that $\overrightarrow{P_0\ell(P_0)} = r e_0$.

The map ℓ is called a Bäcklund transformation from M to M^* and is easily seen to preserve asymptotic coordinates. The angle ϕ between two asymptotic curves on a hyperbolic surface in $\mathbf{R}^3$ satisfies the Sine-Gordon equation

SGE $$\phi_{xt} = \sin \phi$$

w.r.t. asymptotic coordinate x, t, thus the above theorem gives a "Bäcklund transformation" for SGE. In an attempt to find new equations having Bäcklund transformation, S. S. Chern and the author obtained the following analogue of Bäcklund theorem for affine surfaces:

Let M, M^* be two surfaces in A^3, and $\ell: M \to M^*$ a diffeomorphism. Suppose that

(1) The line joining P and $\ell(P)$ is a common tangent line for M and M^* for all $P \in M$.

(2) The affine normals at P and $\ell(P)$ are parallel.

(3) ℓ is a W-congruence, i.e. the asymptotic curves correspond under ℓ.

Then both M and M^* are affine minimal. Moreover, if M is an affine minimal surface in A^3, given any $v_0 \in TM_{P_0}$, there exists a surface M^* and ℓ as above such that $\overrightarrow{P_0 \ell(P_0)} = v_0$.

CHUU-LIAN TERNG
SCHOOL OF MATHEMATICS
INSTITUTE FOR ADVANCED STUDY
PRINCETON, NEW JERSEY 08540

REFERENCES

[1] W. Blaschke, Vorlesungen über Differentialgeometrie, II, Berlin 1923.

[2] E. Calabi, Improper affine hyperspheres of convex type and a generalization of a theorem by K. Jörgens. Michigan Math. J.S. (1958), 105-126.

[3] ———, Notes on affine differential geometry. Institute for Advanced Study. Fall 1979.

[4] S. Y. Cheng and S. T. Yau, Complete affine hyperspheres, to appear.

[5] S. S. Chern, Affine minimal hypersurfaces, Proc. of the Japan-United States Seminar on Minimal submanifold and geodesics 1977, 17-30.

[6] S. S. Chern and C. L. Terng. An analogue of Bäcklund's theorem in affine geometry. Rocky Mountain J. of Math. vol. 10, no. 1, 1979, 105-124.

[7] K. Jörgens, "Uber die Lösungen der Differentialgleichung $rt - s^2 = 1$," Math. Ann. 127 (1954), 130-134.

THE MINIMAL VARIETIES ASSOCIATED TO A CLOSED FORM

F. Reese Harvey and H. Blaine Lawson, Jr.

1. Introduction

How do you prove that a given submanifold M of a Riemannian space X is volume-minimizing? The classical argument goes as follows: Suppose M is oriented and of dimension p, and suppose one can find a closed exterior p-form ϕ such that

$$\phi|_M = \mathrm{vol}_M \tag{1}$$

and with the additional property that

$$\phi|_\xi \leqq \mathrm{vol}_\xi \tag{2}$$

for all oriented tangent p-planes ξ at all points of X. *Then* M *is homologically volume-minimizing in* X, that is, $\mathrm{vol}(M) \leqq \mathrm{vol}(M')$ for all oriented p-manifolds M' with $\partial M' = \partial M$ and $[M-M'] = 0$ in $H_p(X;R)$. To see this, note that $\mathrm{vol}(M) = \int_M \phi = \int_{M'} \phi \leqq \mathrm{vol}(M')$, (where the first equality comes from (1), the second from the homology assumption, and where the inequality comes from (2)). This argument generalizes to show that M minimizes mass among all compactly supported deRham currents homologous to M (cf. [HL_1]).

Seminar on Minimal Submanifolds
0-691-08324-X/83/217-07 $0.85/0 (cloth)
0-691-08319-3/83/217-07 $0.85/0 (paperback)

Taking this process in reverse, one associates to a given p-form ϕ on X, a collection of p-dimensional submanifolds, which we call the ϕ-geometry of X. In particular, we assume that ϕ satisfies condition (2), which is to say in the language of Whitney and Federer, that ϕ has *comass* $\leqq 1$. Then any compact oriented p-dimensional submanifold M which satisfies (1), will be called a *ϕ-submanifold* of X. More generally, an integral p-current $\mathfrak{M}$ in X such that $\mathfrak{M}(\phi) = \mathrm{Mass}(\mathfrak{M})$ will be called a *ϕ-subvariety* of X. The collection of ϕ-subvarieties constitutes the *ϕ-geometry* of X.

The above argument shows that if ϕ is closed, then every ϕ-variety is homologically mass minimizing (over $\mathbf{R}$) in X. In particular, every ϕ-submanifold is a minimal submanifold. We call such a closed form ϕ of comass-one a *calibration*, and the pair (X, ϕ) is called a *calibrated manifold.*

For generic forms ϕ the ϕ-geometry is completely uninteresting. Nevertheless, this process recaptures most of the standard geometric structures on manifolds. For example, if X is a complex hermitian manifold and ω is its associated Kähler form, then $(\omega^p/p!)$-geometry consists precisely of the complex p-dimensional subvarieties (with boundary) in X. If ω is closed, i.e., if X is Kählerian, then every complex subvariety is homologically mass minimizing. This is the classical result of H. Federer [F].

2. *Tight foliations and the Kähler question*

Suppose now that X is a smooth manifold equipped with an oriented, p-dimensional foliation $\mathcal{F}$ of class C^1 (cf. [L_2]). For any Riemannian structure on X there are many p-forms ϕ such that the ϕ-geometry is generated by domains of finite volume in the leaves. Such forms must have comass-one and the property that

$$\phi|_{\mathcal{F}} = \mathrm{vol}_{\mathcal{F}} . \tag{3}$$

In this light, it is natural to ask the following.

QUESTION 1. Can one find a Riemannian metric on X for which there exists a calibration ϕ which satisfies (3)?

Note that in such a metric, every leaf would be a minimal submanifold. Moreover, every relatively open subset of finite volume in a leaf, and, in fact, every positive $\mathbb{R}$-linear combination of such things, would be a current of least mass for its boundary. This would have the following purely topological consequence.

PROPERTY 2. *Every compact leaf, and in fact, every positive integral chain of compact oriented leaves from* $\mathfrak{F}$, *represents a non-zero class in* $H_p(X; \mathbb{R})$.

However there is an even stronger topological consequence which follows easily from the discussion above. By a *foliation cycle* or $\mathfrak{F}$*-cycle*, we mean a d-closed current T of finite mass and compact support in X, such that $\vec{T}_x = \vec{\mathfrak{F}}_x$ for $\|T\|$ – a.a.x. Foliation cycles appear, for example, in constructing the asymptotic homology classes of leaves with polynomial growth (cf. [P]).

PROPERTY 2′. *Every non-zero foliation cycle represents a non-zero class in* $H_p(X; \mathbb{R})$.

Interestingly, this condition is sufficient.

THEOREM 3 (Harvey and Lawson [HL_2]). *Let* $\mathfrak{F}$ *be an oriented* p-*dimensional foliation of class* C^1 *on a compact manifold* X. *If* $\mathfrak{F}$ *has Property 2′, then the answer to Question 1 is "yes".*

If the codimension of $\mathfrak{F}$ *is one and* $\mathfrak{F}$ *simply satisfies Property 2, then the answer to Question 1 is "yes".*

This result is closely related to the work of D. Sullivan [S], and part of the proof uses techniques presented there.

One can formulate an analogue to Question 1 in the context of complex geometry. This boils down to asking: *When does a given complex manifold*

X *admit a Kähler metric*? When X is compact, the authors have established necessary and sufficient conditions for an affirmative answer. (See [HL_3].)

3. Special Langrangian geometry

Recently the authors have conducted a systematic investigation of the geometries associated to parallel forms in Euclidean space. We shall briefly sketch here the principal ideas and results. Complete details will appear in [HL_1].

The first new and interesting geometry is associated to the form

$$\phi = \mathrm{Re}\{dz_1 \wedge \cdots \wedge dz_n\}$$

in $\mathbb{C}^n = \mathbb{R}^n \oplus i\mathbb{R}^n$ where, as usual, we write $z = x + iy$. The corresponding ϕ-submanifolds are Lagrangian submanifolds of "constant phase", and the geometry is therefore called *special Lagrangian.* To be specific, we recall that the Lagrangian n-planes in $\mathbb{C}^n$ are just the U_n-images of the x-axis. Accordingly, the special Lagrangian n-planes (those where attains its maximum) are just the SU_n-images of the x-axis. If we let Lag and SLag denote the Grassmannians of oriented Lagrangian and special Lagrangian planes, then there are diffeomorphisms: $\mathrm{Lag} = U_n/SO_n$ and $\mathrm{SLag} = SU_n/SO_n$, and there exists a "Maslov" fibration

$$\mathrm{SLag} \longrightarrow \mathrm{Lag} \xrightarrow{\det} S^1 .$$

given by the complex determinant.

It is an easy and well-known fact that, after a possible unitary change of coordinates, every Lagrangian submanifold can be locally represented as the graph of a function: $y = f(x)$ where $f = \mathrm{grad}(F)$ for some real-valued potential function $F(x)$. Accordingly, after a possible special-unitary change of coordinates, every special Lagrangian submanifold can be locally represented as the graph of a gradient: $y = (\nabla F)(x)$ where the potential function F satisfies a certain non-linear elliptic differential equation. When $n = 3$, this equation is simply

$$\Delta F = \det(\mathrm{Hess}F) . \tag{4}$$

It follows that the graph of the gradient of any solution of (4) is an absolutely volume-minimizing 3-fold in $\mathbf{R}^6$. In particular, any C^2 solution is real analytic.

In dimension 3, special Lagrangian geometry bears an intimate relation to the work of Hans Lewy on harmonic gradient maps [Le] and explains the mysterious appearance there of the minimal surface equation.

The geometry of special Lagrangian varieties in all dimensions is richly endowed and constitutes a large new class of minimizing currents in $\mathbf{R}^n$. Among the many interesting special cases are the following. Let $\Sigma \subset S^3 \subset \mathbf{R}^4$ be a compact oriented minimal surface in the 3-sphere, and for $x \in \Sigma$, let x^* denote the unit normal to Σ at x in S^3. Then

$$V_\Sigma = \{(sx, tx^*) \in \mathbf{R}^4 \oplus \mathbf{R}^4 : x \in \Sigma \text{ and } s, t \geqq 0\}$$

is a special Lagrangian 4-fold in $\mathbf{R}^8$. Since we may choose Σ to be of arbitrary genus (cf. [L_1]) the topological type of the cone V_Σ can be quite complicated.

4. *Exceptional geometries*

One of our earliest discoveries in this investigation was the appearance of certain beautiful "exceptional" geometries in low dimensions. There is a geometry of 3-folds (and a dual geometry of 4-folds) in $\mathbf{R}^7$, which is invariant under the standard representation of the group G_2. It is associated to the 3-form $\psi(x, y, z) = \langle x, yz \rangle$ where $x, y, z \in \mathbf{R}^7$ are considered as imaginary Cayley numbers. A 3-manifold $M \subset \mathbf{R}^7 = \mathrm{Im}\,\mathbf{O}$ belongs to this geometry iff each of its tangent planes is a (canonically oriented) imaginary part of a quaternion subalgebra of the Cayley numbers $\mathbf{O}$. The Grassmannian of all such 3-planes is naturally diffeomorphic to the homogeneous space G_2/SO_4.

The local system of differential equations for this geometry is essentially deduced from the vanishing of the *associator* $[x,y,z] = x(yz) - (xy)z$, and thus the geometry is called *associative*. This system of equations has a striking and elegant form: Write $\mathrm{Im}\,\mathbf{O} = \mathrm{Im}\,\mathbf{H} \oplus \mathbf{H}$, where $\mathbf{H}$ denotes the quaternions, and consider a function $f : U \subset \mathrm{Im}\,\mathbf{H} \to \mathbf{H}$. Then the graph of f is an associative submanifold if and only if f satisfies the equation

$$(5) \qquad Df = \sigma f$$

where D is the Dirac operator and σ is a certain first-order "Monge-Ampere" operator. Specifically,

$$Df = -\Sigma(\nabla_{e_j} f)e_j \qquad \text{and} \qquad \sigma f = (\nabla_{e_1} f) \times (\nabla_{e_2} f) \times (\nabla_{e_3} f)$$

where $e_1, e_2, e_3 = e_1 e_2$ is any oriented orthonormal basis of $\mathrm{Im}\,\mathbf{H}$, and where the triple-cross-product on $\mathbf{H}$ is defined by $x \times y \times z = (1/2)(x\bar{y}z - z\bar{y}x)$. Implicit Function Theorem techniques show that (5) has many solutions. Moreover, an application of the Cartan-Kähler Theorem shows that any real analytic surface in $\mathbf{R}^7$ is contained in a uniquely determined associative 3-fold. The associative geometry is definitely quite large.

The dual geometry of 4-folds in $\mathbf{R}^7$ is called *coassociative*, and has a local system of differential equations similar to (5). This geometry includes the "cone on the Hopf map", that is, the graph of the Lipschitz (and not C^1) solution to the minimal surface system constructed in [LO]. It follows that this singular graph is actually volume-minimizing in $\mathbf{R}^7$.

The richest and most intriguing of the exceptional geometries is the family of Cayley 4-folds in $\mathbf{R}^8 \cong \mathbf{O}$. This is the collection of subvarities associated to the 4-form $\Psi(x,y,z,w) = \frac{1}{2}\langle x(\bar{y}z) - z(\bar{y}x), w\rangle$. It is invariant under the 8-dimensional representation of Spin_7; and the Cayley 4-planes constitute a homogeneous space of the form Spin_7/H where $H = SU_2 \times SU_2 \times SU_2/\mathbf{Z}_2$. This geometry contains the coassociative 4-folds. It also contains both the complex and the special Lagrangian subvarieties for a seven-dimensional family of complex structures on $\mathbf{R}^8$. In each of these complex structures the Cayley submanifolds can be characterized geometrically as the oriented 4-manifolds $M \subset \mathbf{R}^8 = \mathbf{C}^4$ which have the property that $\omega|_M$ is *self-dual* and which satisfy a "constant phase" condition analogous to that occurring in the special Lagrangian case.

The associated system of equations for a function $f: U \subset \mathbf{H} \to \mathbf{H}$ (guaranteeing that the graph of f in $\mathbf{O} = \mathbf{H} \oplus \mathbf{H}$ is a Cayley 4-fold) is again of the form $Df = \sigma f$. Here D is the Dirac operator (or more suggestively, the quaternion analogue of the operator $\partial/\partial z$ for functions $f: U \subset \mathbf{C} \to \mathbf{C}$). The operator σ is a homogeneous cubic expression in the first derivatives if f, constructed using the 3-fold cross-product in $\mathbf{H}$. This equation may be solved by Implicit Function Theorem techniques. Further examples can be constructed by application of the Cartan-Kähler Theorem.

F. REESE HARVEY
DEPARTMENT OF MATHEMATICS
RICE UNIVERSITY
HOUSTON, TEXAS 77001

BLAINE LAWSON, JR.
DEPARTMENT OF MATHEMATICS
STATE UNIVERSITY OF NEW YORK
STONY BROOK, NEW YORK 11790

REFERENCES

[F] H. Federer, *Some theorems on integral currents*, Trans. A.M.S., *117* (1965), 43-67.

[HL_1] R. Harvey and H. B. Lawson, Jr., *Calibrated geometries*, Acta Math., 148 (1982), 47-157.

[HL_2] ———, *Calibrated foliations*, Amer. J. Math., 103 (1981), 411-435.

[HL_3] ———, An intrinsic characterization of Kähler manifolds, (to appear).

[L_1] H. B. Lawson, Jr., *Complete minimal surfaces in* S^3, Ann. of Math., *92* (1970), 335-374.

[L_2] ———, *Foliations*, Bull. A.M.S., *80* (1974), 369-418.

[LO] H. B. Lawson, Jr. and R. Osserman, *Non-existence, non-uniqueness and irregularity of solutions to the minimal surface system*, Acta Mathematics, *139* (1977), 1-17.

[Le] H. Lewy, *On the non-vanishing of the Jacobian of a homeomorphism by harmonic gradients*, Ann. of Math., *88* (1968), 529-578.

[P] J. Plante, *Asymptotic properties of foliations*, Comm. Math. Helv. *47* (1972), 449-456.

[S] D. Sullivan, *A homological characterization of foliations consisting of minimal surfaces*, (to appear).

NECESSARY CONDITIONS FOR SUBMANIFOLDS AND CURRENTS WITH PRESCRIBED MEAN CURVATURE VECTOR

Robert Gulliver*

Previous investigations have considered the nonhomogeneous problem of Plateau, in which a Riemannian manifold M^n is given and an oriented submanifold Σ^k is sought, having a prescribed mean curvature vector and a given closed (k–1)-dimensional submanifold Γ as boundary. Here the mean curvature vector is prescribed as a function of the point in M and of the tangent space to Σ. In the present paper we derive necessary conditions on the magnitude of the prescribed mean curvature vector, depending on Γ and the geometry of M, for this Plateau problem to have a solution (see Theorem 2.1 below). In the Euclidean space E^7, for example, these necessary conditions imply that a plane circle Γ of radius $R > 1$ cannot be the boundary of any surface Σ whose mean curvature vector is the Cayley cross product of an orthonormal pair of tangent vectors. On the other hand, if $R \leq 1$, then Γ is the boundary of such a surface, namely, a spherical cap in a three-dimensional subspace of E^7 closed under the Cayley cross product. More generally, it has been shown that such a surface exists bounded by any closed curve Γ lying in the unit ball of E^7. These and other existence results are now seen to be the best possible.

---

*Research partially supported by National Science Foundation grant MCS 79-02032, and by a grant to the Institute for Advanced Study.

Seminar on Minimal Submanifolds
0-691-08324-X/83/225-18 \$1.40/0 (cloth)
0-691-08319-3/83/225-18 \$1.40/0 (paperback)

The first necessary conditions for parametric surfaces were given by Heinz for surfaces of constant mean curvature in Euclidean space E^3 ([7]). We then derived a necessary condition in codimension one for the more general case of prescribed mean curvature, relative to an infinitesimal isometry V of the Riemannian manifold M, assuming that the prescribed mean curvature function is constant along the integral curves of V ([5]).

If H is a tensor of type $(k,1)$ on M, and if H is alternating in its k arguments, with values orthogonal to each of its arguments, then an oriented k-dimensional submanifold Σ of M is said to have *prescribed mean curvature vector* H if its mean curvature vector at each point $p \in \Sigma$ is equal to $H(u_1, \cdots, u_k)$ for any positively oriented, orthonormal basis $\{u_1, \cdots, u_k\}$ for the tangent space to Σ at p. In codimension $n-k=1$, if M is oriented, then the (k,1)-tensor H is usually represented by a real-valued function h on M, where $|h(p)| = |H(u_1, \cdots, u_k)|$ for any orthonormal set $\{u_1, \cdots, u_k\}$ of tangent vectors to M at p, and the sign of $h(p)$ is determined by the orientation of $\{u_1, \cdots, u_k, H(u_1, \cdots, u_k)\}$. In this case, one speaks of a hypersurface of prescribed mean curvature h.

Submanifolds with prescribed mean curvature vector H may be characterized as stationary submanifolds for the functional

$$E(\Sigma) = \text{Area}(\Sigma) + \int_\Sigma \alpha \tag{1.1}$$

where α is a differential k-form such that

$$d\alpha(v_0, \cdots, v_k) = \langle v_0, kH(v_1, \cdots, v_k)\rangle$$

for any tangent vectors $v_0, \cdots, v_k$ to M. Hence, an alternating (k–1)-tensor H, whose values are orthogonal to each of its k arguments, may be related to a variational problem only if the associated differential (k+1)-form β, defined by

$$\beta(v_0, v_1, \cdots, v_k) = \langle v_0, kH(v_1, \cdots, v_k)\rangle ,$$

satisfies the integrability condition

$$d\beta = 0 . \tag{1.2}$$

This integrability condition was assumed for the existence theorems in [4] and appears again as a hypothesis in the necessary conditions of the present paper. It may be observed that (1.2) holds automatically in co-dimension one.

A further hypothesis to the necessary conditions below, as in [5], is that H be invariant under a one-parameter family of isometries $\phi_t : M \to M$, or equivalently, that the Lie derivative $L_V H = 0$, where V is the infinitesimal isometry $V = \frac{\partial}{\partial t}\big|_{t=0} \phi_t$. In the most familiar case, V would be the last coordinate vector e_n in Euclidean E^n, and it is required that the coefficients of the tensor H depend on $x^1, \cdots, x^{n-1}$ only (see Example 3.1 below). The relevance of such a hypothesis, in addition to a lower bound on the magnitude of H, may be seen by considering the curve $(x^1(t), x^2(t)) = (\sin t, \cos t + bt)$ in E^2 : Its curvature may be written as a "prescribed" function of x^1, and has one sign if $|b| < 1$. The curve connects points arbitrarily far apart in the x^2-direction. After rescaling, this example shows that two given points may be connected by curves with an arbitrarily large lower bound for curvature.

When $k > 2$, a variational problem for k-dimensional submanifolds has solutions which are not submanifolds, in general, but belong to a wider class of k-dimensional objects, for example, the class of integral currents of dimension k. It would therefore be desirable to have necessary conditions for the existence of integral currents Σ with prescribed mean curvature vector and having a given boundary Γ. In Theorem 2.4 below, we indicate necessary conditions, analogous to those for smooth submanifolds, for the existence of an integral current which minimizes the functional corresponding to (1.1).

The general necessary condition for submanifolds with prescribed mean curvature vector is derived in Section 2. In this differentiable context, there is no need to assume that the submanifold minimizes the functional (1.1). Section 3 is devoted to a series of examples in which the ambient manifold M has a high degree of symmetry.

NOTATION. The metric tensor of the Riemannian manifold M is written in either of two ways: $g(V, W) = \langle V, W \rangle$. The associated Levi-Civita connection is denoted ∇. The volume k-form of a k-dimensional Riemannian manifold Σ is written as ν_Σ; the volume of Σ is $\mathrm{Vol}(\Sigma) = \int_\Sigma \nu_\Sigma$. For a (k+1)-form β and a vector field V, inner multiplication by V defines the k-form $(i(V)\beta)(v_1, \cdots, v_k) := \beta(V, v_1, \cdots, v_k)$. The norm of β is $|\beta|(p) := \sup\{\beta(u_0, \cdots, u_k) : (u_0, \cdots, u_k)$ orthonormal vectors at $p\}$. The Lie derivative of a tensor H with respect to a vector field V is $L_V H = \lim_{t\to 0} \frac{\phi_t^* H - H}{t}$, where $\{\phi_t\}$ is the flow associated with V: $\frac{\partial}{\partial t} \phi_t = V \circ \phi_t$, $\phi_0 =$ identity. With respect to a particular vector field V, we may define a reduced inner product

$$\tilde{g}(X, Y) := \langle X, Y \rangle - \frac{\langle X, V \rangle \langle Y, V \rangle}{|V|^2}. \tag{1.3}$$

Equivalently, $\tilde{g}(X, Y) = \langle X^0, Y^0 \rangle$, where X^0 denotes the component of X orthogonal to V. Using $\tilde{g}$, we define a reduced volume form $\tilde{\nu}_\Sigma$ for a submanifold Σ^k: $\tilde{\nu}_\Sigma(X_1, \cdots, X_k) := \nu_\Sigma(X_1^0, \cdots, X_k^0)$.

2. *Immersed submanifolds*

A rather natural condition, and a strong one, on a vector field in a Riemannian manifold is that it preserve the Riemannian structure, that is, that the associated flow consist of isometries. Such a vector field V is called an *infinitesimal isometry* (or Killing vector field); this is equivalent to the condition that the Lie derivative $L_V g$ of the metric tensor vanishes, or that $\langle \nabla_W V, W \rangle = 0$ for any vector field W.

2.1. THEOREM. *Let* M *be an* n*-dimensional Riemannian manifold,* H *an alternating* (k,1)*-tensor on* M*, with values orthogonal to its arguments, satisfying the integrability condition (1.2). Suppose* V *is an infinitesimal isometry on* M*, such that the Lie derivative* $L_V H = 0$*. Consider a* C^1 *immersed* (k–1)*-dimensional submanifold* Γ*. If there exists a* C^2 *immersed submanifold* Σ *with prescribed mean curvature vector* H*, of class* C^1 *up to the boundary, so that* $\partial\Sigma = \Gamma$*, then for any* k*-dimensional submanifold* Σ_0 *with* $\partial\Sigma_0 = \Gamma$ *and homologous to* Σ*,*

$$\left| \int_{\Sigma_0} < kH(u_1, \cdots, u_k), V > \nu_{\Sigma_0} \right| \leq \int_{\Gamma} |V| \tilde{\nu}_{\Gamma}, \tag{2.1}$$

where $(u_1, \cdots, u_k)$ *denotes a local choice of orthonormal basis of tangent vectors to* Σ_0*. The right-hand side of (2.1) is unchanged if* Γ *is deformed smoothly along the integral curves of* V*.*

The usefulness of Theorem 2.1 for nonexistence statements follows from the fact that no information is needed regarding a possible submanifold with prescribed mean curvature vector other than its homology class. The result is further strengthened in that Σ_0 need only be homologous to Σ modulo V, and that Γ may be replaced by an immersed (k–1)-dimensional submanifold Γ_0 obtained from Γ by flowing along a vector field ϕV, for any scalar function ϕ. It might be observed that this invariance property of the right-hand side of (2.1) is highly analogous to a property enjoyed by the Hilbert invariant integral associated with a field of extremals in one-dimensional variational problems. The meaning of this invariance property is perhaps clearest in the case treated in Example 3.1 below.

The proof of Theorem 2.1 is based on the following formula, which may be verified directly (cf. the proof of Lemma 3.2 in [5]).

2.2. LEMMA. *Let* Σ *be a* k*-dimensional immersed submanifold of a Riemannian manifold* M*, with mean curvature vector field* $\mathcal{H}$*. Then for* *any vector field* V *on* M, *the orthogonal component* V^t *of* V *tangent to* Σ *has divergence*

$$\mathrm{div}_\Sigma V^t = \langle k\mathcal{H}, V\rangle + \frac{1}{2}\sum_i (L_V g)(u_i, u_i)$$

where $(u_1, \cdots, u_k)$ *is any orthonormal basis for the tangent space to* Σ.

Proof of Theorem 2.1. Let N denote the outward unit normal vector to $\Gamma = \partial\Sigma : N$ is tangent to Σ. The divergence theorem in Σ, as a Riemannian manifold-with-boundary, implies

$$\int_\Sigma \mathrm{div}_\Sigma V^t \nu_\Sigma = \int_\Gamma \langle V^t, N\rangle \nu_\Gamma . \tag{2.2}$$

Now since V is an infinitesimal isometry, we have $L_V g = 0$, and since Σ has prescribed mean curvature vector H, its mean curvature vector $\mathcal{H} = H(u_1, \cdots, u_k)$ for any oriented orthonormal basis $(u_1, \cdots, u_k)$ for the tangent space to Σ. Thus Lemma 2.2 implies

$$\mathrm{div}_\Sigma V^t = \langle kH(u_1, \cdots, u_k), V\rangle . \tag{2.3}$$

Now $\tilde{\nu}_\Gamma = c\nu_\Gamma$, where c^2 is the determinant of the $(k-1)\times(k-1)$ matrix with entries $\tilde{g}(v_i, v_j)$, and where $(v_1, \cdots, v_{k-1})$ is an orthonormal basis for the tangent space of Γ. Writing $e = V/|V|$, we have $\tilde{g}(v_i, v_j) = \delta_{ij} - \langle v_i, e\rangle\langle v_j, e\rangle$; therefore, via an eigenvalue computation, $c^2 = 1 - \Sigma\langle v_i, e\rangle^2$. Now $(v_1, \cdots, v_{k-1}, N)$ is an orthonormal set, so that $\langle e, N\rangle^2 + \Sigma\langle v_i, e\rangle^2 \le |e|^2 = 1$ and hence

$$|\langle V^t, N\rangle| = |\langle V, N\rangle| \le |V|(1 - \Sigma\langle v_i, e\rangle^2)^{1/2} = |V|c .$$

Therefore,

$$\left|\int_\Gamma \langle V^t, N\rangle \nu_\Gamma\right| \le \int_\Gamma |\langle V^t, N\rangle| \nu_\Gamma \le \int_\Gamma |V| \tilde{\nu}_\Gamma . \tag{2.4}$$

The desired inequality (2.1), with Σ in place of Σ_0, follows from (2.2), (2.3) and (2.4).

It remains to prove the invariance properties of both sides of (2.1). Consider the (k+1)-form β defined by

$$\beta(w_0, w_1, \cdots, w_k) := \langle w_0, kH(w_1, \cdots, w_k)\rangle . \tag{2.5}$$

Since H satisfies the integrability condition (1.2), we have $d\beta = 0$. Further, since β is the constant k times a contraction of the (k,1)-tensor H with the metric tensor g, the assumptions $L_V g = 0$ and $L_V H = 0$ imply $L_V \beta = 0$. Therefore, $d(i(V)\beta) = L_V\beta - i(V)d\beta = 0$. Note that the left-hand side of (2.1) is nothing more than the integral of the k-form $i(V)\beta$ over Σ_0. So suppose that an immersed k-dimensional submanifold Σ_0 is homologous to Σ modulo V; that is, there exists a differentiable k-chain Σ_1 and a (k+1)-chain R in M so that $\partial R = \Sigma - \Sigma_0 + \Sigma_1$, and so that V lies in the tangent space to Σ_1 whenever this tangent space is k-dimensional. Then the restriction of $i(V)\beta$ to Σ_1 vanishes. Therefore,

$$\int_\Sigma \langle kH(f_1, \cdots, f_k), V\rangle \, d\nu_\Sigma - \int_{\Sigma_0} \langle kH(u_1, \cdots, u_k), V\rangle \, d\nu_{\Sigma_0} =$$

$$= \int_{\partial R} i(V)\beta = \int_R d(i(V)\beta) = 0$$

where $(f_1, \cdots, f_k)$ is a local orthonormal basis of tangent vector fields to Σ. This proves the invariance property asserted for the left-hand side of (2.1).

The invariance property of the right-hand side of (2.1) may be stated as follows. Let $\phi_t : M \to M$, $t \in \mathbf{R}$, be the flow generated by the vector field V, that is: $\partial\phi_t/\partial t = V \circ \phi_t$, and $\phi_0 = id_M$. If τ is a C^1 real-valued function on Γ, then $\pi(p) = \phi_{\tau(p)}(p)$ defines a diffeomorphism π

of Γ with another immersed (k–1)-submanifold Γ_0. We assert that

(2.6)
$$\int_{\Gamma_0} |V|\,\tilde{\nu}_{\Gamma_0} = \int_{\Gamma} |V|\,\tilde{\nu}_{\Gamma}\,.$$

To see this, consider $p \in \Gamma$ and write $p_0 = \pi(p)$. Then $V(p_0) = (\phi_{\tau(p)})_*(V(p))$ since V is invariant under its own flow. Also, $\phi_{\tau(p)}$ is an isometry, so $|V(p_0)| = |V(p)|$. Now for a tangent vector w to Γ at p, one may compute that $\pi_*(w)$ and $(\phi_{\tau(p)})_*(w)$ differ by a multiple of $V(p_0)$. Their components orthogonal to $V(p_0)$ are therefore the same, and hence they have the same length with respect to the reduced metric $\tilde{g}$. But $\phi_{\tau(p)}$ is an isometry for $\tilde{g}$ as well as for g. This implies that $\pi: \Gamma \to \Gamma_0$ is a $\tilde{g}$-isometry, and in particular $\pi^* \tilde{\nu}_{\Gamma_0} = \tilde{\nu}_{\Gamma}$. This completes the proof of the invariance property (2.6). Q.E.D.

2.3. REMARK. *Branch points.* The existence theory for $k = 2$ in, for example, [4] yields smooth mappings of class C^1 up to the boundary ([8]), which are not necessarily immersions but only *branched immersions.* That is, for an appropriate coordinate system (u, v) at any singular point of f in its domain, and an appropriate coordinate system $(x^1, \cdots, x^n)$ in M at its image, the mapping satisfies an asymptotic relation

(2.7)
$$(x^1 + ix^2)(f(u,v)) = (u+iv)^{m+1} + \sigma^1(u,v) + i\sigma^2(u,v)$$
$$x^r(f(u,v)) = \sigma^r(u,v)\,, \qquad 3 \le r \le n\,,$$

as (u, v) tends to the singular point $(0, 0)$, where $\sigma^r(0,0) = 0$ and the first partial derivatives of $\sigma^r(u,v)$ tend to zero faster than $(u^2+v^2)^{m/2}$, $1 \le r \le n$. The natural number m is called the order of branching, and the singular point is called a branch point. It may be seen from (2.7) that branch points are isolated.

It has been shown in the codimension-one case $n = 3$ that simply-connected minimizing surfaces are immersions in the interior ([3]), and are immersions up to the boundary, provided that Γ is an analytic Jordan curve ([6]). In both cases, it is assumed that the functional E is positive definite, that is $|\alpha| \leq a < 1$. However, in the case of higher codimension the minimizing surfaces may well have branch points, as is shown by the example of a holomorphic mapping from a plane domain into $C^n = R^{2n}$, $n \geq 2$, which is always an area-minimizing minimal surface ([9], p. 33). It is therefore desirable to extend Theorem 2.1 to allow branched immersions for $k = 2$.

The proof of Theorem 2.1 required Σ to be an immersed submanifold in applying the divergence theorem; see equation (2.2). However, the divergence theorem continues to hold when Σ is the image of $f: M' \to M$, M' is a differentiable 2-manifold with boundary, f is a C^1 branched immersion up to the boundary, and V is a C^1 vector field on M. The proof is similar to arguments carried out on pp. 22-23 of [5], and shows that the contribution from a small neighborhood of each branch point is negligible. The remainder of the proof holds without change. We conclude that Theorem 2.1 is valid in the case $k = 2$ when Σ is the image of a branched immersion.

A rather more general class of geometric objects is comprised of the *integral currents* of geometric measure theory (see [1]). With a somewhat stronger hypothesis than for submanifolds and branched immersions, one may show that the analogous necessary conditions obtain for integral currents.

2.4. THEOREM. *Let* V *be an infinitesimal isometry on* M, *and suppose* H *is a* (k,1)-*tensor with* $L_V H = 0$, *and satisfying the integrability condition (1.2). If a* k-*dimensional integral current* T *is homologically minimizing for the functional* $E(T) = \mathbf{M}(T) + T(\alpha)$ *with respect to its boundary* $\partial T = \Gamma$, *then*

$$|T(i(V)\,d\alpha)| \leq \int |V|\,|\vec{\Gamma}|^{\sim}\, d\|\Gamma\| . \tag{2.8}$$

Moreover, the left-hand side of (2.8) is an invariant of the homology class of T *modulo* V, *while the right-hand side is unchanged if* Γ *is deformed along the integral curves of* V. *Here* $|\vec{\Gamma}|^{\sim}$ *denotes the length with respect to the reduced inner product* $\tilde{g}$ *of the unit* (k–1)*-vector* $\vec{\Gamma}$; $\|\Gamma\|$ *is the measure associated with* Γ *as a current;* $\mathbf{M}(T)$ *denotes the mass of* T ([1], p. 349); *and* α *is a* k*-form satisfying* $d\alpha = \beta$ *as above.*

The proof of Theorem 2.4 involves the direct comparison of T with a current obtained by flowing along V for small time, plus a small piece at the boundary. The theorem holds, by the same proof, for currents satisfying a hypothesis weaker than homological minimization but stronger than stationarity, namely: that $E(T) \le E(T') + \varepsilon F(T-T')$ for any given $\varepsilon > 0$, and for all currents T' with $F(T-T')$ sufficiently small. Here F is the flat seminorm ([1], p. 367). We conjecture that the necessary condition (2.8) holds also for a stationary current T, but present knowledge of regularity at $\Gamma = \partial T$ appears insufficient to permit generalization of Lemma 2.2.

The extension of our results to integral currents is of particular interest owing to the absence of an even remotely satisfactory existence theory in the context of k-dimensional immersed submanifolds when $k > 2$. By contrast, if M is a compact manifold and the k-form α may be found with $|\alpha| \le 1-\varepsilon < 1$, then for every closed (k–1)-dimensional rectifiable current Γ, there is an integral current minimizing E in each homology class with Γ as boundary ([1], p. 521). The solution is a smooth submanifold with prescribed mean curvature vector H on an open dense subset of the support of its measure ([1], pp. 466, 471). It should be pointed out that the two-dimensional existence theory in [4], for example, is carried out in a manifold-with-boundary M on which α may be constructed; we conjecture that the existence of minimizing integral currents holds also in an ambient manifold-with-boundary, provided its boundary allows an appropriate maximum principle.

3. *Examples*

It was assumed in the hypotheses of Theorem 2.1 that V is an infinitesimal isometry. Although this is a strong assumption, it is nonetheless satisfied in a number of interesting cases. In light of Theorem 2.4 above, it should be observed that all of these examples apply to homologically minimizing currents as well as to submanifolds and branched immersions with prescribed mean curvature vector.

3.1. *Euclidean space.* The most familiar infinitesimal isometry is a parallel vector field in Euclidean space E^n. For clarity, let us assume that $V = e_n$, the last coordinate vector field in standard orthogonal coordinates $(x_1, \cdots, x_n)$, and that Γ is the graph $\{(x_1, \cdots, x_n) : x_n = \phi(x_1, \cdots, x_{n-1}), (x_1, \cdots, x_{n-1}) \in \Gamma_0\}$ of a C^1 function ϕ over an immersed (k–1)-submanifold Γ_0 of E^{n-1}. The hypothesis $L_V H = 0$ in this context simply means that the components of the tensor H do not depend on x^n. We identify E^{n-1} with the hyperplane $\{x^n = 0\}$ in E^n. Now choose any k-chain Σ_0 in E^{n-1} with Γ_0 as boundary. A necessary condition for Γ to be the boundary of a submanifold with prescribed mean curvature vector H is, according to Theorem 2.1, that

$$(3.1) \qquad \int_{\Sigma_0} < kH(u_1, \cdots, u_k), e_n > \nu_{\Sigma_0} \leq \mathrm{Vol}(\Gamma_0)$$

where $(u_1, \cdots, u_k)$ is an orthonormal basis of tangent vectors to Σ_0. The invariance property of the left-hand side of inequality (2.1) shows further that the choice of Σ_0 is immaterial, that is, the necessary condition is on Γ_0 alone, and moreover holds equally for all graphs Γ over Γ_0. This may be realized explicitly by finding a k-form α whose components do not depend on x^n, with $d\alpha = \beta$. Then $i(e_n)\beta = -d(i(e_n)\alpha)$, and the necessary condition (3.1) may be rewritten

$$\left| \int_{\Gamma_0} i(e_n)\alpha \right| \leq \mathrm{Vol}(\Gamma_0) .$$

One might observe that this is not unrelated to the condition $|\alpha| \leq 1-\varepsilon < 1$ which, in the company of an appropriate maximum principle, is sufficient for the existence of solutions to the variational problem in the case $k = 2$ (cf. Lemma 4.2 of [4]). Condition (3.1) itself resembles the necessary condition given by Giaquinta for the existence of a nonparametric minimizing hypersurface ([2]).

3.2. *Codimension two.* For $k = n-2$, where M is oriented, a $(k,1)$-tensor H satisfying the integrability condition (1.2) corresponds to a divergence-free vector field B, with the correspondence defined by $\langle v_0, H(v_1, \cdots, v_k)\rangle = \nu_M(B, v_0, v_1, \cdots, v_k)$. In this case, the necessary condition (2.1) may be written

$$k \left| \int_{\Sigma} \langle B, \xi \rangle \right| V^n |\nu_\Sigma| \leq \int_{\Gamma} |V| \tilde{\nu}_\Gamma \tag{3.2}$$

where V^n is the orthogonal component of V normal to Σ, and ξ is a unit normal to Σ orthogonal to V.

In the Euclidean context with $V = e_n$, as above, we may consider $B = e_1$ and let Γ be a graph $\Gamma = \{x : x^n = \phi(x^2, \cdots, x^{n-1}), (x^2, \cdots, x^{n-1}) \in \Gamma_0, x^1 = 0\}$ over the $(k-1)$-dimensional sphere Γ_0 of radius R in the k-plane $\{x^1 = x^n = 0\}$. Σ_0 may be chosen as the ball interior to Γ_0 in this k-plane. Using the invariance properties shown in Theorem 2.1, we may write (3.2) for Σ_0 and Γ_0, obtaining the necessary condition $k\,\mathrm{Vol}(\Sigma_0) \leq \mathrm{Vol}(\Gamma_0)$. But $\mathrm{Vol}(\Gamma_0) = k\,\mathrm{Vol}(\Sigma_0)/R$, so that Γ can be the boundary of a submanifold of prescribed mean curvature vector H only if $R \leq 1$. On the other hand, Γ_0 itself is the boundary of a k-dimensional hemisphere Σ_2 of radius 1 in the hyperplane $L := \{x : x^1 = 0\}$, which has prescribed mean curvature vector H. This shows that our necessary condition is sharp.

In this example Σ_2 is a hypersurface of constant mean curvature 1 in the hyperplane L, which implies that Σ_2 has prescribed mean curvature vector H as a submanifold of E^n. In fact, on any submanifold Σ with prescribed mean curvature vector H, the coordinate function x^1 is harmonic. Hence by the maximum principle, if $\partial\Sigma$ lies in L, then Σ lies in L, and Σ is a hypersurface of constant mean curvature 1 in L. That is, the use of the maximum principle reduces the Plateau problem to codimension one.

This nonexistence statement for $B = e_1$ may be further compared to existence results given in [4] for $k = 2$: Since $|H| \leq 1$, any Jordan curve Γ lying in a closed ball $D^4(1)$ of radius one, or in any closed cylinder $D^3\left(\frac{1}{2}\right) \times R$ of radius one-half, is the boundary of a smooth (possibly branched) surface with prescribed mean curvature vector H, of the type of the disk. For the case at hand $(B = e_1)$, however, existence holds if Γ lies in the specific cylinder of radius one $D^3(1) \times R = \{x : (x^2)^2 + (x^3)^2 + (x^4)^2 \leq 1\}$, with axis parallel to B, or in certain bicylinders such as $D^2\left(\frac{1}{2}\right) \times R^2 = \{x : (x^2)^2 + (x^3)^2 \leq 1/4\}$, whose flat directions include the direction of B. In fact, these regions satisfy the hypotheses of Lemma 4.2 and Lemma 5.1 of [4] for this choice of tensor H.

3.3. *Charged particle in a stationary magnetic field.* The case $k = 1$, $n = 3$ is of physical interest. If H is defined as in Section 3.2 from the magnetic induction vector field B, then a curve of prescribed curvature vector H is the path of a particle of unit mass and charge, travelling with unit speed. In E^3, if the components B^1, B^2, B^3 of B depend only on (x^1, x^2), then the necessary condition of Theorem 2.1 gives a lower bound S_0 for the (constant) speed of a unit particle which passes through $x_0 = (x_0^1, x_0^2, x_0^3)$ and $\tilde{x} = (\tilde{x}_1, \tilde{x}_2, \tilde{x}_3)$:

$$2S_0 = \int B^1\,dx^2 - B^2\,dx^1$$

where the integral is over any path in the (x^1,x^2)-plane from (x_0^1,x_0^2) to $(\tilde{x}^1,\tilde{x}^2)$. The integrability condition (1.2) holds since the magnetic induction is divergence-free.

3.4. *Manifolds of constant sectional curvature.* Let M be the complete, simply-connected Riemannian n-manifold of constant sectional curvature b^2, where b is either real or imaginary. We would like to construct an alternating (k,1)-tensor H on M in such a way that our necessary conditions come as close as possible to the sufficient conditions provided by the existence theory. Among many natural constructions, the following appears most satisfying for the case $b^2 \leq 0$.

Given $p_0 \in M$, we may choose a complete, totally geodesic hypersurface P through p_0 and an infinitesimal isometry V on M which is everywhere orthogonal to P, and with length $|V(q)| = \cos(bd(p_0,q))$ for all $q \in P$. Here $d(p_0,q)$ denotes the Riemannian distance in M from p_0 to q. In general, $|V(q)| = \cos(br(q))$, where $r(q)$ is the distance from q to the complete geodesic tangent to $V(p_0)$. Let Q be a complete, totally geodesic (k+1)-submanifold through p_0 to which V is tangent. We define the mapping $\pi : M\setminus Q^* \to Q$ by letting $\pi(p)$ be the point in Q nearest to p; π is not defined on the focal set Q^* of Q, which is an (n–k–2)-dimensional sphere in the case $b^2 > 0$, and which is empty for $b^2 \leq 0$.

Consider the (k+1)-form $\pi^*\nu_Q$ on $M\setminus Q^*$ pulled back from the volume form of Q. Observe that $\pi^*\nu_Q$ is a closed (k+1)-form. One may compute the norm of $\pi^*\nu_Q$ by observing that there is an orthogonal basis of Jacobi fields $(u_1,\cdots,u_n)$ along the minimizing unit-speed geodesic σ from Q to any point $p \in M\setminus Q^*$, such that $u_n = \sigma'$, $u_1,\cdots,u_{k+1}$ all have length $\cos bt$ at $\sigma(t)$, and $u_{k+2},\cdots,u_{n-1}$ all have length $(\sin bt)/b$ at $\sigma(t)$. It follows that $|\pi^*\nu_Q(p)| = (\cos b\rho(p))^{-k-1}$, where $\rho(p)$ denotes the distance from p to Q.

We may define an alternating (k,1)-tensor H on $M\setminus Q^*$ as in equation (2.5) from the (k+1)-form $\beta = kh\,\pi^*\nu_Q$, for any positive constant h; note $|H| = h$. Let Γ_0 be the (k–1)-sphere at distance R from p_0 in the totally geodesic k-submanifold $P \cap Q$, and let its interior in $P \cap Q$ be denoted Σ_0. Then since V is orthogonal to Σ_0, we have

$$\int_{\Sigma_0} \langle V, kH(u_1, \cdots, u_k) \rangle \nu_{\Sigma_0} = \int_{\Sigma_0} i(V)\beta = \pm h\alpha_k b^{-k} \sin^k(bR) ,$$

where α_k denotes the volume of the unit ball in E^k. Further,

$$\int_{\Gamma_0} |V| \tilde{\nu}_{\Gamma_0} = \int_{\Gamma_0} |V| \nu_{\Gamma_0} = \alpha_k \cos(bR)\, b^{1-k} \sin^{k-1}(bR) .$$

Therefore, according to Theorem 2.1, a necessary condition for the existence of a k-dimensional submanifold in $M \setminus Q^*$ with prescribed mean curvature vector H and having Γ_0 as boundary is that $h \leq b \cot(bR)$. In fact, this necessary condition is sharp, since Γ_0 is the boundary of one hemisphere Σ of the k-dimensional sphere in Q at distance R from p_0; Σ is a submanifold of M with prescribed mean curvature vector H, defined using $h = b \cot(bR)$. The necessary condition $h \leq b \cot(bR)$ applies equally to any (k–1)-dimensional submanifold Γ whose points are in one-to-one correspondence with points of Γ_0 lying on the same integral curve of V.

By contrast, the existence results of [4] for $k = 2$ show that any closed curve lying in a closed ball D_R of radius R in M is the boundary of a (branched) surface of prescribed mean curvature vector H, provided that H satisfies the integrability condition (1.2) and $|H| \leq b \cot bR$ in D_R. In the case of nonpositive sectional curvature $b^2 = -a^2 \leq 0$, the sufficient condition $|H| \leq b \cot bR = a \coth aR$ is equivalent in the above construction of H to $h \leq b \cot bR$, which we have just shown to be a necessary condition as well (note that Γ_0 lies in $D_R(p_0)$). Thus the existence results are sharp for $b^2 \leq 0$.

For $b^2 > 0$, however, in order to ensure $|H| \leq b \cot bR$ everywhere in D_R, we must require $h \leq b \cot bR (\cos bR)^{k+1}$. This gap between the necessary and sufficient conditions may be closed as in the next example.

3.5. *Product manifolds.* For $b > 0$, let M be constructed as the Riemannian product $M^n = Q^{k+1} \times E^{n-k-1}$, where Q is the sphere of radius $\frac{1}{b}$ in E^{k+2}. Let $\pi : M \to Q$ be the projection onto the first factor. Given a constant $h > 0$, we define the (k+1)-form β by $\beta = hk\pi^*\nu_Q$. Then the corresponding (k,1)-tensor H, defined as in equation (2.5), has constant norm $|H| = h$. Choose a totally geodesic hypersurface P^k in Q^{k+1}, and a point $p_0 \in P$. Then there is an infinitesimal isometry V_0 of Q, with V_0 orthogonal to P and having length $|V_0(q)| = \cos(bd(p_0,q))$ for $q \in P$, where $d(p_0, q)$ is the Riemannian distance from p to q. We shall work with the unique infinitesimal isometry V of M with $\pi_* V = V_0$ and with V tangent to $Q \times \{x\}$ for each $x \in E^{n-k-1}$.

Identifying Q with $Q \times \{0\} \subset M$, we may choose Γ_0 to be the (k–1)-dimensional sphere at distance R from p_0 in P, and let Σ_0 be the interior of Γ_0 in P. Then

$$\int_{\Sigma_0} < kH(u_1, \cdots, u_k), V > \nu_{\Sigma_0} = \int_{\Sigma_0} i(V)\beta = \pm\, kh\alpha_k b^{-k} \sin^k(bR) ,$$

where $\{u_1, \cdots, u_k\}$ is an orthonormal basis of tangent vectors to Σ_0. A further computation yields

$$\int_{\Gamma_0} |V| \nu_{\Gamma_0} = k\alpha_k \cos(bR) b^{1-k} \sin^{k-1}(bR) .$$

Now according to Theorem 2.1, Γ_0 can be the boundary of a k-dimensional submanifold of prescribed mean curvature vector H only if $h \leq b \cot(bR)$.

On the other hand, Γ_0 is the boundary of a k-dimensional hemisphere in Q of radius R, which is a submanifold of prescribed mean curvature vector H defined as above with $h = b \cot(bR)$. Further, since $b^2 > 0$ is an upper bound for the sectional curvatures of M, the existence results of [4] (cf. Theorem 6.3) for $k = 2$ show that any closed curve lying in a ball of radius R is the boundary of a surface of prescribed mean curvature vector H, provided that $|H| \leq b \cot(bR)$. It is now shown that this existence result is sharp.

It might be pointed out that with this choice of H, the existence results of [4] hold for any curve lying in a cylindrical region $D_R \times E^{n-k-1}$, for any ball $D_R \subset Q^{k+1}$ of radius R, where $|H| \leq b \cot(bR)$ (compare Lemmas 4.2 and 5.1 of [4]).

3.6. *Cayley cross product.* There is an alternating (2,1)-tensor H on E^7 with the special property that $|H(u,v)| = 1$ for any orthonormal pair (u,v). In fact, let $\mu: E^8 \times E^8 \to E^8$ be the multiplication of Cayley numbers, $i: E^7 \to E^8$ an isometric linear transformation whose image is orthogonal to the scalars, and $\pi: E^8 \to E^7$ the orthogonal projection onto the image of i. Then H is defined by $H(u,v) = \pi(\mu(i(u), i(v)))$. Similar examples exist for a (k,1)-tensor on E^{2k+3}.

Let an orthonormal basis $(e_1, \cdots, e_7)$ for E^7 be chosen so that $H(e_1, e_2) = e_3$. Let $V = e_3$: V is an infinitesimal isometry. Since H has constant components, it satisfies the integrability condition (1.2) as well as $L_V H = 0$. If $\Gamma = \{(r\cos\theta, r\sin\theta, \phi(\theta), 0, 0, 0, 0) : \theta \in [0, 2\pi]\}$ is the graph of a smooth function ϕ over the circle Γ_0 of radius r in the plane spanned by e_1 and e_2, then according to Theorem 2.1, Γ can be the boundary of a surface of prescribed mean curvature vector H only when $r \leq 1$. The existence results of [4] show that any curve lying in a ball of radius 1 bounds a surface of prescribed mean curvature vector H; the existence results are now seen to be sharp in this case.

With this choice of H, a surface with prescribed mean curvature vector H will have a mean curvature vector of constant length 1. It would be of interest to know if there are necessary conditions, in codimension greater than one, for a curve to be the boundary of a surface satisfying this weaker condition, that its mean curvature vector has length 1 .

ROBERT GULLIVER
SCHOOL OF MATHEMATICS
UNIVERSITY OF MINNESOTA
MINNEAPOLIS, MINNESOTA 55455

REFERENCES

[1] Federer, H., *Geometric Measure Theory*, Springer-Verlag, New York, 1969.

[2] Giaquinta, M., On the Dirichlet problem for surfaces of prescribed mean curvature, *manuscripta math.*, 12 (1974), 73-86.

[3] Gulliver, R., Regularity of minimizing surfaces of prescribed mean curvature, *Annals of Math.*, 97 (1973), 275-305.

[4] ———, Existence of surfaces with prescribed mean curvature vector, *Math. Z.*, 131 (1973), 117-140.

[5] ———, On the nonexistence of a hypersurface of prescribed mean curvature with a given boundary, *manuscripta math.*, 11 (1974), 15-39.

[6] Gulliver, R., and F. D. Lesley, On boundary branch points of minimizing surfaces, *Arch. Rat. Mech. Anal.*, 52 (1973), 20-25.

[7] Heinz, E., On the nonexistence of a surface of constant mean curvature with finite area and prescribed rectifiable boundary, *Arch. Rat. Mech. Anal.*, 35 (1969), 249-252.

[8] Heinz, E., and S. Hildebrandt, Some remarks on minimal surfaces in Riemannian manifolds, *Comm. Pure Appl. Math.*, 23 (1970), 371-377.

[9] Lawson, B., *Lectures on Minimal Submanifolds*, IMPA, Rio de Janeiro, 1973.

APPROXIMATION OF RECTIFIABLE CURRENTS BY LIPSCHITZ Q VALUED FUNCTIONS

F. J. Almgren, Jr.[1]

0. *Introduction*

We show how to approximate in mass a rectifiable current of small excess within a cylinder by the "graph" of a Lipschitz function with values in $\mathbf{Q}$, a space of unordered Q tuples of points. Such approximation is a central theme of the manuscript [AF3] where nearly stationary integral varifolds with small tilt excess are so approximated as a first step in the study of the generalized branching type behavior of mass minimizing integral currents as discussed in [AF2]. The approximation there which extends [AW2 8.12] seems to depend in an essential way on the monotonicity formula of [AW2 5.1(3)] which, in view of [AW1], has no apparent generalization to surfaces which are stationary or minimizing for integrals other than area. The present method of approximation (Corollary 7) requires no such extremal properties for the current being approximated although slightly better estimates are obtained for minimizing currents (Corollary 6). Estimates on the size of Lipschitz constants depend on both the excess and the degree of mass approximation required. We conclude (Corollary 8) by showing how to approximate a rectifiable current of small multiplicity excess by the "graph" of a Lipschitz $\mathbf{Q}$ valued func-

[1] This research was supported in part by grants from the National Science Foundation.

Seminar on Minimal Submanifolds
0-691-08324-X/83/243-17 $1.35/0 (cloth)
0-691-08319-3/83/243-17 $1.35/0 (paperback)

tion, possibly with large Lipschitz constant, up to a set having small projected area. The proof of our basic estimate (Theorem 5) was suggested by the proof of [SS Lemma 3 (Lipschitz Approximation Theorem)] which in turn was based in part on the clever utilization of Besicovitch's covering theorem in [AW2 8.12].

1. *Definitions and terminology*

Except when otherwise indicated we will follow the terminology of [FH 669-671].

(1) m, n, Q denote positive integers with $m \geqq 2$. The results are of interest primarily when $n \geqq 2$ and $Q \geqq 2$ also.

(2) $\{R^m \times \{0\}\}_\natural : R^m \times R \to R^m$,

$\{R^m \times \{0\}\}_\natural : R^m \times R^n \to R^m$,

$\{R^{m+1} \times \{0\}\}_\natural : R^{m+n} = R^{m+1} \times R^{n-1} \to R^{m+1}$

all denote projections onto the first factor. In mapping currents we abbreviate $\{R^m \times \{0\}\}_\# = \{R^m \times \{0\}\}_{\natural\#}$, etc. Also for $0 < \rho < \infty$, $\mu(\rho): R^n \to R^n$ maps $p \in R^n$ to ρp.

(3) For $k \in \{1, m, n, m+n\}$, $p \in R^k$, $0 < r < \infty$ we set

$$B^k(p,r) = R^k \cap \{x : |x-p| \leqq r\},$$

$$U^k(p,r) = R^k \cap \{x : |x-p| < r\},$$

$$\partial B^k(p,r) = R^k \cap \{x : |x-p| = r\}.$$

We also set $\alpha(m) = \mathcal{L}^m(B^m(0,1))$ and let $\beta(m)$ denote the constant of Besicovitch's covering theorem in R^m; in particular, whenever C is a family of closed balls in R^m with $\sup\{\operatorname{diam} B : B \in C\} < \infty$ and A is the union of the centers of members of C then there are $\beta(m)$ disjointed subfamilies $C_1, \cdots, C_{\beta(m)}$ of C such that $A \subset \cup\{\cup C_i : i = 1, \cdots, \beta(m)\}$.

(4) $\mathcal{R}_m(R^{m+n})$ denotes the group of m-dimensional rectifiable currents in R^{m+n} and

$$EX, MEX, TEX : \mathcal{R}_m(R^{m+n}) \to R$$

are defined by setting for each $T \in \mathcal{R}_m(R^{m+n})$

$EX(T) = M(T) - M(\{R^m \times \{0\}\}_\# T)$ (the excess of T),

$MEX(T) = M(\{R^m \times \{0\}\}_\# |T|) - M(\{R^m \times \{0\}\}_\# T)$ (the multiplicity excess of T),

$TEX(T) = M(T) - M(\{R^m \times \{0\}\}_\# |T|)$ (the tilt excess of T);

here $|T|$ denotes the integral varifold associated with T.

(5) Whenever $\Phi : R^{m+n} \times \Lambda_m R^{m+n} \to R$ is a continuous parametric integrand of degree m on R^{m+n} and $T \in \mathcal{R}_m(R^{m+n})$,

$$<\Phi, T> = \int_{p \in R^{m+n}} \Phi(p, \vec{T}(p))\, d\|T\|p .$$

(6) For each $p \in R^n$, $[\![p]\!] \in I_0(R^n)$ denotes the zero-dimensional integral current in R^n defined by setting

$$[\![p]\!](\phi) = \phi(p) \text{ for each } \phi \in \mathcal{E}^0(R^n) .$$

(7) $Q = Q(R^n) = I_0(R^n) \cap \{[\![p_1]\!] + [\![p_2]\!] + \cdots + [\![p_Q]\!] : p_1, \cdots, p_Q$ are (not necessarily distinct) points in $R^n\}$.

(8) Q carries the metric $\mathcal{G}$ defined by requiring

$$\mathcal{G}\Big(\sum_i [\![p_i]\!], \sum_j [\![q_j]\!]\Big) =$$
$$= \inf \Big\{\Big(\sum_{i=1}^{Q} |p_i - q_{\sigma(i)}|^2\Big)^{1/2} : \sigma \text{ is a permutation of } \{1, \cdots, Q\}\Big\},$$

whenever $p_1, \cdots, p_Q,\ q_1, \cdots, q_Q \in R^n$.

2. *Two basic facts about Lipschitz* $\mathbf{Q}$ *valued functions.*

The two general facts about $\mathbf{Q}$ valued functions which are required in the present paper are the following.

2.1. LIPSCHITZ EXTENSION THEOREM [AF3 1.3(2)]. *There is a constant* $1 < C_2 < \infty$ *depending only on* m *and* n *with the following property. Suppose* $A \subset \mathbf{R}^m$ *and* $f : A \to \mathbf{Q}$ *with* $\mathrm{Lip}(f) < \infty$. *Then there exists* $g : \mathbf{R}^m \to \mathbf{Q}$ *such that* $g|A = f$ *and* $\mathrm{Lip}(g) \leqq C_2 \mathrm{Lip}(f)$.

2.2. *Mapping currents by* $\mathbf{Q}$ *valued functions* [AF3 1.5, 1.6]. Corresponding to each bounded open subset A of $\mathbf{R}^m$ and each Lipschitz map $f : A \to \mathbf{Q}$. There is a naturally defined rectifiable current

$$T = ([\![1_A]\!] \bowtie f)_{\#}(\mathbf{E}^m \llcorner A) \in \mathcal{R}_m(\mathbf{R}^{m+n})$$

characterized by the conditions that $\vec{T}(p) \bullet e_1 \wedge \cdots \wedge e_m > 0$ for $\|T\|$ almost all $p \in \mathbf{R}^{m+n}$ and

$$<T, \{\mathbf{R}^m \times \{0\}\}_{\natural}, x> = [\![x]\!] \times f(x) \in \mathbf{I}_0(\mathbf{R}^{m+n})$$

for each $x \in A$. Here $<T, \{\mathbf{R}^m \times \{0\}\}_{\natural}, x>$ denotes the slice of T by the mapping $\{\mathbf{R}^m \times \{0\}\}_{\natural}$ at the point x as in [FH 4.3].

For each $B \subset A$ which is $\mathcal{L}^m$ measurable,

$$([\![1_B]\!] \bowtie f|B)_{\#}(\mathbf{E}^m \llcorner B) = T \llcorner B \times \mathbf{R}^n \in \mathcal{R}_m(\mathbf{R}^{m+n}) .$$

In case ∂B is a smooth compact $m-1$-dimensional submanifold of A then

$$([\![1_{\partial B}]\!] \bowtie f|\partial B)_{\#}(\partial(\mathbf{E}^m \llcorner B)) = \partial(T \llcorner B \times \mathbf{R}^n) \in \mathcal{R}_{m-1}(\mathbf{R}^{m+n}) .$$

Whenever B is any bounded $\mathcal{L}^m$ measurable subset of $\mathbf{R}^m$ and $g : B \to \mathbf{Q}$ with $\mathrm{Lip}(g) < \infty$, fact 2.1 implies the existence of $h : \mathbf{R}^m \to \mathbf{Q}$ with $\mathrm{Lip}(h) \leqq C_2 \mathrm{Lip}(g)$ which can be used to define $([\![1_B]\!] \bowtie g)_{\#}(\mathbf{E}^m \llcorner B)$ as above; the definition is independent of the particular extension chosen and satisfies the inequality

$$\mathbf{M}[([\![1_B]\!] \bowtie g)_{\#}(\mathbf{E}^m \llcorner B)] \leqq Q(1 + [\mathrm{Lip}(g)]^2)^{m/2} \mathcal{L}^m(B) .$$

3. *Lemma*

There is a constant $0 < C_3 < \infty$ *depending only on* m *with the following property. Suppose* $T \in \mathcal{R}_m(R^{m+1})$ *with* $\operatorname{spt} T \subset B^m(0,1) \times R$, $\operatorname{spt} \partial T \subset \partial B^m(0,1) \times R$, $\{R^m \times \{0\}\}_\# T = E^m \llcorner B^m(0,1)$, *and*

$$EX(T) < EX(T)^{1/2}/2 < \alpha(m)/6 .$$

Then there is $s \in R$ *such that*

$$\|T\|(B^m(0,1) \times \{r : |r-s| > C_3 EX(T)^{1/2m}\}) \leqq 3EX(T)^{1/2} .$$

Proof. Use [AF1 4.8] with $n=1$ and $E = EX(T)^{1/2}$.

4. *Lemma*

HYPOTHESES

(a) $T \in \mathcal{R}_m(R^{m+n})$ *with* $\operatorname{spt} T \subset B^m(0,1) \times R^n$, $\operatorname{spt} \partial T \subset \partial B^m(0,1) \times R^n$, *and* $\{R^m \times \{0\}\}_\# T = Q\, E^m \llcorner B^m(0,1)$.

(b) $0 < \Gamma < \infty$ *such that* $\|T\|\, U^{m+n}(p,r) \geqq \Gamma\, r^m$ *whenever* $p \in \operatorname{spt} T$ *and* $0 < r < \operatorname{dist}(p, \partial B^m(0,1) \times R^n)$.

(c) $E = EX(T) < EX(T)^{1/2}/2 < \alpha(m)/6$.

(d) $\delta = [(3Q+1)/\Gamma]^{1/m} E^{1/2m} < 1/2$,

(e) $\rho = 1 - 2\delta \leqq 1$.

(f) $T_* = T \llcorner U^m(0,\rho) \times R^n$.

(g) $C_4(\Gamma) = 2n^{1/2} Q^{3/2} (C_3 + [(3Q+1)/\Gamma]^{1/m})$.

CONCLUSIONS

(1) *These are numbers* $N_1, \cdots, N_n \in \{1, 2, \cdots, Q\}$ *and closed intervals* $L(k,i) \subset R$ *for* $k = 1, \cdots, n$ *and* $i = 1, \cdots, N_k$ *with the following properties.*

(i) $\operatorname{dist}(L(k,i), L(k,j)) > 0$ *whenever* $k \in \{1, \cdots, n\}$ *and* $i,j \in \{1, \cdots, N_k\}$ *with* $i \neq j$.

(ii) $\operatorname{diam} L(k,i) \leqq 2Q(C_3 + [(3Q+1)/\Gamma]^{1/m}) E^{1/2m}$ *for each* $k \in \{1, \cdots, n\}$ *and each* $i \in \{1, \cdots, N_k\}$.

(iii) $\operatorname{spt} T_* \subset \cup \{B^m(0,\rho) \times L(1,i_1) \times L(2,i_2) \times \cdots \times L(n,i_n) :$ $i_k \in \{1, \cdots, N_k\}$ *for each* $k = 1, \cdots, n\}$.

(2) *There exist* $J \in \{1,\cdots,Q\}$, $Q_1,\cdots,Q_J \in \{1,\cdots,Q\}$ *with* $Q_1+\cdots+Q_J=Q$, $p_1,\cdots,p_J \in \mathbf{R}^n$, *and* $T_1,\cdots,T_J \in \mathcal{R}_m(\mathbf{R}^{m+n})$ *with the following properties.*

(i) $\operatorname{spt} T_j \subset \mathbf{B}^m(0,\rho)\times\mathbf{B}^n(p_j, n^{1/2}Q(C_3+[(3Q+1)/\Gamma]^{1/m})E^{1/2m})$ *for each* $j=1,\cdots,J$.

(ii) $\operatorname{spt} \partial T_j \subset \partial\mathbf{B}^m(0,\rho)\times\mathbf{B}^n(p_j, n^{1/2}Q(C_3+[(3Q+1)/\Gamma]^{1/m})E^{1/2m})$ *for each* $j=1,\cdots,J$.

(iii) $\operatorname{spt} T_i \cap \operatorname{spt} T_j = \emptyset$ *for each* $i,j \in \{1,\cdots,J\}$ *with* $i\neq j$.

(iv) $\{\mathbf{R}^m\times\{0\}\}_\# T_j = Q_j\,\mathbf{E}^m \llcorner \mathbf{B}^m(0,\rho)$ *for each* $j=1,\cdots,J$.

(v) $T_* = T_1+T_2+\cdots+T_J$.

(3) *Suppose* $x,z \in \mathbf{U}^m(0,\rho)$ *and* $p_1,p_2,\cdots,p_Q, q_1,q_2,\cdots,q_Q \in \mathbf{R}^n$ *are not necessarily distinct. In case*

$$<T,\{\mathbf{R}^m\times\{0\}\}_\natural, x> = [\![p_1]\!]+\cdots+[\![p_Q]\!],$$
$$<T,\{\mathbf{R}^n\times\{0\}\}_\natural, z> = [\![q_1]\!]+\cdots+[\![q_Q]\!],$$

then

$$\mathcal{G}\Big(\sum_{i=1}^{Q}[\![p_i]\!], \sum_{j=1}^{Q}[\![q_j]\!]\Big) \leqq C_4(\Gamma)E^{1/2m}.$$

(4) *Suppose* $x,z \in \mathbf{U}^m(0,\rho)$, $K,L \in \{0,1,2,\cdots\}$, *and*

$$p_1,p_2,\cdots,p_{Q+K},\ r_1,\cdots,r_K,\ q_1,\cdots,q_{Q+L},\ s_1,\cdots,s_L \in \mathbf{R}^n$$

are not necessarily distinct. In case

$$<T,\{\mathbf{R}^m\times\{0\}\}_\natural, x> = [\![p_1]\!]+\cdots+[\![p_{Q+K}]\!]-[\![r_1]\!]-\cdots-[\![r_K]\!],$$
$$<T,\{\mathbf{R}^m\times\{0\}\}_\natural, z> = [\![q_1]\!]+\cdots+[\![q_{Q+L}]\!]-[\![s_1]\!]-\cdots-[\![s_L]\!].$$

Then

$$\mathcal{F}\Big(\sum_{i=1}^{Q+K}[\![p_i]\!]-\sum_{i=1}^{K}[\![r_i]\!]-\sum_{i=1}^{Q+L}[\![q_i]\!]+\sum_{i=1}^{L}[\![s_i]\!]\Big)$$
$$\leqq 2n^{1/2}Q(C_3+[(3Q+1)/\Gamma]^{1/m})E^{1/2m}(Q+K+L) \quad \text{[FH 4.1.24]}.$$

Proof.

STEP 1. We set $S = \{R^m \times R \times \{0\}\}_{\#} T \in \mathcal{R}_m(R^m \times R)$. One readily checks the existence of $S_1, \cdots, S_Q \in \mathcal{R}_m(R^m \times R)$ such that $S = S_1 + \cdots + S_Q$, $\mathbf{M}(S) = \mathbf{M}(S_1) + \cdots + \mathbf{M}(S_Q)$, $\{R^m \times \{0\}\}_{\#} S_i = \mathbf{E}^m \llcorner \mathbf{B}^m(0,1)$ for each $i = 1, \cdots, Q$, and $\operatorname{spt} \partial S_i \subset \partial \mathbf{B}^m(0,1) \times R$ for each $i = 1, \cdots, Q$. One infers from 3 the existence of $s_1, \cdots, s_Q \in R$ such that for each $i = 1, \cdots, Q$,

$$\|S_i\|(R^m \times \{y : |y - s_i| > C_3 E^{1/2m}) \leqq 3E^{1/2}$$

so that

$$\|S_i\|(R^m \times \{y : |y - s_i| \leqq C_3 E^{1/2m}) \geqq \boldsymbol{\alpha}(m) - 3E^{1/2}$$

which implies

$$\|S\|(R^m \times \{y : |y - s_j| \leqq C_3 E^{1/2m} \text{ for some } j \in \{1, \cdots, Q\}\}) \geqq Q\,\boldsymbol{\alpha}(m) - 3Q\,E^{1/2}$$

which further implies (since projections cannot increase mass)

$$\|T\|[R^m \times (R^n \cap \{y : |y_1 - s_j| \leqq C_3 E^{1/2m} \text{ for some } j \in \{1, \cdots, Q\}\})] \geqq Q\boldsymbol{\alpha}(m) - 3QE^{1/2}.$$

In case $\delta < r < 1$ and $(x,y) \in \operatorname{spt} T \cap \mathbf{B}^m(0, 1-r) \times R^n$ with $|y_1 - s_i| \geqq C_3 E^{1/2m} + r$ for each $i = 1, \cdots, Q$ one infers from hypotheses (b) and (c) that

$$\begin{aligned} \mathbf{M}(T) &\geqq \Gamma r^m + Q\boldsymbol{\alpha}(m) - 3Q\,E^{1/2} \\ &> \Gamma \delta^m - 3Q\,E^{1/2} + Q\,\boldsymbol{\alpha}(m) \\ &= E^{1/2} + Q\,\boldsymbol{\alpha}(m) \\ &\geqq E + Q\,\boldsymbol{\alpha}(m) \end{aligned}$$

which contradicts the definition of E in hypothesis (c). One therefore concludes $\operatorname{spt} T_* \subset R^m \times (R^n \cap \{y : |y_1 - s_i| \leqq C_3 E^{1/2m} + \delta$ for some $i \in \{1, \cdots, Q\}\})$.

STEP 2. We suppose $s_1, \cdots, s_Q \in R$ are as in Step 1 and partition $\{s_1, \cdots, s_Q\}$ into distinct equivalence classes $\Sigma_1, \cdots, \Sigma_{N_1}$ by defining s_i to be equivalent to s_j if and only if there exist (not necessarily distinct) $t_1, t_2, \cdots, t_Q \in \{s_1, \cdots, s_Q\}$ such that $s_i = t_1$, $s_j = t_Q$, and $|t_k - t_{k+1}| \leqq 2(C_3 E^{1/2m} + \delta)$ for each $k = 1, \cdots, Q-1$. It follows that the sets $L(1,i) = R \cap \{y ; |y - s_j| \leqq C_3 E^{1/2m} + \delta$ for some $s_j \in \Sigma_i\}$ for $i = 1, 2, \cdots, N_1$ are closed intervals, no two of which overlap. Clearly also

$$\operatorname{spt} T_* \subset B^m(0,\rho) \times (\cup\{L(1,i) : i = 1, \cdots, N_1\}) \times R^{n-1}$$

and

$$\operatorname{diam} L(1,i) \leqq 2Q(C_3 E^{1/2m} + \delta)$$

for each $i = 1, \cdots, N_1$. The remainder of the proof of conclusion (1) is left to the reader. Conclusion (2) is a straightforward consequence of conclusion (1). Conclusion (3) is a straightforward consequence of conclusion (2) as is conclusion (4).

5. *Basic approximation theorem for rectifiable currents with small excess and density ratios bounded from below.*

HYPOTHESES

(a) $T \in \mathcal{R}_m(R^{m+n})$ *with* $\operatorname{spt} T \subset B^m(0,1) \times R^n$, $\operatorname{spt} \partial T \subset \partial B^m(0,1) \times R^n$ *and* $\{R^m \times \{0\}\}_\# T = Q\, E^m \llcorner B^m(0,1)$.

(b) $0 < \Gamma < \infty$ *such that* $\|T\|\, U^{m+n}(p,r) \geqq \Gamma\, r^m$ *whenever* $p \in \operatorname{spt} T$ *and* $0 < r < \operatorname{dist}(p, \partial B^m(0,1) \times R^n)$.

(c) $0 \leqq \sigma < 1$, $0 < \tau < \infty$, *and* $E = EX(T)$.

(d) $E < E^{1/2}/2 < \boldsymbol{\alpha}(m)/6$.

(e) $[(3Q+1)/\Gamma]^{1/m} E^{1/2m} < 1/16$.

(f) $E/\boldsymbol{\alpha}(m) \leqq \tau E^\sigma < 1$.

(g) $\tau E^\sigma < (\tau E^\sigma)^{1/2}/2 < \boldsymbol{\alpha}(m)/6$.

(h) $C_5(\Gamma) = (64/7)\, \boldsymbol{\alpha}(m)^{1/2m} C_2 C_4(\Gamma)$.

CONCLUSION. *There exist* $F : B^m(0, 7/8) \to \mathbf{Q}$ *together with a partitioning of* $B^m(0,7/8)$ *into* $\mathcal{L}^m$ *measurable sets* A *and* B *with the following properties.*

(1) $\operatorname{Lip} F \leqq C_5(\Gamma)\, \tau^{1/2m} E^{\sigma/2m}$.

(2) *For each* $x \in A$,

$$[\![x]\!] \times F(x) = \langle T, \{R^m \times \{0\}\}_\natural, x \rangle = \Sigma\{\theta^m(\|T\|, p)[\![p]\!] : p \in \operatorname{spt} T \cap \{x\} \times R^n\},$$

(3) $\mathcal{L}^m(B) \leqq \boldsymbol{\beta}(m)\,\tau^{-1} E^{1-\sigma}$.

(4) $\|T\| B \times R^n \leqq E + Q\,\boldsymbol{\beta}(m)\,\tau^{-1} E^{1-\sigma}$.

(5) $T \llcorner A \times R^n = ([\![1_A]\!] \bowtie F|A)_\#(E^m \llcorner A)$.

(6) $\mathbf{M}[([\![1_{B^m(0,7/8)}]\!] \bowtie F)_\#(E^m \llcorner B)] \leqq Q[1 + (\operatorname{Lip} F)^2]^{m/2} \mathcal{L}^m(B)$.

Proof. We set

$$\eta = \tau E^\sigma,$$

$$W = B^m(0,7/8) \cap \{x : \text{for some } 0 < r < 1 - |x|,\ \|T\| B^m(x,r) \times R^n \geqq (\eta + Q)\,\boldsymbol{\alpha}(m)\,r^m\},$$

$$X = B^m(0,7/8) \cap \{x : \langle T, \{R^m \times \{0\}\}_\natural, x \rangle \text{ does not exist}\},$$

$$Y = [B^m(0,7/8) \sim X] \cap \{x : \mathbf{M}(\langle T, \{R^m \times \{0\}\}_\natural, x \rangle) > Q\},$$

$$Z = [B^m(0,7/8) \sim (X \cup Y)] \cap \{x : \langle T, \{R^m \times \{0\}\}_\natural, x \rangle \neq \Sigma\{\theta^m(\|T\|, p)[\![p]\!] : p \in \operatorname{spt} T \cap \{x\} \times R^n\},$$

$$B = W \cup X \cup Y \cup Z,$$

$$A = B^m(0,7/8) \sim B.$$

Using Besicovitch's covering theorem one infers the existence of $x_1, x_2, x_3, \cdots \in W$ and associated $r_1, r_2, r_3, \cdots > 0$ such that $W \subset \cup\{B^m(x_k, r_i) : i = 1, 2, 3, \cdots\}$ and

$$\begin{aligned} \mathcal{L}^m(W) &\leqq \sum_i \boldsymbol{\alpha}(m)\, r_i^m \\ &\leqq \sum_i [\|T\| B^m(x_i, r_i) \times R^n - Q\,\boldsymbol{\alpha}(m)\, r_i^m]/\eta \\ &\leqq \boldsymbol{\beta}(m)\,[\|T\| \cup_i B^m(x_i, r_i) \times R^n - Q\,\mathcal{L}^m(\cup_i B^m(x_i, r_i))]/\eta . \end{aligned}$$

One infers from [FH 4.3.2, 4.3.6] that $\mathcal{L}^m(X) = 0$. One readily checks that

$$\|T\| \cup_i B^m(x_i, r_i) \times R^n - Q\,\mathcal{L}^m(\cup_i B^m(x_i, r_i)) + \mathcal{L}^m(Y - \cup_i B^m(x_i, y_i)) \leqq E$$

and then infers from hypothesis (f) and our estimate above that $\mathcal{L}^m(\cup_i B^m(x_i, r_i) \cup Y) \leqq \boldsymbol{\beta}(m) E/\eta$. Using hypothesis (b) and [FH 4.1.28] one verifies that

$$\|T \llcorner U^m(0,1) \times R^n\| = \mathcal{H}^m \llcorner \operatorname{spt} T \cap U^m(0,1) \times R^n \wedge \theta^m(\|T\|, \cdot)$$

and additionally that $\mathcal{L}^m(Z) = 0$. It follows that $\mathcal{L}^m(B) \leqq \boldsymbol{\beta}(m) E/\eta$ which is conclusion 3.

We now define $F(x) = \langle T, \{R^m \times \{0\}\}_\natural, x\rangle$ for each $x \in A$. Whenever $x, w \in A$ with $|x-w| \geqq 7/64$ we infer from Lemma 4(3) and hypothesis (f) that

$$\mathcal{G}(F(x), F(w)) \leqq C_4(\Gamma) E^{1/2m} \leqq C_4(\Gamma)(64/7)(\boldsymbol{\alpha}(m)\tau E^\sigma)^{1/2m}|x-w| .$$

Whenever $x, w \in A$ with $|x-w| < 7/64$ we consider $B^m(x, (8/7)|x-w|) \subset U^m(0,1)$ and infer from Lemma 4(3), hypothesis (g), and the fact that $x \notin W$ that

$$\mathcal{G}(F(x), F(w)) \leqq C_4(\Gamma)(\boldsymbol{\alpha}(m)\eta)^{1/2m}[(8/7)|x-w|] .$$

Conslusion (1) follows with use of 2.1. Conclusion (2) is clear and conclusion (4) is readily checked. Conclusions (5) and (6) follow with the use of 2.2.

6. COROLLARY. (*Approximation of* Φ *minimizing rectifiable currents with small* excess.)

HYPOTHESES

(a) $0 < \gamma < \infty$.

(b) $\Gamma = 1/2m^m(m+n)^{2m^2}\gamma^{m-1}$.

(c) $\Phi : R^{m+n} \times \Lambda_m R^{m+n} \to R$ *is a continuous parametric integrand of degree* m *on* R^{m+n} *with* $0 < \inf \Phi \leqq \sup \Phi \leqq \gamma \inf \Phi < \infty$; *here we have abbreviated* $\inf \Phi = \inf\{\Phi(p, \lambda) : p \in R^{m+n}, \lambda \in \Lambda_m R^{m+n}$ *is simple with* $|\lambda| = 1\}$, *etc.*

(d) $T \in \mathcal{R}_m(R^{m+n})$ *with* $\operatorname{spt} T \subset B^m(0,1) \times R^n$, $\operatorname{spt} \partial T \subset \partial B^m(0,1) \times R^n$ *and* $\{R^m \times \{0\}\}_\# T = Q\, E^m \llcorner B^m(0,1)$.

(e) T *is absolutely* Φ *minimizing with respect to* $U^m(0,1) \times R^n$, *i.e* $<\Phi, T> \leqq <\Phi, T+S>$ *whenever* $S \in \mathcal{R}_m(R^{m+n})$ *with* $\operatorname{spt} S \subset U^m(0,1) \times R^n$ *and* $\partial S = 0$.

(f) $0 \leqq \sigma < 1$, $0 < \tau < \infty$, *and* $E = EX(T)$.

(g) E *is sufficiently small so that inequalities (d), (e), (f) of the hypotheses of Theorem 5 hold, and*

$$C_5(\Gamma) = (64/7)\,\alpha(m)^{1/2m} C_2 C_4(\Gamma)\,.$$

CONCLUSIONS

(1) $\|T\| U^{m+n}(p,r) \geqq \Gamma\, r^m$ *whenever* $p \in \operatorname{spt} T$ *and*

$$0 < r < \operatorname{dist}(p, \partial B^m(0,1) \times R^n)\,.$$

(2) *Conclusions (1), (2), (3), (4), (5) of Theorem 5 hold.*

Proof. Conclusion (2) follows readily from conclusion (1) which is hypothesis (b) of Theorem 5. We will verify conclusion (1).

Suppose $p \in \operatorname{spt} T \cap U^m(0,1) \times R^n$, $r_0 = \operatorname{dist}(p, \partial B^m(0,1) \times R^n)$, and $\mu = \|T\| U^{m+n}(p,\cdot) : (0, r_0) \to R$. One infers from [FH 4.2.1, 4.3.6] that for $\mathcal{L}^1$ almost all $0 < r < r_0$, $\mu'(r)$ exists and $T \llcorner U^{m+n}(0,r) \in I_{m-1}(R^{m+n})$ with $M(\partial[T \llcorner U^{m+n}(p,r)]) \leqq \mu'(r)$. Furthermore, for each such r, [FH 4.2.10] implies the existence of $W \in I_m(R^{m+n})$ with $\operatorname{spt} W \subset B^{m+n}(p,r)$, $\partial W = \partial[T \llcorner U^{m+n}(p,r)]$, and $M(W)^{(m-1)/m} \leqq 2(m+n)^{2m} \mu'(r)$. Setting $S = W - T \llcorner U^{m+n}(p,r)$ in hypothesis (e) one uses hypotheses (c) and (e) to infer

$$\begin{aligned}\mu(r) \inf \Phi = M(T \llcorner U^{m+n}(p,r)) \inf \Phi &\leqq <\Phi, T \llcorner U^{m+n}(p,r)> \leqq <\Phi, W> \leqq M(W) \sup \Phi \\ &\leqq 2(m+n)^{2m^2/(m-1)} \mu'(r)^{m/(m-1)} \sup \Phi\end{aligned}$$

so that

$$\mu'(r)/\mu(r)^{(m-1)/m} \geqq 1/2(m+n)^{2m} \gamma^{(m-1)/m}\,.$$

Integration of this differential inequality yields conclusion (1).

7. COROLLARY. *(Approximation of general rectifiable currents with small excess.)*

HYPOTHESES

(a) $T \in \mathcal{R}_m(R^{m+n})$ *with* spt $T \subset B^m(0,1) \times R^n$, spt $\partial T \subset \partial B^m(0,1) \times R^n$, *and* $\{R^m \times \{0\}\}_\# T = Q\, E^m \llcorner B^m(0,1)$.

(b) $0 \leqq \sigma < 1$, $0 < \tau < \infty$, $0 < \varepsilon < 1$, $E = EX(T)$.

(c) $E < E^{1/2}/2 < \alpha(m)/6$.

(d) $\Gamma = \varepsilon^{m-1}/2^m m^m (m+n)^{2m^2}$.

(e) $[(3Q+1)/\Gamma]^{1/m} E^{1/m} < 1/16$.

(f) $E/\alpha(m) \leqq \tau E^\sigma < 1$.

(g) $\tau E^\sigma < (\tau E^\sigma)^{1/2}/2 < \alpha(m)/6$.

(h) $C_5(\Gamma) = (64/7)\, \alpha(m)^{1/2m} C_2 C_4(\Gamma)$.

CONCLUSION. *There exist* $F : B^m(0,7/8) \to Q$ *together with a partitioning of* $B^m(0,7/8)$ *into* $\mathcal{L}^m$ *measurable sets* A *and* B *with the following properties.*

(1) $\mathrm{Lip}\, F \leqq C_5(\Gamma)\, \tau^{1/2m} E^{\sigma/2m}$.

(2) *For each* $x \in A$,

$$[\![x]\!] \times F(x) = \langle T, \{R^m \times \{0\}\}_\natural, x \rangle = \Sigma\{\theta^m(\|T\|, p)\, [\![p]\!] : p \in \mathrm{spt}\, T \cap \{x\} \times R^n\}.$$

(3) $\mathcal{L}^m(B) \leqq \beta(m)\, \tau^{-1} E^{1-\sigma} + [(1+\varepsilon)/(1-\varepsilon)] E$.

(4) $\|T\| B \times R^n \leqq 2E + Q\beta(m)\, \tau^{-1} E^{1-\sigma}$.

(5) $T \llcorner A \times R^n = ([\![1_A]\!] \bowtie F|A)_\# (E^m \llcorner A)$.

(6) $M[([\![1_{B^m(0,7/8)}]\!] \bowtie F)_\# (E^m \llcorner B)] \leqq Q[1 + (\mathrm{Lip}\, F)^2]^{m/2} \mathcal{L}^m(B)$.

Proof. The proof is in several parts.

Part 1. *There exists* $S \in \mathcal{R}_m(R^{m+n})$ *with the following properties.*

(i) spt $S \subset B^m(0,1) \times R^n$,

(ii) $\partial S = \partial T$,

(iii) $\mathbf{M}(S-T) \leqq [(1+\varepsilon)/(1-\varepsilon)]E$.

(iv) *For each* $p \in \operatorname{spt} S$ *and each* $0 < r < \operatorname{dist}(p, \partial \mathbf{B}^m(0,1) \times \mathbf{R}^n)$,
$\varepsilon \|S\| \mathbf{B}^{m+n}(p,r) \leqq \inf\{\|V\| \mathbf{B}^{m+n}(p,r) : V \in \mathcal{R}_m(\mathbf{R}^{m+n})$ *with* $\partial V = \partial T$ and $\operatorname{spt}(V-T) \subset \mathbf{B}^{m+n}(p,r)\}$.

(v) $\mathbf{EX}(S) \leqq \mathbf{EX}(T) - [(1-\varepsilon)/(1+\varepsilon)]\mathbf{M}(S-T)$.

Proof. Corresponding to each $V \in \mathcal{R}_m(\mathbf{R}^{m+n})$ with $\operatorname{spt} V \subset \mathbf{B}^m(0,1) \times \mathbf{R}^n$ and $\partial V = \partial T$, to each $p \in \mathbf{U}^m(0,1) \times \mathbf{R}^n$, and to each

$$0 < r < \operatorname{dist}(p, \partial \mathbf{B}^m(0,1) \times \mathbf{R}^n)$$

we set $e(V,p,r) = \sup\{\mathbf{M}(V) - \mathbf{M}(W) : W \in \mathcal{R}_m(\mathbf{R}^{m+n})$ with $\partial W = \partial T$ and $\operatorname{spt}(W-V) \subset \mathbf{B}^{m+n}(p,r)\}$, $e(V) = \sup\{0, \sup\{e(V,q,s) : q \in \mathbf{U}^m(0,1) \times \mathbf{R}^n$, $0 < s < \operatorname{dist}(q, \partial \mathbf{B}^m(0,1) \times \mathbf{R}^n)$, and $e(V,q,s) > (1-\varepsilon)\|V\| \mathbf{B}^{m+n}(q,s)\}\}$. We now sequentially select $p_i \in \mathbf{U}^m(0,1) \times \mathbf{R}^n$, $0 < r_i < \operatorname{dist}(p_i, \partial \mathbf{B}^m(0,1) \times \mathbf{R}^n)$, and $T_i \in \mathcal{R}_m(\mathbf{R}^{m+n})$ with $\operatorname{spt} T_i \subset \mathbf{B}^m(0,1) \times \mathbf{R}^n$ and $\partial T_i = \partial T$ for each $i = 1, 2, 3, \cdots$, as follows.

In case $e(T) = 0$ we set $T = T_1 = T_2 = T_3 = \cdots$ and set $0 = p_1 = p_2 = p_3 = \cdots$ and $1/2 = r_1 = r_2 = r_3 = \cdots$.

In case $e(T) > 0$ we choose p_1, r_1, T_1 so that $p_1 \in \mathbf{U}^m(0,1) \times \mathbf{R}^n$, $0 < r_1 < \operatorname{dist}(p_1, \partial \mathbf{B}^m(0,1) \times \mathbf{R}^n)$, and $T_1 \in \mathcal{R}_m(\mathbf{R}^{m+n})$ with $\operatorname{spt} T_1 \subset \mathbf{B}^m(0,1) \times \mathbf{R}^n$, $\partial T_1 = \partial T$, $\operatorname{spt}(T_1 - T) \subset \mathbf{B}^{m+n}(p_1, r_1)$, $\mathbf{M}(T) - \mathbf{M}(T_1) \geqq e(T)/2$, and $\mathbf{M}(T) - \mathbf{M}(T_1) > (1-\varepsilon)\|T\| \mathbf{B}^{m+n}(p_1, r_1)$.

Assuming $i \in \{2, 3, 4, \cdots\}$ and $T_1, \cdots, T_{i-1}$, $p_1, \cdots, p_{i-1}$, and $r_1, \cdots, r_{i-1}$ have been selected we select T_i, p_i, r_i as follows.

In case $e(T_{i-1}) = 0$ we set $T_{i-1} = T_i = T_{i+1} = T_{i+2} = \cdots$ and set $0 = p_i = p_{i+1} = p_{i+2} = \cdots$ and $1/2 = r_i = r_{i+1} = r_{i+2} = \cdots$.

In case $e(T_{i-1}) > 0$ we choose p_i, r_i, T_i so that $p_i \in \mathbf{U}^m(0,1) \times \mathbf{R}^n$, $0 < r_i < \operatorname{dist}(p_i, \partial \mathbf{B}^m(0,1) \times \mathbf{R}^n)$, and $T_i \in \mathcal{R}_m(\mathbf{R}^{m+n})$ with $\operatorname{spt} T_i \subset \mathbf{B}^m(0,1) \times \mathbf{R}^n$, $\partial T_i = \partial T$, $\operatorname{spt}(T_i - T_{i-1}) \subset \mathbf{B}^{m+n}(p_i, r_i)$, $\mathbf{M}(T_{i-1}) - \mathbf{M}(T_i) \geqq e(T_{i-1})/2$, and $\mathbf{M}(T_{i-1}) - \mathbf{M}(T_i) > (1-\varepsilon)\|T\| \mathbf{B}^{m+n}(p_i, r_i)$.

Setting $T_0 = T$ we estimate for each $i = 1, 2, 3, \cdots$

$$\|T_{i-1}\| B^{m+n}(p_i, r_i) - \|T_i\| B^{m+n}(p_i, r_i)$$
$$= M(T_{i-i}) - M(T_i)$$
$$> (1-\varepsilon) \|T_{i-1}\| B^{m+n}(p_i, r_i)$$

so that

$$\|T_i\| B^{m+n}(p_i, r_i) < \varepsilon \|T_{i-1}\| B^{m+n}(p_i, r_i)$$

and hence

$$\begin{aligned} M(T_i - T_{i-1}) &= \|T_i - T_{i-1}\| B^{m+n}(p_i, r_i) \\ &\leqq \|T_i\| B^{m+n}(p_i, r_i) + \|T_{i-1}\| B^{m+n}(p_i, r_i) \\ &< (1+\varepsilon) \|T_{i-1}\| T_{i-1}\| B^{m+n}(p_i, r_i) \\ &< [(1+\varepsilon)/(1-\varepsilon)] [M(T_{i-1}) - M(T_i)] . \end{aligned}$$

One notes for each $j \in \{1, 2, 3, \cdots\}$

$$\begin{aligned} \sum_{i=1}^{j} M(T_i - T_{i-1}) &< [(1+\varepsilon)/(1-\varepsilon)] \sum_{i-1}^{j} M(T_{i-1}) - M(T_i) \\ &= [(1+\varepsilon)/(1-\varepsilon)] [M(T) - M(T_j)] \\ &\leqq [(1+\varepsilon)/(1-\varepsilon)] \mathrm{EX}(T) ; \end{aligned}$$

with regard to the last inequality one recalls $\partial T_j = \partial T$ and concludes

$$\{R^m \times \{0\}\}_{\#} T_j = \{R^m \times \{0\}\}_{\#} T = Q\, E^m \llcorner B^m(0,1)$$

from [FH 4.1.7 (constancy theorem)] so that $M(T_j) \geqq M(\{R^m \times \{0\}\}_{\#} T)$ by [FH 4.1.14]. In particular, the sequence $T, T_1, T_2, T_3, \cdots$ is a Cauchy sequence in the mass metric. We set $S = \lim_{i\to\infty} T_i \in \mathcal{R}_m(R^{m+n})$ and assert that $e(S) = 0$. If this were not true there would exist $p \in U^m(0,1) \times R^n$, $0 < r < \operatorname{dist}(p, \partial B^m(0,1) \times R^n)$, and $W \in \mathcal{R}_m(R^{m+n})$ with $\operatorname{spt} W \subset B^m(0,1) \times R^n$, $\partial W = \partial T$, $\operatorname{spt}(W - S) \subset B^{m+n}(p, r)$, and $M(S) - M(W) >$

$(1-\varepsilon)\,\|S\|B^{m+n}(p,r) > 0$. This is not possible, as the following argument shows. We define $G : R^{m+n} \to R^{m+n}$ by setting

$$G(q) = q \quad \text{in case } q \in B^{m+n}(p,r)$$
$$= p + r(q-p)/|q-p| \quad \text{in case } q \in R^{m+n} \sim B^{m+n}(p,r)\,.$$

Additionally for each $i \in \{1,2,3,\cdots\}$ we set

$$W_i = T_i \llcorner (R^{m+n} \sim B^{m+n}(p,r))$$
$$+ G_\#[(S-T_i) \llcorner (R^{m+n} \sim B^{m+n}(p,r))]$$
$$+ W \llcorner B^{m+n}(p,r)\,,$$

noting that $\partial W_i = \partial T_i = \partial T$ and $\operatorname{spt}(W_i - T_i) \subset B^{m+n}(p,r)$. The mass convergence of the $T_1, T_2, T_3, \cdots$ to S implies for all sufficiently large $i \in \{1,2,3,\cdots\}$,

$$M(T_i) - M(W_i) > (1-\varepsilon)\,\|S\|B^{m+n}(p,r) > 0\,;$$

this is not possible since each W_i, $i = 1,2,3,\cdots$, is a candidate for T_{i+1} and $\lim_{i\to\infty}(M(T_i) - M(T_{i+1})) = 0$.

The assertions of part 1 follow readily.

Part 2. *Suppose* S *is as in part 1 and* $p \in U^m(0,1) \times R^n$.

(i) *For each* $0 < r < \operatorname{dist}(p, \partial B^m(0,1) \times R^n)$, [FH 4.2.10] *readily implies the existence of* $W \in \mathcal{R}_m(R^{m+n})$ *such that* $\operatorname{spt} W \subset B^{m+n}(p,r)$, $\partial W = \partial[S \llcorner B^{m+n}(p,r)]$, *and*

$$M(W) \leqq [2(m+n)^{2m}]^{m/(m-1)} M(\partial[S \llcorner B^{m+n}(p,r)])^{m/(m-1)}\,;$$

furthermore, in view of part 1(iv),

$$\varepsilon\,\|S\|B^{m+n}(p,r) \leqq [2(m+n)^{2m}]^{m/(m-1)} M(\partial[S \llcorner B^{m+n}(p,r)])^{m/(m-1)}\,.$$

(ii) *By estimates virtually identical with those of the proof of Corollary 6 one concludes that if* $p \in \operatorname{spt} S$ *and*

$$0 < r < \operatorname{dist}(p, \partial \mathbf{B}^m(0,1) \times \mathbf{R}^n),$$

$$\|S\| \mathbf{U}^{m+n}(p,r) \geqq [\varepsilon^{m-1}/2^m m^m (m+n)^{2m^2}] r^m .$$

(iii) *One readily checks that the hypotheses of Theorem 5 are satisfied with* S *replacing* T *there. The present theorem then readily follows with "B" of the present theorem equal to the union of "B" of Theorem 5 with*

$$\mathbf{B}^m(0,7/8) \cap \{x : 0 < \limsup_{r \downarrow 0} \|T-S\| \mathbf{B}^m(x,r) \times \mathbf{R}^n / r^m\} .$$

8. COROLLARY. (*Approximation of general rectifiable currents with small multiplicity excess.*)

HYPOTHESES

(a) $T \in \mathcal{R}_m(\mathbf{R}^{m+n})$ *with* $\operatorname{spt} T \subset \mathbf{B}^m(0,1) \times \mathbf{R}^n$, $\operatorname{spt} \partial T \subset \partial \mathbf{B}^m(0,1) \times \mathbf{R}^n$, *and* $\{\mathbf{R}^m \times \{0\}\}_\# T = Q\, \mathbf{E}^m \llcorner \mathbf{B}^m(0,1)$.

(b) $0 \leqq \sigma < 1$, $0 < \tau < \infty$, $0 < \varepsilon < 1$, $0 < \rho \leqq 1$, $E = \rho\, \mathrm{TEX}(T) + \mathrm{MEX}(T)$.

(c) $E < E^{1/2} < \boldsymbol{\alpha}(m)/6$.

(d) $\Gamma = \varepsilon^{m-1}/2^m m^m (m+n)^{2m^2}$.

(e) $[(3Q+1)/\Gamma]^{1/m} E^{1/m} < 1/16$.

(f) $E/\boldsymbol{\alpha}(m) \leqq \tau E^\sigma < 1$.

(g) $\tau E^\sigma < (\tau E^\sigma)^{1/2} < \boldsymbol{\alpha}(m)/6$.

(h) $C_5(\Gamma) = (64/7)\, \boldsymbol{\alpha}(m)^{1/2m} C_2 C_4(\Gamma)$.

CONCLUSION. *There exist* $F : \mathbf{B}^m(0,7/8) \to \mathbf{Q}$ *together with a partitioning of* $\mathbf{B}^m(0,7/8)$ *into* $\mathcal{L}^m$ *measurable sets* A *and* B *with the following properties.*

(1) $\operatorname{Lip} F \leqq C_5(\Gamma) \tau^{1/2m} \rho^{-1} E^{\sigma/2m}$.

(2) *For each* $x \in A$,

$$[\![x]\!] \times F(x) = \langle T, \{\mathbf{R}^m \times \{0\}\}_\natural, x \rangle = \Sigma\{\theta^m(\|T\|, p)\, [\![p]\!] : p \in \operatorname{spt} T \cap \{x\} \times \mathbf{R}^n\}.$$

(3) $\mathcal{L}^m(B) \leqq \boldsymbol{\beta}(m) \tau^{-1} E^{1-\sigma} + [(1+\varepsilon)/(1-\varepsilon)] E$.

(4) $\|T\| B \times \mathbf{R}^n \leqq \rho^{-m}[2E + Q\,\beta(m)\tau^{-1}E^{1-\sigma}]$.

(5) $T \llcorner A \times \mathbf{R}^n = ([\![1_A]\!] \bowtie F|A)_\#(\mathbf{E}^m \llcorner A)$.

(6) $\mathbf{M}[([\![1_{\mathbf{B}^m(0,7/8)}]\!] \bowtie F)_\#(\mathbf{E}^m \llcorner B)] \leqq Q[1+(\operatorname{Lip} F)^2]^{m/2}\mathcal{L}^m(B)$.

Proof. Apply Corollary 7 to $[1_{\mathbf{R}^m} \times \mu(\rho)]_\# T = T'$ checking that $\mathrm{TEX}(T') \leqq \rho\,\mathrm{TEX}(T)$ so that $\mathrm{EX}(T') = \mathrm{MEX}(T') + \mathrm{TEX}(T') \leqq E$.

F. J. ALMGREN, JR.
DEPARTMENT OF MATHEMATICS
PRINCETON UNIVERSITY
PRINCETON, NEW JERSEY 08544

REFERENCES

[AW1] W. K. Allard, *A characterization of the area integrand*, Symposia Mathematica XIV (1974), 429-444.

[AW2] ———, *On the first variation of a varifold*, Ann. of Math. 95 (1972), 417-491.

[AF1] F. J. Almgren, Jr., *Existence and regularity almost everywhere of solutions to elliptic variational problems among surfaces of varying topological type and singularity structure*, Ann. of Math. 87 (1968), 321-391.

[AF2] ———, *Dirichlet's problem for multiple valued functions and the regularity of mass minimizing integral currents*, Minimal Submanifolds and Geodesics, edited by M. Obata, Kaigai Publications, Ltd., Tokyo, 1978, 1-6.

[AF3] ———, *Q valued functions minimizing Dirichlet's integral and the regularity of area minimizing rectifiable currents up to codimension two* (preprint).

[FH] H. Federer, Geometric Measure Theory, Springer-Verlag, New York, 1969.

[FW] W. H. Fleming, *Flat chains over a finite coefficient group*, Trans. Amer. Math. Soc. 121 (1966), 160-186.

[SS] R. Schoen and L. Simon, *A new proof of the regularity theorem for rectifiable currents which minimize parametric elliptic functionals* (preprint).

SIMPLE CLOSED GEODESICS ON OVALOIDS AND THE CALCULUS OF VARIATIONS

Melvyn S. Berger*

Global differential geometry and the calculus of variations in the large are closely linked. Nonetheless, until recently, many interesting geometric problems that could be naturally expressed in terms of the calculus of variations have remained unresolved. In this article I shall describe some joint work with E. Bombieri [1] on the partial resolution of a classic seventy-five year old problem of Poincaré [2] on simple closed geodesics for a two-dimensional smooth ovaloid, M. This problem requires the joint utilization of the calculus of variations and of global geometry. I had tried to resolve this basic problem for a number of years via Hilbert space techniques associated with spaces of parametrized closed curves on M. Actually, the resulting Hilbert manifold has an interesting topology (as utilized by Klingenberg [4] for example) but even this rich structure did not seem adequate to carry through Poincaré's approach. Indeed Poincaré's approach awaited the thorough development of a relatively new field of analysis: the geometric measure theory of integral currents and varifolds for isoperimetric variational problems.

I believe the problem discussed here is the simplest example of a whole hierarcy of geometric problems that can be understood and resolved by further development of the techniques I describe below.

*Research partially supported by the National Science Foundation.

Seminar on Minimal Submanifolds
0-691-08324-X/83/261-10 $1.00/0 (cloth)
0-691-08319-3/83/261-10 $1.00/0 (paperback)

The basic problems

If (M, g) is a compact two-dimensional simply-connected Riemannian manifold, the determination of even one simple closed geodesic form M is a subtle problem. Indeed $\pi_1(M) = \{0\}$, so the usual approach of minimizing arc length over curves in a given homotopy class breaks down. The attempt to use the topology of closed curves over M, as described above, also has its difficulties, since the simplicity of possible saddle points of the arc length functional often requires ingenious geometric arguments. A different approach, suggested by Poincaré [2], utilized isoperimetric problems in the calculus of variations.

1. *Poincaré's idea and program*

Poincaré noted that the Gauss-Bonnet theorem yield an alternate obstruction of a differential geometry nature to the problem of simple closed geodesics on ovaloids. Indeed if C is a simple closed curve on an ovaloid (M^2, g) (embedded in R^3 say) then if $\Sigma(C)$ is the two-dimensional portion of M bounded by C

(1) $$\int_{\Sigma(C)} K + \oint_C K_g = 2\pi \quad \text{where} \quad \begin{cases} K & \text{denotes Gauss curvature} \\ K_g & \text{denotes geodesic curvature.} \end{cases}$$

If, in addition, C is a geodesic for (M, g) then $K_g = 0$ and so

(2) $$\int_{\Sigma(C)} K = 2\pi .$$

Thus every simple closed geodesic C on (M, g) must satisfy (2). Conversely if C is a simple smooth extremal of arc length *restricted* to the set of curves C satisfying (2) then on C, the relevant Euler equation is easily shown to be

(3) $$K_g = \lambda K$$

where λ is an undetermined constant, a Lagrange multiplier. But in this case the constraint (2) is so well chosen that $\lambda = 0$ and so (3) reduces to $K_g = 0$ (i.e. C is a geodesic). In other words the imposition of the constraint (2) does not affect the Euler equation for geodesics. Thus, in this case, the Gauss-Bonnet formula (1) acts as a "natural" substitute for the topological homotopy class!

Proof that $\lambda = 0$ *in* (3) *for a simple closed extremal* C. Integrate (3) over C, to find

$$\oint_C K_g = \lambda \oint_C K .$$

Apply the Gauss-Bonnet theorem (1) to the left side of (3) to find

$$2\pi - \int_{\Sigma(C)} K = \lambda \oint K .$$

But since C satisfies (2) we find

$$\lambda \oint K = 0 .$$

Since (M, g) is an ovaloid, $K > 0$ on M so this last equation implies $\lambda = 0$.

Now Poincaré's program to study simple closed geodesics on ovaloids is contained in the following outline:

Step 1. Find the extremal for the isoperimetric problem.

(II) : Minimize arclength among all smooth closed curves $\{C\}$ satisfying (2).

Step 2. Prove the resulting extremal is smooth.

Step 3. Prove the resulting extremal is nonself-intersecting.

Step 4. Study the stability of this extremal, regarded possibly as a simple periodic orbit of a dynamical system.

To discuss steps 1 and 2, Poincaré refers to the work of Hilbert. For step 3, Poincaré devises a physical experiment to prove that at least intuitively that the extremal is nonself-intersecting. For the stability of step 4, Poincaré carries out an intuitive discussion based on his ideas of stability of periodic orbits of mechanical systems. Poincaré's ideas and program were never carried out however. We next show how all four steps can be *reformulated* and *resolved* by using differential geometry and the notion of integral currents.

A.2. *Reformulation of Poincaré's ideas in terms of integral currents*

We first review the notions leading up to integral currents on a subset U of R^2. Let $\mathcal{S}$ be a differential 2-form defined on U with compact support, and T be a distribution on $\mathcal{S}$ (i.e. bounded linear functional on $\Lambda^2(U)$). The boundary of T, ∂T, can be easily defined on one forms $\mathcal{S}$ by $(\partial T)(\mathcal{S}) = T(d\mathcal{S})$, where d denotes exterior derivative. If T and ∂T are continuous with respect to the L^∞ norm on forms, we obtain the normal currents; these can be made into a Banach space using duality. Indeed let $M(\mathcal{S})$ be any norm on 2 forms and let $M(T)$ be the dual norm

$$M(T) = \sup_{M(\mathcal{S}) \leq 1} T(\mathcal{S}) .$$

Let the normal current norm $N(T)$

$$N(T) = M(T) + M(\partial T) . \tag{4}$$

A special norm of forms, equivalent to the L^∞ norm, called comass (Federer [3]) has a dual norm called *mass* in which 2-dimensional area (in case $T(\mathcal{S}) = \int_T \mathcal{S}$) coincides with the two-dimensional area of a simplex T. Not all normal currents can be obtained by integration on sets however. Thus it is necessary for our purpose to restrict attention to integral currents (i.e. those currents that can be approximated in mass norm by finite polyhedral chains and their deformations under Lipschitz continuous mappings). By this restriction, we shall be able to establish the compactness, closure and regularity results necessary to carry through the steps in Poincaré's program.

First however we are now in a position to reformulate Poincaré's isoperimetric variational principle in terms of integral currents. First we note that all the above definitions can be carried over to Riemannian manifolds invariantly. In this way, Poincaré's isoperimetric problem can be written

($\tilde{\text{II}}$) Find the current $T \in I_2(M)$ such that

$$\inf_{T \in \Delta} M(\partial T) = \min \tag{5}$$

where $\Delta = \{T | T \in I_2(M), 2\pi = \int_T K d\|T\|\}$. Here $I_2(M)$ denotes the set of integral 2-currents on M.

Now we list the advantages of utilizing the formulation (II) of Poincaré's problem.

1) The Mass norm in $I_2(M)$ is lower semicontinuous, with respect to "flat" convergence.
2) With respect to this very weak notion of convergence ("flat" convergence of Whitney) standard closure and compactness theorems allow us to prove the extremal of ($\tilde{\text{II}}$) above is actually attained.
3) Smoothness and simpleness of the extremal can be established in one blow by applying standard regularity techniques of integral currents.
4) A new notion of stability based on perturbation of metrics can be established, for simple closed geodesics on M of minimal nonzero length.

First we separate the problem into two distinct parts: existence and smoothness except that for integral currents *smoothness also implies simpleness of the extremal.* The existence part is solved by compactness closure and semicontinuity arguments plus in this case: topological arguments based on homology groups. Indeed since cohomology enters analysis naturally via differential forms, homology groups arise naturally by duality for integral currents.

3. *The results and their proof*

The idea of our proof of the steps 1)-4) of Poincaré's program is now easily described: First we show the infimum of the variational problem $(\tilde{\Pi})$ defined by equation (5) is attained by an integral current $T \in \Delta$. We shall outline this step below since it can easily be described. Then we mention an a priori estimate that is needed for the extremal, since the extremal is not an absolute minimizer but rather an "almost" minimizing current in the sense of Bombieri [5] and Almgren [6]. This estimate enables us to conclude regularity of the extremal almost everywhere, together with a criterion for distinguishing smooth points of the exceptional set. It turns out that in this case this criterion can be used to prove regularity everywhere of the extremal. However our proof is still not completed at this stage: the *connectivity* of the extremal, essential for Poincaré's idea is lost in the limiting arguments with integral currents. The proof of connectivity of the extremal requires a computation of the second variation for the variational problem $(\tilde{\Pi})$ defined by (5) and this in turn restricts the Gaussian curvature of the ovaloid mentioned in *Theorem* 1 below. A virtue of our treatment of the problem of closed simple geodesics on ovaloids is our ability to prove a result on the stability of the closed simple geodesic of minimal nonzero length characterized by $(\tilde{\Pi})$. Thus completing step 4 in Poincaré's program outlined above. We state our result in Theorem 2 below. It involves a new type of stability in that we study the derivation of a simple closed geodesic from a great circle under perturbation of the metric. This gives a new differential geometric notion of stability for closed simple geodesics.

We now state our main results as follows:

THEOREM 1. *Poincaré's isoperimetric variational problem* $(\tilde{\Pi})$ *defined by (5) has a smooth solution* $T \in I_2(V)$ *without self-intersections for all* C^2 *ovaloids* (M^2, g). *Provided* $\left|\frac{\partial K}{\partial n}\right| < 2K^{3/2}$ *along* ∂T, ∂T *is a connected one-dimensional manifold and thus a simple closed geodesic of shortest nonzero length on* (M^2, g).

With the aid of Theorem 1 and geometric measure theory we prove

STABILITY THEOREM. *There is a function* $\eta(t)$ *tending to zero as* $t \to 0$ *with the following property.*

Let (S^2, g) *be the standard sphere in* R^3 *and* $(S^2, \tilde{g})$ *be a small* C^3 *perturbation of the metric* g *in the sense that*

$$\|g - \tilde{g}\|_{C^3} = \sup_{i,j} |g_{ij}(x) - \tilde{g}_{ij}(x)|_{C^3} \leq \varepsilon_0$$

for a suitable absolute constant $\varepsilon_0 > 0$.

Whenever $\|g - g\|_{C^3} \leq \varepsilon_0$, *there is a closed simple geodesic* $\tilde{C}$ *of minimum length on* (S^2, g) *and a great circle* C *on* (S^2, g) *such that the Hausdorff distance of* C *and* $\tilde{C}$ *satisfies*

$$\operatorname{dist}(\tilde{C}, C) \leq \eta(\|g - \tilde{g}\|_{C^3}) .$$

Some proofs

Proof that $\inf_{T \in \Delta} M(\partial T) > 0$, where

$$\Delta = \left\{ T \mid T \in I_2(M), \int Kd\, \|T\| = 2\pi \right\} .$$

This result does not follow from classical isoperimetric inequalities. For example in Huber [6], we find on an obvious notation,

$$L^2(\partial T) \geq 2\pi A(T) \left\{ 1 - \frac{1}{2\pi} \int_T Kd\, \|T\| \right\} .$$

So that $\inf M(\partial T) \geq 0$.

We prove our inequality using an argument by contradiction, assuming the following compactness and closure theorems for integral currents: (a) if a flat limit of integral currents is normal, then the limit is also integral; (b) bounded sets of integral currents are precompact in the flat topology.

Now both theorems utilize the notion of flat convergence. There are various equivalent definitions of the notion of flat norm for integral currents; for our purpose the flat norm of $T \in I_k(V)$

$$F(T) = \inf_{S \in I_{k+1}(V)} \{M(T - \partial S) + M(S)\} .$$

Notice that if $T \in I_2(V)$ with V a two-dimensional manifold $I_3(V) = \{0\}$ so the flat norm coincide with mass norm. We now utilize these facts. Let $T_n \in \Delta$ be a minimizing sequence for (II) and assume $0 = \inf_{\Delta} M(\partial T)$. Then $N(T_n) = M(T_n) + M(\partial T_n)$ is uniformly bounded. We have $M(\partial T_n) \to 0$ while

$$M(T_n) = \int d\,\|T_n\| \le \int \left[\frac{K}{\inf_V K}\right] d\,\|T_n\|$$

$$\le \left[\frac{1}{\inf_V K}\right] \int K\,d\,\|T_n\| = \left(\frac{1}{\inf_V K}\right)(2\pi) .$$

Here we have used the facts that $\inf_V K$ is a strictly positive constant (as V is an ovaloid) and that $T_n \in \Delta$. Now by the compactness theorem $T_n \to T$ in flat norm (i.e. $T_n \to T$ in mass norm since T_n have maximal dimension) where $T \in \Delta$. (Here we utilize the lower semicontinuity of mass under weak convergence and the Closure Theorem.) Now by lower semicontinuity $M(\partial T) = 0$ hence by homology theory for integral currents we have $[T] = N[V]$ for some integer N. Then by the Gauss-Bonnet theorem, since $T \in \Delta$, we obtain

$$2\pi = \int K\,d\,\|T\| = \int K\,d\,\|NV\| = N \int K\,d\,\|V\| = 4\pi N .$$

But this implies $N = \frac{1}{2}$ (a contradiction, since N must be a integer). This is the desired contradiction so the proof is completed.

Proof that $\inf_{T\in\Delta} M(\partial T)$ *is attained.* Again this proof uses the lower semi-continuity of mass and the compactness and closure theorem for integral currents. Let $T_n \in \Delta$ with $M(\partial T_n) \to \inf_{\Delta} M(\partial T) > \alpha$ (say). Again $N(T_n) = M(\partial T_n) + M(T_n)$ is uniformly bounded. By utilizing the argument in the above paragraph we deduce $T_n \to T$ first in flat norm and then in mass norm (since T_n is maximal dimensional). Moreover, by lower-semicontinuity

$$M(\partial T) = \alpha$$

and by utilizing closure results we get $T \in \Delta$. Here we have also used the fact that Δ is closed under mass convergence.

Idea of the regularity proof

The basic idea is to show that the extremal $T \in I_2(M)$ for $(\tilde{\Pi})$ is such that ∂T is almost minimal in the sense of Bombieri [5]. For this it suffices to prove the "local" a priori estimate

$$(*) \qquad M(\partial T) \leq M(\partial(T+X)) + c_4 M(X)$$

for any $X \in I_2(M)$ with support in a sufficiently small neighborhood of a point $a \in V$ such that $M(X)$ and $M(\partial X) \leq 1$. The constant $c_4 = c_4(M)$ in $(*)$ must depend only on M. This is achieved by a deformation argument and the variational characterization of T by $(\tilde{\Pi})$.

The results of Bombieri [5], then apply and show that ∂T is smooth apart from a small exceptional set E, which can be eliminated by density and tangent cone properties of ∂T. Results of Fleming [7] and Federer [3] show that E is empty and so ∂T is smooth everywhere.

MELVYN S. BERGER
DEPARTMENT OF MATHEMATICS
UNIVERSITY OF MASSACHUSETTS
AMHERST, MASSACHUSETTS 01003

BIBLIOGRAPHY

[1] M. S. Berger and E. Bombieri, On Poincaré's Isoperimetric Problem for Simple Closed Geodesics, Journal of Functional Analysis, Vol. 42, 1981, 274-298.

[2] H. Poincaré, Sur les lignes géodésiques des surfaces convexes, Trans. Am. Math. Soc. 6 (1905), 237-274.

[3] H. Federer, Geometric Measure Theory, Springer-Verlag, New York, 1969.

[4] W. Klingenberg, Lectures on Closed Geodesics, Springer-Verlag, Berlin, 1978.

[5] E. Bombieri, Regularity theory for almost minimal currents, to appear in Arch. Rat. Mech. Analysis.

[6] H. Huber, On the isoperimetric inequality on surfaces of variable Gaussian curvature, Ann. of Math. 60 (1954), 237-247.

[7] W. Fleming, On the oriented Plateau problem, Rend. Circ. Mat. (Palermo) vol. 11 1962, 1-22.

ON THE GEHRING LINK PROBLEM

Enrico Bombieri and Leon Simon

1. A conjecture of Gehring asserts that if we have a (n–1)-sphere and a k-sphere differentiably embedded and linked in R^{n+k} and at a distance at least 1 from each other, then their volumes are not less than the volumes of the corresponding standard unit spheres. Proofs of Gehring's conjecture have been obtained in special cases ($n=2$, $k=1$ or 2) in [E-S], [Ga], [O]; the existence of positive lower bounds for the volumes has been obtained in [Ge] using techniques from geometric measure theory. In this paper we prove Gehring's conjecture in the case in which the linking number of the two manifolds is non-zero.

2. Let us denote by A and B the S^{n-1} and S^k so embedded in R^{n+k} and let X be the (n–1)-dimensional integral current in R^{n+k} determined by the oriented manifold A; let also $r = \text{dist}(A, B)$.

Since $\partial X = 0$ we can find an absolutely area minimizing integral current $Y \in I_n(R^{n+k})$ with $\partial Y = X$. We claim that $\text{spt}(Y) \cap B$ is not empty. Indeed, if it were empty then Y would have compact support in $R^{n+k} \sim B$, hence $Y \in I_n(R^{n+k} \sim B)$ and thus $X = \partial Y \in \mathcal{B}_{n-1}(R^{n+k} \sim B)$ would be an integral (n–1)-dimensional boundary. This contradicts the fact that the homology class of A in

Seminar on Minimal Submanifolds
0-691-08324-X/83/271-04 $0.70/0 (cloth)
0-691-08319-3/83/271-04 $0.70/0 (paperback)
For copying information, see copyright page.

$$H_{n-1}(R^{n+k} \sim B, Z) \cong \mathcal{Z}_{n-1}(R^{n+k} \sim B)/\mathcal{B}_{n-1}(R^{n+k} \sim B)$$

is non-zero because the linking number of A with B is non-zero. (For the above isomorphism of singular homology with the homology of the complex of currents, see Federer, Geometric Measure Theory, Springer-Verlag, New York 1969, 4.4.1, p. 464.)

Let us take $p \in \operatorname{spt} Y \cap B$ and note that $\operatorname{dist}(p, \operatorname{spt} \partial Y) \geqq \operatorname{dist}(A,B) = r$. Now everything comes from the following isoperimetric inequality, to be proved in the next section.

LEMMA. *Let* $Y \in I_n(R^{n+k})$ *be absolutely minimizing and let* $p \in \operatorname{spt} Y$. *Then we have*

$$n\alpha_n[\operatorname{dist}(p, \operatorname{spt} \partial Y)]^{n-1} \leqq M(\partial Y),$$

where α_n *is the volume of the* n*-dimensional unit ball.*

If we apply this lemma to Y as defined before, we obtain

$$\operatorname{vol}(A) = M(\partial Y) \geqq n\alpha_n r^{n-1},$$

which is the conclusion of Gehring's conjecture for A. If we interchange the role of A and B we obtain the full Gehring conjecture. We remark that the above proof does not require A, B to be spheres, but only smooth oriented compact manifolds without boundary in R^{n+k}. Furthermore, it suffices that A, B are immersed rather than embedded.

3. Let $g(x)$ be a piecewise smooth Lipschitz vector field on R^{n+k} and let $h_t(x) = x + tg(x)$. The first variation of the current Y is a linear functional δY whose value on the vector field g is

$$(\delta Y)(g) = \frac{d}{dt} M(h_{t\#}Y)\Big|_{t=0}$$

$$= \int \operatorname{div}(g)\, d\,\|Y\|;$$

here $\operatorname{div}(g) = \sum_{i=1}^{n} \langle e_i, \nabla_{e_i} g\rangle$, where $e_1, \cdots, e_n$ is an orthonormal frame in the approximate tangent space $\operatorname{Tan}(Y, x)$ and where ∇_{e_i} is the covariant derivative, in the direction e_i, in the ambient Riemannian manifold (in this case R^{n+k}). Since Y is absolutely area minimizing with smooth boundary, comparison of Y with $\tilde{Y} = h_{t\#}Y - h_{\#}([0,t] \times \partial Y)$ (here $h = h(t,x) = x + tg(x)$ is considered as a map from $R \times R^{n+k}$ to R^{n+k}) yields

$$|(\delta Y)|(g) \leqq \int |g \wedge \overrightarrow{\partial Y}|\, d\, \|\partial Y\| \leqq \int |g|\, d\, \|\partial Y\|$$

(for more details, see [A, (4), p. 440]). We have shown that

$$\int \operatorname{div}(g)\, d\, \|Y\| \leqq \int |g|\, d\, \|\partial Y\| .$$

For the proof of the lemma, we assume that $p = 0$ is the origin of R^{n+k} and choose for $g(x)$ the vector field

$$g(x) = \min\left(\frac{1}{\rho^n}, \frac{1}{|x|^n}\right) x .$$

A simple calculation in local coordinates shows that $\operatorname{div}(g) \geqq 0$ and $\operatorname{div}(g) = \frac{n}{\rho^n}$ if $|x| < \rho$, hence

$$n\rho^{-n} M(Y \llcorner \{|x| < \rho\}) \leqq \int \frac{d\, \|\partial Y\|}{|x|^{n-1}} .$$

Letting $\rho \to 0$, we deduce

$$n\alpha_n \Theta^*(\|Y\|, 0) \leqq \int \frac{d\, \|\partial Y\|}{|x|^{n-1}}$$

where $\Theta^*(\|Y\|, 0)$ is the upper density of Y at $0 \in \operatorname{spt} Y$. Since $\Theta^*(\|Y\|, x) \geqq 1$, $\|Y\|$ – a.e., we have shown

$$n\alpha_n \leqq \int \frac{d\,\|\partial Y\|}{|x-x_0|^{n-1}}$$

whenever $x_0 \in \operatorname{spt}(Y) \sim \operatorname{spt}(\partial Y)$; our lemma follows at once from this inequality.

ENRICO BOMBIERI
SCHOOL OF MATHEMATICS
INSTITUTE FOR ADVANCED STUDY
PRINCETON, NEW JERSEY 08540

LEON SIMON
DEPARTMENT OF MATHEMATICS
RESEARCH SCHOOL OF PHYSICAL SCIENCES
AUSTRALIAN NATIONAL UNIVERSITY
CANBERRA, A.C.T. 2600
AUSTRALIA

REFERENCES

[A] W. K. Allard, On the first variation of a varifold, Annals of Math. 95 (1972), 417-491.

[E-S] M. Edelstein and B. Schwarz, On the length of linked curves, Israel J. Math. 23 (1976), 94-95.

[Ga] M. E. Gage, On Gehring's linked sphere problem. Stanford 1978.

[Ge] F. W. Gehring, The Hausdorff measure of sets which link in Euclidean space. Contribution to Analysis, A collection of papers dedicated to Lipman Bers, Academic Press 1974, 159-167.

[O] R. Osserman, Some remarks on the isoperimetric inequality and a problem of Gehring, J. Analyse Math. 30 (1976), 404-410.

CONSTRUCTING CRYSTALLINE MINIMAL SURFACES

Jean E. Taylor*

1. *Introduction*

The object of this note is to give two examples of the application of a general algorithm (outlined in the seminar of this title) for constructing certain polyhedral surfaces having a given boundary and minimizing the integral of a crystalline integrand. That the surfaces of the examples minimize the integral locally in space will be proven here. The complete description of the algorithm and the description and proof of the full nature of the minimization are beyond the scope of this note.

Crystalline integrands arose from the consideration of the surface tension functions of crystalline materials. They are of interest when restricted to that context, but are also of interest because they are not elliptic, and as a result have properties quite different from the area integrand. One of these is the possession of a family of simple but extraordinarily useful "barriers" (see Section 3 below).

A major tool in the construction is an association between polyhedral surfaces and certain labelled graphs (see Section 2 below). Using graphs one can modify the local structure of polyhedral surfaces without immediate regard for the global feasibility of constructing such surfaces.

[1] This work was partially supported by a Sloan Fellowship and an NSF grant.

Seminar on Minimal Submanifolds
0-691-08324-X/83/275-18 $1.40/0 (cloth)
0-691-08319-3/83/275-18 $1.40/0 (paperback)

Much of the algorithm has been programmed on a microcomputer; in particular, the examples of this paper have been run from start to finish. I am indebted to Robert F. Almgren for assistance in the computer programming.

2. *Labelled graphs, surfaces, integrands, and integrals*

(1) Integrands and their crystals. A general reference for this section is [TJ].

A *two-dimensional constant-coefficient integrand* F *on* R^3 is a continuous function from the Grassmannian of oriented two-planes through the origin in R^3 to the positive real numbers; by duality, it can and will in this paper be regarded as a function on the unit sphere in R^3. The *crystal* of such an integrand is defined to be the set

$$w = \{x \in R^3 : x.v \leq F(v) \text{ for each } v \in R^3 \text{ with } |v| = 1\},$$

oriented positively. An integrand is called *crystalline* if and only if its crystal is a polyhedron. (Elliptic integrands, on the other hand, have uniformly convex crystals.) The unit oriented normals to the faces of the crystal of a crystalline integrand are called *crystalline directions*.

For any convex region containing the origin in its interior and oriented positively, there is a unique convex integrand having this as its crystal (see [TJ]).

The *integral* F(S) *of* F over an oriented polyhedral surface S is

$$F(S) = \sum_i F(n_i)\,\text{Area}(P_i),$$

where the sum is over all plane segments P_i in S and, for each i, n_i is the unit oriented normal to the plane segment P_i. By an appropriate extension of this, one can obtain the definition of integrating F over integral currents (or see [F].

For the examples of this paper, we consider the particular convex crystalline integrand F whose crystal W is obtained by taking the unit cube

$$\{x = (x_1, x_2, x_3) : |x_i| \leq 1 \text{ for } i = 1, 2, 3\}$$

and cutting off the corners by planes with normals $(\pm1, \pm1, \pm1)$, each at distance $(\sqrt{3})/2$ from the origin. (In the region of the unit sphere defined in spherical coordinates by

$$0 \leq \theta \leq 1/\sqrt{2}, \ |\cos \phi| \leq (\sqrt{(2/3)}) \sin \theta ,$$

F is given explicitly by

$$F(\theta, \phi) = \sin \phi (\cos \theta + (1/2) \sin \theta) ,$$

and its value in the regions obtained by reflecting this region around is that induced by the reflections.) A sketch of this crystal W (and a simple closed curve on it relevant to example 2 below) appears in Figure 2(a). I do not know if any physical crystalline material has precisely W as its infinitesimal equilibrium shape (W is the surface of least surface integral among all integral currents with the same mass as W –see [TJ]) but a mineral called Hauerite has approximately this habit [DM].

(2) Labelled graphs associated dually to polyhedral surfaces. Let S be a polyhedral surface, with or without a (polyhedral) boundary. For each plane segment in the surface, put a vertex in the graph (labelled a surface vertex). For each intersection of plane segments along an edge, put an edge in the graph connecting the corresponding vertices (if the polyhedral surface is being regarded as an immersion of a disk with handles, then put in an edge if and only if the preimages of the plane segments are adjacent). For each line segment of the boundary, put a vertex in the graph, labelled a boundary vertex. To each plane segment of the surface containing a portion (of positive length) of a boundary line segment as part of its boundary, put an edge in the graph between the corresponding surface and boundary vertices. Finally, for each boundary corner, put an edge in the graph connecting the corresponding boundary vertices. Such a graph will be planar if in particular the surface giving rise to it is an immersion of a disk. Since the surface is oriented, the

graph is also oriented and there is a notion of "counterclockwise" around each vertex.

The vertices are labelled with two additional labels—that is, we define the function

$$L : \{\text{vertices of graph}\} \to \text{unit sphere} \times \mathbf{R} ,$$

where for each surface vertex v, $(Lv)_1$ (the "direction" of the vertex) is the unit normal to that plane segment and $(Lv)_2$ (the "distance" of the vertex) gives the translation of the oriented plane containing the plane segment from the origin. For each boundary vertex, the unit vector is chosen arbitrarily among all unit vectors normal to the corresponding boundary line segment and not equal to the direction of an adjacent vertex, and the real number is chosen so that a plane with that normal translated from the origin by that distance contains the boundary line segment; in what follows, it is convenient to take the unit vector to be a noncrystalline direction.

For a crystalline integrand F, the graph associated to the boundary of its crystal is called its *crystal graph*; if the integrand is convex, its crystal graph contains all the information about the integrand.

(3) Polyhedral surfaces associated to certain labelled graphs. Given a graph G embedded in a sphere with k handles (the examples of this paper have $k = 0$), with each vertex v labelled with a unit vector ("direction") $(Lv)_1$ and a real number ("distance") $(Lv)_2$, one can associate an oriented plane P(v) to each vertex v, the plane with the unit vector as its oriented normal and translated from the origin by the given real number. Suppose suitable faces of G are designated as boundary faces and all their vertices as boundary vertices. We designate as surface vertices all other vertices.

We define such a labelled graph to be *surface-compatible* if and only if for each face f of the graph other than a boundary face, all the planes corresponding to the vertices of the face have a unique point CRNR(f) in common. If all the faces of a graph which do not contain a boundary vertex are triangular and if the three corresponding planes meet in general position, then such a unique point automatically exists for those faces, and the above is a condition only on the remaining faces.

For such a surface compatible graph, say with one boundary face, one can construct an oriented polyhedral boundary curve by forming the line segments linking sequentially the points CRNR(f) corresponding to faces f of the graph adjacent to the boundary face, in clockwise order around the boundary face. Given any surface vertex v, the faces adjacent to it, in counterclockwise order, determine an oriented polyhedral curve in the plane associated to that vertex. In case this curve is in one-to-one correspondence with the boundary of a polyhedral region of positive orientation, that plane segment can be associated to that vertex. Other behavior is possible but will not be treated in this note. In this way one constructs a surface corresponding to the labelled graph.

If all the faces of the graph which do not contain a boundary vertex are indeed triangular, with their corresponding planes in general position, then the graph is also *variation-compatible*: each surface vertex which is not adjacent to a boundary vertex can have its distance varied and the labelled graph remains surface compatible.

(4) Integrating an integrand over a labelled graph. Given a surface compatible graph G, one can define the function AREA associating a real number to each surface vertex as follows:

$$\mathrm{AREA}(v) = \left[(Lv)_1 \cdot \left(\sum_i (\mathrm{CRNR}(f(i)) - C) \times (\mathrm{CRNR}(f(i{+}1)) - C)\right)\right]/2 \;;$$

here "$\times$" means the cross product, "$\cdot$" means the dot product, the sum runs from $i = 2$ to $i = n(v)-1$, where $n(v)$ is the number of faces adjacent to v and for each i, f(i) is the ith such face (the faces are numbered in counterclockwise order, starting anywhere), and $C = \mathrm{CRNR}(f(1))$. The integral of F over G is the sum over all surface vertices v of the numbers $\mathrm{AREA}(v)\,F((Lv)_1)$.

Observe that if G is the labelled graph corresponding to an oriented embedded polyhedral surface, then the integral of F over G equals F(S); however, if negative orientations occur, then this integral does not correspond to any of the standard notions of integration over the resulting surface.

3. *Some basic* F*-minimizing surfaces*

The following results are proved in [TJ] for any convex crystalline integrand F. (Here F-minimizing means in the class of all integral currents with the same boundary.)

Any oriented plane segment is F-minimizing.

An oriented surface which is a subset of the boundary of the crystal of F and which is composed of one or two plane segments is uniquely F-minimizing.

An oriented surface which is a subset of the boundary of the crystal of F, whose segments have more than two distinct directions, and which contains only plane segments which share a common point, is F-minimizing but not uniquely so.

If W' is the central inversion of the crystal of F (oriented negatively) then the above results are also true for similar subsets of the boundary of W'.

Since F is constant coefficient, any translation or homothety of an F-minimizing surface is also F-minimizing. Thus the uniquely minimizing above surfaces, translated and uniformly dilated or contracted, form a collection of barriers. Such a surface composed of one plane segment with its unit normal a crystalline direction is called a *crystal plane barrier*, if instead it comes from two adjacent plane segments in ∂W it is called a *"down" edge barrier*, if it comes from two adjacent plane segments in $\partial W'$ it is called an *"up" edge barrier*. By a slight abuse of language, we also refer to *crystal corner* or *inverse crystal corner "barriers."*

4. *Example 1*

Let F be the integrand defined in 2(1), and let B be the oriented closed curve

$$[B_0, B_1] + [B_1, B_2] + [B_2, B_3] + [B_3, B_0],$$

where $B_0 = (1,0,1)$, $B_1 = (1,1,0)$, $B_2 = (0,1,1)$, and $B_3 = (0,0,0)$. Consider first the oriented surface S′ having this boundary which is obtained by "applying crystal plane barriers" to B from behind; it consists of two planar surfaces with oriented normals $n_0 = (-a, a, a)$ and $n_2 = (a, -a, a)$, where $a = 1/\sqrt{3}$. (These planes span the line segments of B containing B_0 and B_2 respectively.) This surface is sketched in Figure 1(a) and its graph G in Figure 1(b). The two surface vertices of

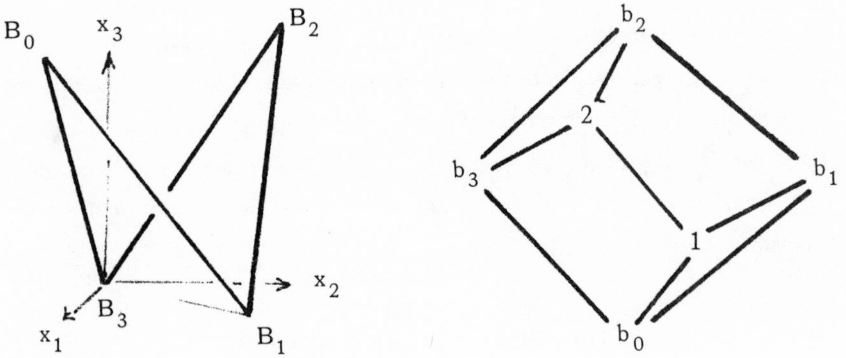

Figure 1(a) Figure 1(b)

G are adjacent in G but their directions are not adjacent in the crystal graph—rather, a vertex of direction (0, 0, 1) intervenes. One might therefore suspect that S′ is not F-minimizing, and that a smaller integral might be achieved by replacing a neighborhood of the intersection of the planes with a plane segment having unit normal (0, 0, 1) and hanging at some height together with patches in the boundary corners as necessary;

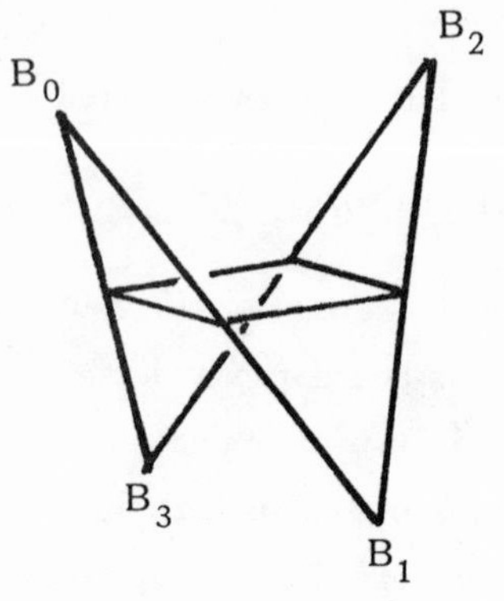

Figure 1(c)

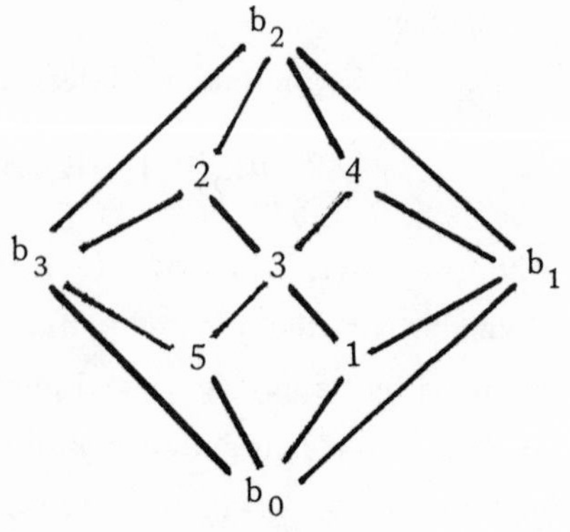

Figure 1(d)

Directions of vertices: 1 : $(-a, a, a)$ 2 : $(a, -a, a)$ 3 : $(0, 0, 1)$ 4 : (a, a, a)
5 : $(-a, -a, a)$ where $a = 1/\sqrt{3}$.

the resulting surface is sketched in Figure 1(c) and its graph G' is shown in Figure 1(d). Note that all pairs of adjacent surface vertices in G' have their directions adjacent in the crystal graph. If the hanging plane has its x_3 coordinate equal to t (i.e. the corresponding surface vertex has distance t), then one calculates that the integral of F over this surface is

$$t^2 - t + 3/2 .$$

Barrier considerations restrict the possible values of t to $0 \leq t \leq 1$. One calculates immediately that the integral of F is indeed less for $0 < t < 1$ than it is for $t = 0$; in fact, there is a unique minimum for the integral of F in the family of surfaces parameterized by t and it occurs at $t = 1/2$. Moreover, we have the following proposition:

PROPOSITION. *The surface of the above example with* $t = 1/2$ *is uniquely* F*-minimizing among all surfaces which are graphs of Lipschitz functions taking boundary* B.

Proof. Given M greater than or equal to 2, let S be any surface of least F integral in the class of all graphs of Lipschitz functions with Lipschitz constant bounded by M. We will show that S is the surface of the above example. It then follows that S is the unique surface of least integral among all graphs of Lipschitz functions.

The proof consists of repeated barrier arguments. Let P_i, for $i = 0, 1, 2, 3$, be the plane containing the boundary line segments adjacent to boundary corner i, oriented by the unit vector n_i where n_0 and n_2 are as above and $n_1 = (a, a, a)$, $n_3 = (-a, -a, a)$, $a = 1/\sqrt{3}$, and let d_i be such that $B_i \cdot n_i = d_i$. Note that each oriented plane P_i is a barrier for S from one side or the other.

Let

$$t = \inf\{x_3 : (x_1, x_2, x_3) \in S \sim (P_1 \cup P_3)\},$$

and let

$$T = \{x \in S : x_3 \le t\},$$

$$T' = T \sim (P_1 \cup P_3).$$

Observe that T is closed.

The first step is to see that T' is nonempty. If this were not the case, then for some $\varepsilon > 0$,

$$\{x \in S \sim (P_1 \cup P_3) : x_3 \le t + \varepsilon\}$$

lies within thin neighborhoods of $P_1 \cup P_3$, and a "down" edge barrier consisting of half planes with normals $(0, 0, 1)$ and n_1, at heights $t + \varepsilon$ and d_1 respectively, shows that in fact none of this set is near P_1. A similar barrier near P_3 then contradicts the definition of t.

We then see, by the same argument, that

$$\{x \in S : x_3 < t\} \subset P_1 \cup P_3 .$$

We similarly check, using the same "down" edge barrier argument, that each connected component of T' contains some point in P_1 and some point in P_3. Thus there is at least one simple curve C in T' joining P_1 to P_3. But now we can apply "up" edge barriers from above: use C, the appropriate line segments in P_1 and P_3, and the part of $B \cap P_2$

which lies above the plane $x_3 = t$ to obtain a simple closed curve C_2; now "up" edge barriers applied from above to the part of S bounded by C_2 show that this part of S is contained in the planes P_2 and $x_3 = t$. (If $t = -1$, a minor rewording of the above is required and shows that that part of S is contained in P_2.)

Finally, the remainder of S is shown to lie in the planes P_0 and $x_3 = t$ by applying an "up" edge barrier to that side.

We now have that S consists of a single hanging plane segment with triangular patches as needed in the corners of the boundary; the computation of the integral of F over such surfaces as a function of t, with constraints $0 \leq t \leq 1$, shows that t must be $1/2$.

REMARK. *The surface is in fact* F*-minimizing among all integral currents with boundary* B: if one takes a sequence $\{F_i\}$ of uniformly convex integrands converging to F, and a corresponding sequence $\{S_i\}$ of integral currents minimizing respectively the integral of F_i with boundary B, then (since B has convex projection onto the $x_1 x_2$ plane) for each 1, S_i is unique and is a graph of a function and the above proof shows that if it is deformed to S then its F-integral can only decrease. We therefore conclude, by standard arguments, that S is F-minimizing.

5. *Example 2*

Suppose F remains the crystalline integrand defined in Section 2(1), but B is now the extreme oriented simple closed polyhedral boundary curve whose sixteen corners are (in order): (.5, 0, 1) (0, .5, 1) (0, 1, .5) (.5, 1, 0) (1, .5, 0) (1, 0, –.5) (.5, 0, –1) (0, –5, –1) (–.5, 0, –1) (–1, 0, –.5) (1, –.5, 0) (–.5, –1, 0) (0, –1, –.5) (.5, –1, 0) (0, –1, .5) (0, –.5, 1). B lies on ∂W, and B and W are sketched in Figure 2(a). The surface produced by applying all possible crystal plane and edge barriers from behind B is sketched in Figure 2(b) and its graph is given in Figure 2(c). For convenience in giving the first labels of the surface vertices, we number the vertices of the crystal graph and give one of these numbers in place of

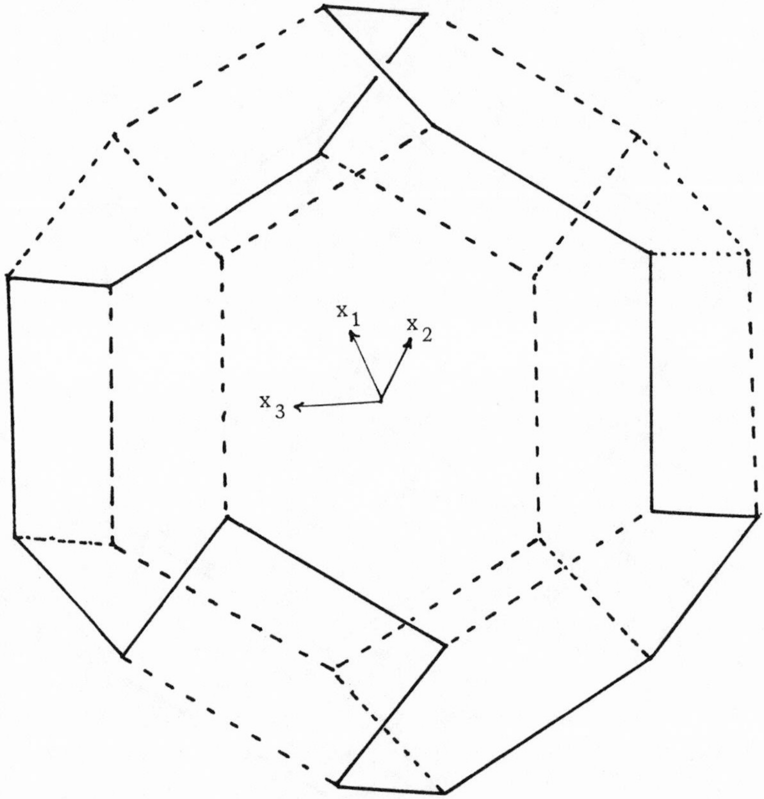

Figure 2(a)

the corresponding unit normals. This numbering of seven of the unit normals is (with $a = 1/\sqrt{3}$):

$$1:(0,0,1) \quad 2:(0,1,0) \quad 3:(0,0,1) \quad 4:(a,a,a) \quad 5:(a,a,-a)$$
$$6:(-a,a,-a) \quad 7:(-a,a,a)$$

and directions 8 through 14 are the negatives of directions 7 through 1 respectively.

The directions of the surface vertices in the graph of Figure 2(c) are then as follows:

$$1:11 \quad 2:6 \quad 3:7 \quad 4:4 \quad 5:6 \quad 6:14\,.$$

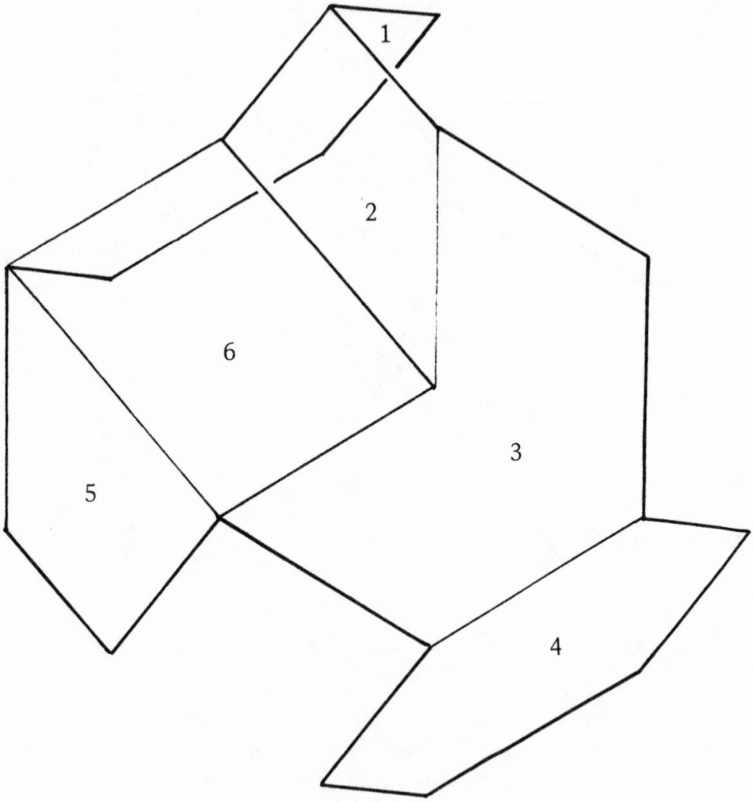

Figure 2(b). The initial surface as seen from behind.

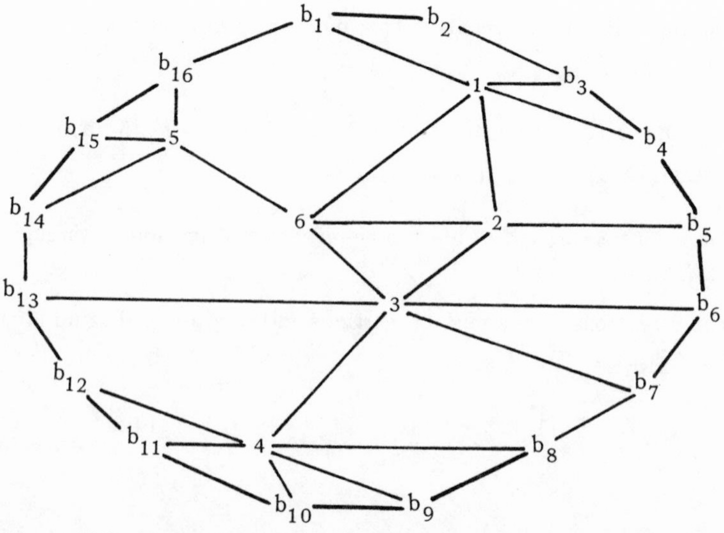

Figure 2(c). This and the following graphs are drawn looking at them from behind, so the orientation is reversed.

All pairs of adjacent surface vertices in this graph indeed have their directions adjacent in the crystal graph, but the vertices are not in "general position": some vertices which "should" be free to have their distance be variable have edges to the boundary, and the surface at two boundary corners (corresponding to the face containing boundary vertices labelled b_{16} and b_1 and to the one containing b_{13} and b_{14} in the Figure 2(c)) is not locally F-minimizing. The algorithm "pulls vertices off the boundary" and then works on the resulting faces, inserting a vertex of the appropriate direction in each boundary corner and filling in the rest of the faces using parts of the crystal graph, in reverse order. The result is the graph of Figure 2(d). The directions of the additional vertices are:

$$7:7 \quad 8:2 \quad 3:5 \quad 10:10 \quad 11:13 \quad 12:10 \quad 13:3 \quad 14:7 \quad 15:5$$
$$16:12 \quad 17:6 \quad 18:9 \quad 19:3 \quad 20:11 \quad 21:12 \quad 22:2 \,.$$

The distances of all the vertices which are adjacent to a boundary vertex can be determined from the boundary corners; the remaining vertices (2, 3, 6, 8, 10, 19) are all variable in distance.

Note that all surface vertices which are not adjacent to the boundary, except for vertex 6, are adjacent to vertices whose directions circle the

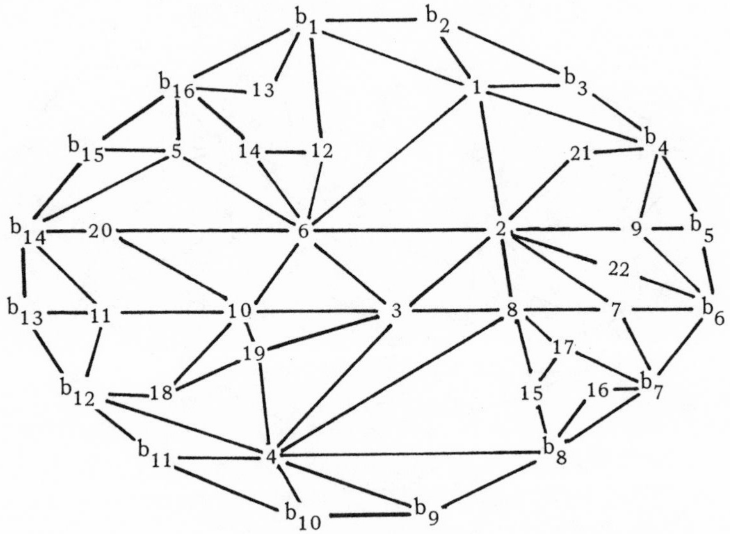

Figure 2(d)

direction of the given vertex exactly once, except possibly for a doubling back by exactly one vertex; the direction of circling in the graph (with its natural orientation) is the opposite from that in the crystal graph. Vertex 6 is circled this way twice. One must consider the possibility of "splitting" vertex 6; there are eight theoretically possible ways of splitting it (not all of which can yield surfaces with no negative orientations), one of which is shown in Figure 2(e).

One computes the integral of F over the graph in 2(d) as a function of the variable distances, subject to constraints of no negative orientation regions; a local minimum is found when the above six variable planes have as their distances:

$$d_2 = 0, \; d_3 = 1/7\sqrt{3}), \; d_6 = 1/7, \; d_4 = 1/28, \; d_{10} = 0, \; d_{19} = 1/28.$$

None of the splittings of vertex 6 which do not create regions of negative orientation decrease integral.

A projection of the surface S corresponding to the graph of Figure 2(d) with these variable values for the distances is shown in Figure 2(f). Com-

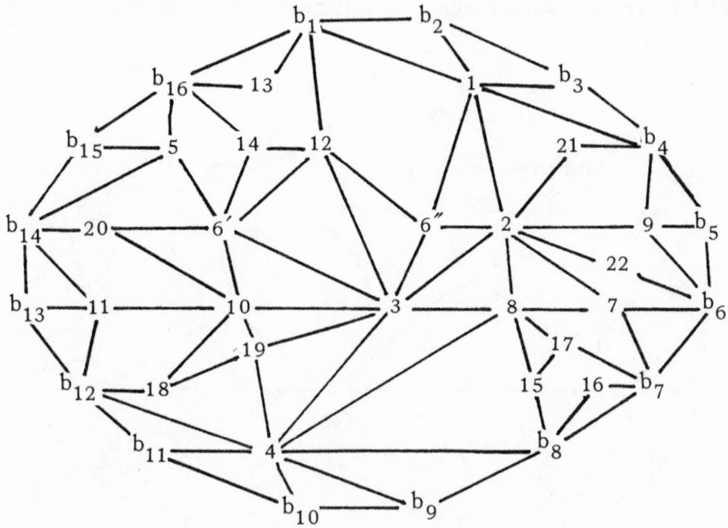

Figure 2(e)

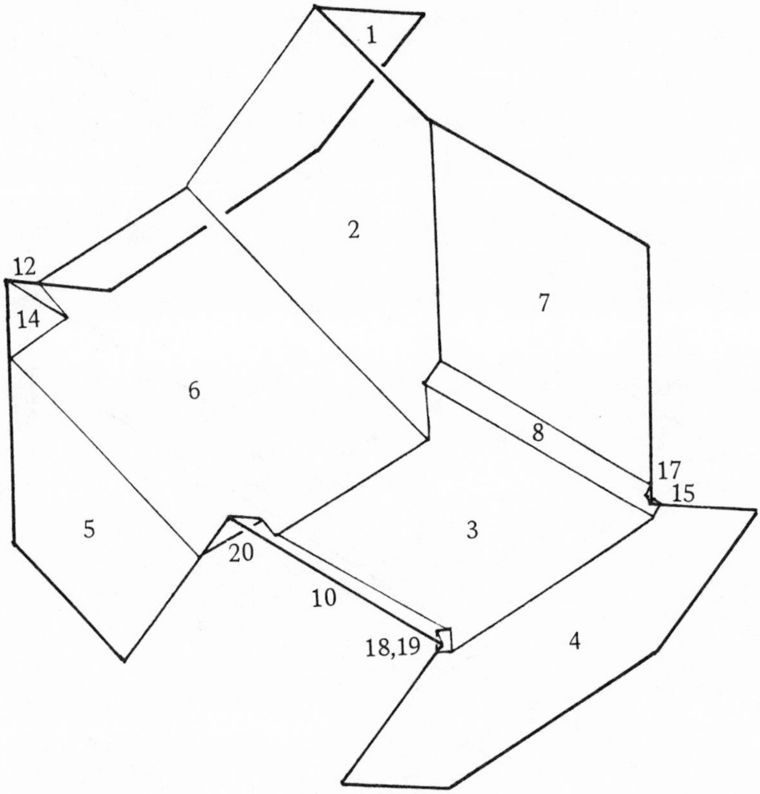

Figure 2(f)

puter generated cutouts (the lines were drawn by hand) of the plane segments are given in Figure 2(g); the numbers around the edges of the plane segments are the planes adjacent to the segments along those edges. A "+" in front of such a number indicates that the intersection is "down" and a "−" indicates it is "up."

PROPOSITION. S *is locally* F*-minimizing among polyhedral surfaces with the same unit normals and adjacency relations; it is also locally* F*-minimizing in space in the family of all integral currents with boundary* B *—i.e. there exists a* $d > 0$ *such that if an integral current* I *agrees with* S *except possibly within some ball of radius* d, *then* $F(S) \leq F(I)$.

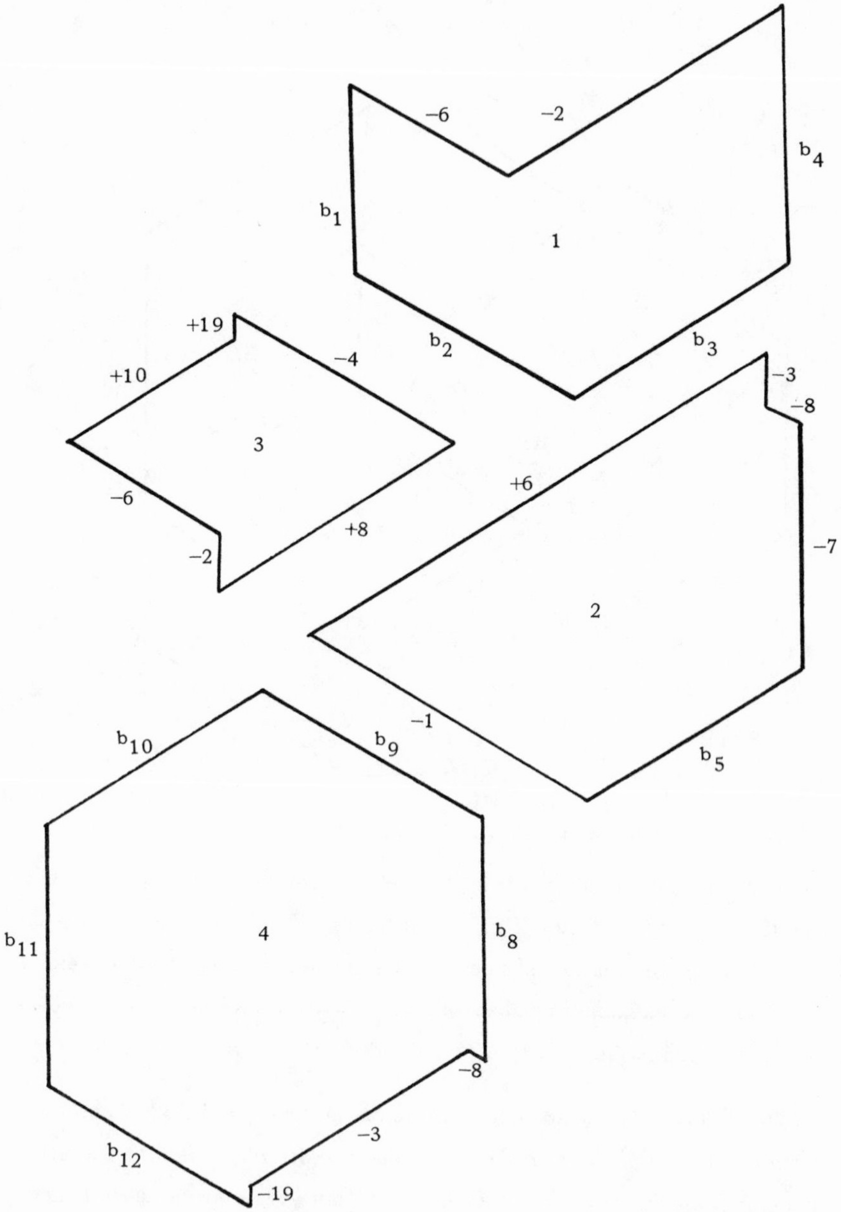

Figure 2(g)

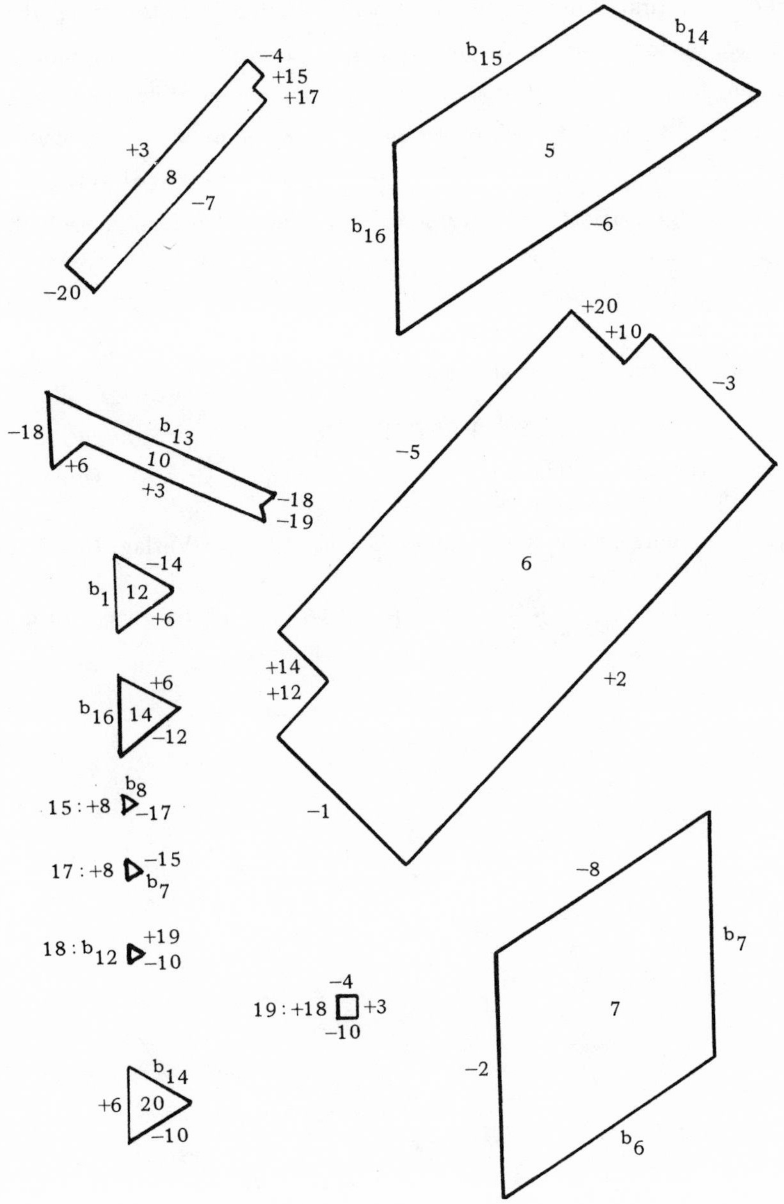

Figure 2(g) (Continued)

Proof. The first statement follows from the choice of distances for the variable planes. For the second statement, d is chosen small enough that any ball of radius d contains at most one corner of any plane segment in S. A neighborhood of radius d of any point of S can now readily be shown to be F-minimizing by the application of crystal plane barriers, edge barriers, and inverse crystal corner "barriers" from both sides.

JEAN E. TAYLOR
DEPARTMENT OF MATHEMATICS
RUTGERS UNIVERSITY
NEW BRUNSWICK, NEW JERSEY 08903

REFERENCES

[DM] V. De Michele (photographs by C. Beviacqua), Crystals, Orbis Publishing Ltd., London, 1973, plate 16.

[FH] H. Federer, Geometric Measure Theory, Springer-Verlag, New York, 1969.

[TJ] J. E. Taylor, *Crystalline variational problems*, Bull. Amer. Math. Soc. 84 (1978), 568-588.

Regularity of Area-Minimizing Hypersurfaces at Boundaries with Multiplicity

Brian White*

Introduction

Let T be an m-dimensional area-minimizing integral current in R^{m+1} with $\partial T = S$, where S is a smoothly embedded, connected manifold with multiplicity 1. Robert Hardt and Leon Simon have recently shown that, in some neighborhood U of spt S, T is a smooth manifold with boundary [HS]. But what if $\partial T = \nu S$ for some multiplicity $\nu > 1$? Our Global Boundary Regularity Theorem (Corollary 1) shows that $T = \Sigma_{i=1}^{\nu} T_i$ where each T_i is area-minimizing with $\partial T_i = S$ and hence is regular at its boundary. More generally, spt ∂T may be the union of several embedded manifolds, which ∂T takes with different multiplicities. This situation is analyzed in our Local Boundary Regularity Theorem (Corollary 2).

Both of these theorems are corollaries of a general decomposition theorem, which essentially says: let T be a codimension 1 rectifiable cycle in an oriented Riemannian manifold such that T(mod ν) is a boundary. Then T can be decomposed geometrically into ν homologous cycles: $T = \Sigma_{i=1}^{\nu} T_i$. Furthermore, T minimizes a parametric integral Φ if and only if each T_i does.

A third corollary answers a question raised by Federer in [F2, p. 396]. The problem is basically the following. Let A be an m+1 dimensional

*Supported by an NSF graduate fellowship.

Seminar on Minimal Submanifolds
0-691-08324-X/83/293-09 $0.95/0 (cloth)
0-691-08319-3/83/293-09 $0.95/0 (paperback)

oriented Riemannian manifold, let Φ be a parametric integrand of degree m on A, and let α be a member of the mth homology group $H_m(A)$. Then clearly:

$$\inf_{T \epsilon k\alpha} <\Phi, T> \leq k \inf_{T \epsilon \alpha} <\Phi, \alpha>$$

since if $T \epsilon \alpha$, then $kT \epsilon k\alpha$ and $<\Phi, kT> = k<\Phi, T>$. The question is: does equality hold? Federer [F2, 5.10] proved that it does if Φ is an even integrand. Here we show it for arbitrary Φ.

This seems to be the first time rectifiable currents mod ν have been used to study integral currents.

I am very grateful to F. J. Almgren, Jr. for encouragement, suggestions, —and patience.

Notations and terminology

In general we follow the conventions of [F1]. In particular, if $B \subset A$ are subsets of Euclidean space:

$$\mathcal{F}_m(A) = \{\text{integral flat m-chains } T \text{ with spt } T \subset A\}$$

$$\mathcal{R}_m(A) = \{T \epsilon \mathcal{F}_m(A) : T \text{ is rectifiable}\}$$

$$\mathcal{Z}_m(A, B) = \{T \epsilon \mathcal{F}_m(A) : \text{spt } \partial T \subset B\}$$

$$\mathcal{B}_m(A, B) = \{T + \partial R : T \epsilon \mathcal{F}_m(B), R \epsilon \mathcal{F}_{m+1}(A)\}.$$

$\mathcal{Z}^\nu_m(A, B)$ and $\mathcal{B}^\nu_m(A, B)$ are the corresponding groups of flat chains modulo ν.

As in [W], if T is a locally flat chain, we denote by TAN(T, x) the oriented tangent cone to T at x, provided it exists and is unique.

Results

DECOMPOSITION THEOREM. *Let* A *be an orientable* m+1 *dimensional* C^1 *submanifold of a Euclidean space. Let* $B \subset A$ *be a closed local Lipschitz neighborhood retract. Suppose*:

$$T \in \mathcal{Z}_m(A,B) \cap \mathcal{R}_m$$
$$T = T \llcorner (A \sim B)$$
$$T \,(\mathrm{mod}\ \nu) \in \mathcal{B}_m^\nu(A,B)\,.$$

Then there exist $T_1, T_2, \cdots, T_\nu$ *such that:*

(1) $$T_i \in \mathcal{Z}_m(A,B) \cap \mathcal{R}_m$$

(2) $$T_i = T_i \llcorner (A \sim B)$$

(3) $$T = \sum T_i$$

(4) $$\|T\| = \sum \|T_i\|$$

(5) $$T_i - T_j \in \mathcal{B}_m(A,B)\,.$$

If Φ *is a parametric integrand of degree* m, *then* T *is homologically* Φ*-minimizing with respect to* (A,B) *if and only if each* T_i *is.*

REMARK. The decomposition given below is quite explicit, as the reader will see if he or she draws a few examples where, say, A is a 2-dimensional torus and $B = \phi$.

Proof. The result is clearly true for $\nu = 1$. Assuming the result true for $\nu = p-1$, we shall prove it for $\nu = p$. By hypothesis, there is an $R \in \mathcal{R}_{m+1}(A)$ such that:

(6) $$T \equiv \partial R \llcorner (A \sim B) \quad (\mathrm{mod}\ p)\,.$$

Since R is an $(m+1)$-dimensional rectifiable current in an $(m+1)$-dimensional orientable manifold:

$$R = \Omega \llcorner f$$

where Ω is a density-one current orienting A and f is an integer-valued $\mathcal{H}^{m+1}$ measurable function. Without loss of generality, $-\frac{1}{2}p < f(x) \le \frac{1}{2}p$

for all x (otherwise replace f by $g(x) = \inf\left\{k > -\frac{1}{2}p : k \equiv f(x) \pmod p\right\}$). It then follows from (6) that $\partial R \llcorner (A \sim B)$ is rectifiable [W, 2.3]. Thus by (6) there is a $T_p \in \mathcal{R}_m(A)$ such that:

$$T - \partial R \llcorner (A \sim B) = pT_p\,. \tag{7}$$

We see immediately:

$$T_p \in \mathcal{Z}_m(A,B) \cap \mathcal{R}_m(A)$$

$$T_p = T_p \llcorner (A \sim B)$$

$$T - T_p = (p-1)T_p + \partial R \llcorner (A \sim B)\,. \tag{8}$$

Now $T - T_p$ satisfies the conditions of the theorem with $\nu = p-1$, so by induction there exist $T_1, T_2, \cdots, T_{p-1} \in \mathcal{Z}_m(A,B) \cap \mathcal{R}_m$ such that:

$$T_i = T_i \llcorner (A \sim B)$$

$$T - T_p = \sum_{i=1}^{p-1} T_i \tag{9}$$

$$\|T - T_p\| = \sum_{i=1}^{p-1} \|T_i\| \tag{10}$$

$$T_i - T_j \in \mathcal{B}_m(A,B) \quad \text{for} \quad 1 \le i,j \le p-1\,. \tag{11}$$

By (9) and (11):

$$(T - T_p) - (p-1)T_1 \in \mathcal{B}_m(A,B)\,.$$

Substituting with (8):

$$(p-1)T_p - (p-1)T_1 \in \mathcal{B}_m(A,B)$$

$$\therefore (p-1)(T_p - T_1) \in \mathcal{B}_m(A,B)\,.$$

But $H_m(A,B)$ has no elements of finite order (since A is orientable).

$$\therefore T_p - T_1 \in \mathcal{B}_m(A,B)\,.$$

We have proved (1), (2), (3), and (5). Thus we need now only prove (4), $\|T\| = \Sigma_{i=1}^{p} \|T_i\|$. By (10) it suffices to prove:

$$\|T\| = \|T_p\| + \|T - T_p\| . \tag{12}$$

Now (12) is purely local. If B is a ball, U its interior, ∂B its boundary, and if $U \cap B = \phi$, then everything we have said remains true when we replace the pair (A, B) by the pair $(A, A \sim U)$ and we replace T, R, T_i by $T \llcorner U$, $R \llcorner U$, and $T_i \llcorner U$, respectively.

But then we may as well assume $A = R^{m+1}$, $B = R^{m+1} \sim U^{m+1}(0,1)$. In this case, since R is an $(m+1)$-dimensional integral current in $U^{m+1}(0,1)$ and since T and T_p are rectifiable, we know that for $\mathcal{H}^m$ almost all $x \in U^{m+1}(0,1)$ there is an oriented hyperplane Γ_x such that:

$$\mathrm{TAN}(R, x) = E^{m+1} \llcorner \tau$$

(where $\tau = i$ on one side of Γ_x, $\tau = i{+}j$ on the other side, and $-\frac{1}{2}p < i \leq i{+}j \leq \frac{1}{2}p$. E^{m+1} is the Euclidean current on R^{m+1}).

$$\mathrm{TAN}(\partial R, x) = j\Gamma_x$$

$$\mathrm{TAN}(T_p, x) = t\Gamma_x \text{ for some integer } t .$$

And by (7):

$$\mathrm{TAN}(T, x) = (pt + j)\,\Gamma_x .$$

Therefore:

$$\mathrm{TAN}(T - T_p, x) = ((p-1)t + j)\,\Gamma_x .$$

Thus:

$$\Theta^m(\|T_p\|, x) = |t|$$

$$\Theta^m(\|T\|, x) = |pt + j|$$

$$\Theta^m(\|T - T_p\|, x) = |(p-1)t + j| .$$

Since $|j| < p$, the numbers t, $pt+j$, and $(p-1)t+j$ all have the same sign. Hence:

$$\Theta^m(\|T\|, x) = \Theta^m(\|T_p\|, x) + \Theta^m(\|T-T_p\|, x)$$

for $\mathcal{H}^m$ almost all x. This proves (12), completing the induction argument.

One readily sees from (3) and (4) that:

$$<\Phi, T> = \sum_{i=1}^{\nu} <\Phi, T_i> .$$

The last sentence of the theorem follows immediately.

COROLLARY 1 (Global Boundary Regularity). *Let* T *be an absolutely area-minimizing* m*-dimensional integral current in* R^{m+1} *such that* $\partial T = \nu S$, *where* S *is an embedded, connected* $C^{k,\alpha}$ *manifold (for some* $k \geq 1$, $0 < \alpha < 1$ *) with multiplicity* 1. *Then* $T = \Sigma_{i=1}^{\nu} T_i$, *where each* T_i *is absolutely area-minimizing with* $\partial T_i = S$. *Furthermore,* $\operatorname{spt} \partial S$ *has a neighborhood* U *such that* $\operatorname{spt} T_i \cap U$ *is a* $C^{k,\alpha}$ *manifold with boundary for all* $1 \leq i \leq \nu$.

Proof. Let $A = R^{m+1}$, $B = \operatorname{spt} S$. Then the hypotheses of the Decomposition Theorem hold. Since $T_i - T_j \in \mathcal{B}_m(A, B)$, there is an $R_{ij} \in \mathcal{R}_{m+1}(R^{m+1})$ such that:

$$\operatorname{spt}(T_i - T_j - \partial R_{ij}) \subset B = \operatorname{spt} S .$$

But then $T_i - T_j - \partial R_{ij}$ is an m-dimensional current supported in an (m–1)-dimensional manifold, which implies:

$$T_i - T_j - \partial R_{ij} = 0$$

$$\therefore \partial T_i = \partial T_j .$$

Since $\nu S = \partial T = \Sigma_{i=1}^{\nu} \partial T_i = \nu \partial T_i$ for all i, $\partial T_i = S$.

The rest of the theorem is an immediate consequence of the Hardt-Simon boundary regularity theorem.

COROLLARY 2 (Local Boundary Regularity). *Let* T *be an* m*-dimensional area-minimizing integral current in* R^{m+1}. *Let* $x \in \operatorname{spt} \partial T$ *have a neighborhood in which* ∂T *is a* $C^{k,\alpha}$ *manifold* (*for some* $k \geq 1$, $0 < \alpha < 1$) *with multiplicity* ν. *Then* x *has a neighborhood* U *such that either*:

(i) $T \llcorner U = \Sigma_{i=1}^{\nu} T_i$, *where each* $\operatorname{spt} T_i \cap U$ *is a* $C^{k,\alpha}$ *manifold with boundary* $\operatorname{spt} \partial T \cap U$. *Also,* T_i *and* T_j *are not tangent at any point of* $\operatorname{spt} \partial T \cap U$ *unless* $T_i = T_j$, *or*

(ii) $\operatorname{spt} T \cap U$ *is a real analytic manifold* (*indeed a minimal submanifold*) *containing* $\operatorname{spt} \partial T \cap U$, *such that* T *has some multiplicity* p *on one side of* $\operatorname{spt} \partial T \cap U$ *and multiplicity* $\nu + p$ *on the other side.*

Proof. Let U be an open ball around x such that $\partial T \llcorner U = \nu S$ where S is a $C^{k,\alpha}$ manifold with multiplicity 1. Let $A = R^{m+1}$ and $B = (R^{m+1} \sim U) \cup \operatorname{spt} S$. Then:

$$T \llcorner U \in \mathcal{Z}_m(A, B) .$$

Furthermore:

$$T \llcorner U \in \mathcal{Z}_m^{\nu}(R^{m+1}, R^{m+1} \sim U)$$

$$\therefore T \llcorner U \in \mathcal{B}_m^{\nu}(R^{m+1}, R^{m+1} \sim U)$$

(since $H_m^{\nu}(R^{m+1}, R^{m+1} \sim U) = 0$).

$$\therefore T \llcorner U \in \mathcal{B}_m^{\nu}(A, B) .$$

As in the proof of Corollary 1, it follows from the Decomposition Theorem that $T = \Sigma_{i=1}^{\nu} T_i$ where each T_i is area-minimizing and $\partial T_i \llcorner U = S \llcorner U$ for all i.

By the Hardt-Simon boundary regularity theorem, we can choose U small enough that each $\operatorname{spt} T_i \cap U$ is either a manifold with boundary $\operatorname{spt} S \cap U$ or a manifold containing $\operatorname{spt} S \cap U$. It is fairly easy to see that the tangent cone to T at x must be as described by (i) or (ii). Furthermore, by the Hopf Boundary Point Lemma [HS, 10.1], T is (in a small enough neighborhood of x) diffeomorphic to its tangent cone.

COROLLARY 3. *Let* A *be an orientable* (m+1)*-dimensional submanifold of a Euclidean space,* $B \subset A$ *be a closed local Lipschitz neighborhood retract, and* Φ *be a parametric integrand of degree* m. *Let* $\alpha \in H_m(A, B; \mathbf{Z})$. *Then*

$$\inf_{T \in k\alpha} \langle \Phi, T \rangle = k \inf_{T \in \alpha} \langle \Phi, T \rangle .$$

Proof. Immediate.

REMARK. Note that every homology class $\alpha \in H_m(A, B; \mathbf{Z})$ with integer coefficients is contained in some $\alpha' \in H_m(A, B; \mathbf{R})$. Federer proves in [F2, 5.8] that:

$$\inf_{T \in \alpha'} \langle \Phi, T \rangle = \lim_{k \to \infty} k^{-1} \inf_{T \in k\alpha} \langle \Phi, T \rangle .$$

Hence (as observed there) Corollary 3 implies:

$$\inf_{T \in \alpha'} \langle \Phi, T \rangle = \inf_{T \in \alpha} \langle \Phi, T \rangle .$$

Sharpness of results

L. C. Young has constructed a smooth curve S in $\mathbf{R}^4$ such that:

$$\min \{\mathbf{M}(T) : \partial T = 2S\} < 2 \min \{\mathbf{M}(T) : \partial T = S\} .$$

([Y]; F. J. Almgren, Jr. has described to me a much simpler proof.) Thus the assumption of codimension 1 is necessary for Corollaries 1 and 3 and hence for the decomposition theorem. Holomorphic varieties in $\mathbf{R}^4 (= \mathbf{C}^2)$ give examples of area-minimizing surfaces with smooth boundaries (of multiplicity 1) for which boundary regularity fails: hence codimension 1 is also necessary for Corollary 2.

That A be assumed orientable is necessary for the decomposition theorem and for Corollary 3, since they fail when, say, A is a projective plane and $B = \phi$.

BRIAN WHITE
COURANT INSTITUTE
NEW YORK UNIVERSITY
WASHINGTON SQUARE
NEW YORK, NEW YORK 10003

REFERENCES

[F1] H. Federer, *Geometric Measure Theory*, Springer-Verlag, Berlin-Heidelberg-New York, 1969.

[F2] ———, *Real Flat Chains, Cochains and Variational Problems*, Indiana Univ. Math. J. 24(1974), 351-407.

[HS] R. Hardt and L. Simon, *Boundary regularity and embedded solutions for the oriented plateau problem*, Ann. of Math. 110(1979), 439-486.

[W] B. White, *The Structure of Minimizing Hypersurfaces Mod 4*, Inventiones Math. 53(1979), 45-58.

[Y] L.C. Young, *Some extremal questions for simplicial complexes: V. The relative area of a Klein bottle*, U.S. Army Mathematics Research Center Report #356, Madison, Wisconsin, 1962.

New Methods in the Study of Free Boundary Problems

David Kinderlehrer

In this note we shall consider briefly some new methods in the study of free boundary problems with their applications to minimal surfaces. Weierstrass gave an early example of a free boundary problem in this context when he observed that a minimal surface partially bounded by a line segment could be analytically reflected across it. Similarly, one of the first uses of the Schwarz reflection principle was to extend a conformal mapping whose image contains an analytic arc.

Here we wish to discuss the analysis of the liquid edge by means of certain hodograph and Legendre transforms. This will permit the application of a well-developed regularity theory for elliptic systems to deduce the appropriate smoothness of the free boundary with the caveat that some initial smoothness must be assumed beforehand. We also consider an extremal eigenvalue problem which involves the mean curvature of the boundary of the domain.

Returning for the moment to our historical observations, a substantial stimulus was imparted to the study of free boundary problems for minimal surfaces when Hans Lewy gave profound extensions of the example of Weierstrass and the use of the Schwarz reflection principle mentioned above. He showed that a minimal surface whose boundary contains an analytic arc is extensible (analytically) across that arc [13]. He also

Seminar on Minimal Submanifolds
0-691-08324-X/83/303-19 $1.45/0 (cloth)
0-691-08319-3/83/303-19 $1.45/0 (paperback)

proved that a minimal surface which meets an analytic manifold transversally has an analytic trace on the manifold [14]. It is worthwhile observing that these problems may be considered from the viewpoint we present here once some initial smoothness is demonstrated. We discuss this in a concluding section.

This note is primarily an exposition of [10], which is a joint work with Louis Nirenberg and Joel Spruck. In a second paper, [11], we consider equations and systems of higher order. For the definition of the notion of an elliptic system with coercive boundary conditions we refer to any of [1] Part II, [10], [12] Chapter VI, or [16] Chapter VI.

1. *The liquid edge*

Consider a configuration of three minimal hypersurfaces in R^{n+1} which meet along a smooth $n-1$ dimensional surface γ, the liquid edge, at known angles, say $2\pi/3$. With the origin of a coordinate system on this surface, let us suppose that we may represent the three surfaces as graphs over the tangent space to one of them at the origin and that Γ is the projection of γ onto this tangent space. Two of the height functions, u^1 and u^2, will be defined on one side, Ω^+, of Γ while the third, u^3, will be defined on the other side Ω^-.

Analytically we obtain the system

$$\begin{aligned} Mu^1 &= Mu^2 = 0 \quad \text{in } \Omega^+ \\ &\ Mu^3 = 0 \quad \text{in } \Omega^- \\ u^1 &= u^2 = u^3 \\ \frac{\nabla u^j \cdot \nabla u^3 + 1}{\sqrt{1+|\nabla u^j|^2}\sqrt{1+|\nabla u^3|^2}} &= \frac{1}{2} \quad \text{on } \Gamma,\ j = 1,2, \\ \nabla u^1 &\neq \nabla u^2 \end{aligned}$$

where

$$\nabla u = (u_{x_1}, \cdots, u_{x_n})$$

and

$$Mu = \left(\delta_{ij} - \frac{u_{x_i} u_{x_j}}{1 + |\nabla u|^2}\right) u_{x_i x_j}$$

is the minimal surface operator written in a form convenient for us. Here the summation convention is understood and the indices i, j range from 1 to n.

Although the original problem concerns surfaces meeting at equal angles, there is no greater difficulty in considering a more general situation. Assume that $\mu_1(x, u)$ and $\mu_2(x, u)$ are analytic functions of $(x, u) \in R^{n+1}$, $|\mu_j| < \pi/2$, $j = 1, 2$, and that u^1 and u^2 meet u^3 at the angles $\mu_1(x, u^1)$ and $\mu_2(x, u^2)$ on Γ. In analytical terms we consider

$$\begin{aligned} Mu^1 &= Mu^2 = 0 \quad \text{in } \Omega^+ \\ & \quad\;\; Mu^3 = 0 \quad \text{in } \Omega^- \\ (*) \qquad u^1 &= u^2 = u^3 \end{aligned}$$

$$\frac{\nabla u^j \nabla u^3 + 1}{\sqrt{1 + |\nabla u^j|^2}\,\sqrt{1 + |\nabla u^3|^2}} = \cos \mu_j < 1 \quad \text{on } \Gamma, \; j = 1, 2,$$

$$\nabla u^3(0) = 0 .$$

THEOREM 1. *Suppose that* Γ *is of class* $C^{1,\alpha}$, $u^j \in C^{1,\alpha}(\Omega^+ \cup \Gamma)$, $j = 1, 2$, *and* $u^3 \in C^{1,\alpha}(\Omega^- \cup \Gamma)$, *for some* α, $0 < \alpha \leq 1$, *satisfy* (*). *If* u^1 *and* u^2 *do not meet at zero angle at* $x = 0$, *that is, if* $\mu_1(0) \neq \mu_2(0)$, *then* Γ *is analytic near* $x = 0$.

We offer several remarks about this theorem.

(1) For $n = 2$, the hypotheses may be verified when the surfaces whose graphs are the u^j, $j = 1, 2, 3$, are the solutions of a certain variational problem. This is the well-known theorem of Jean Taylor [19] and it motivated our consideration of (*).

(2) Also for $n = 2$, J.C.C. Nitsche [18] proved that Γ is of class C^∞ provided $-\mu_1 = \mu_2 = 2\pi/3$.

(3) If $u^2 \geqq u^1$ in Ω^+ and $\mu_1 = \mu_2$ at some (x_0, u_0), $x_0 \epsilon \Gamma$, then $u^1 = u^2$ in Ω^+. This follows from a variation of the Hopf boundary point lemma which is due to R. Finn and D. Gilbarg. Also, if μ_1 and μ_2 are constants and $\mu_1 = \mu_2$ on Γ then $u^1 = u^2$ in Ω^+ by unique continuation ([2]). So a corollary of the theorem is the statement:

> *If three distinct minimal surfaces in* R^{n+1} *meet at constant angles on the* $(n-1)$ *dimensional surface* γ, *then* γ *is analytic.*

Although in the statement of our theorem we have constrained the range of angles μ_1, μ_2 this was only for convenience of exposition. In fact the conclusion holds in general, even when all three surfaces project onto Ω^+ near $x = 0$.

(4) However two minimal surfaces may meet at constant non-zero angle on a surface γ which need not be analytic ([10], p. 115).

(5) Questions related to surfaces of assigned mean curvature may also be considered with these methods.

2. *Hodograph transformations*

Our object is to represent the problem (*) in a known smooth domain at the expense of treating a highly coupled system. This entails the use of what we have termed hodograph transformations ([9], [10], [11]). All our considerations are local.

Let $\Omega \subset R^n$ be a domain and suppose that $w(x) \epsilon C^1(\overline{\Omega})$ satisfies $w_{x_n}(x_0) > 0$ for some $x_0 \epsilon \overline{\Omega}$. Then the transformation

$$y = y(x) = (x', w(x)),\ x \epsilon B_\varepsilon(x_0) \cap \overline{\Omega}\,, \tag{2.1}$$

$x' = (x_1, \cdots, x_{n-1})$ is 1:1 for some $\varepsilon > 0$. Suppose that $\Gamma \subset \partial\Omega$ is some open subset of $\partial\Omega$, $x_0 \epsilon \Gamma$, and $w = 0$ on Γ. Then

$$U = y(\Omega \cap B_\varepsilon(x_0)) \subset \{y_n > 0\}$$

and

$$\Sigma = y(\Gamma \cap B_\varepsilon(x_0)) \subset \{y_n = 0\} .$$

In the sequel we omit the $B_\varepsilon(x_0)$ since we are considering a local question. Now define

$$\psi(y) = x_n . \tag{2.2}$$

The inverse mapping of (2.1) is given by

$$g(y) = (y', \psi(y)) \qquad y \in U \cup \Sigma , \tag{2.3}$$

and

$$\Gamma : x_n = \psi(x', 0) , \qquad (x', 0) \in \Sigma . \tag{2.4}$$

In this way the smoothness of Γ becomes a question about the smoothness of $\psi(y)$ in $U \cup \Sigma$.

If w satisfies an equation in Ω, we would like to know that ψ satisfies one in U. Now the property of our transforms (2.1), (2.2) is that

$$dx_n = d\psi = \sum_1^{n-1} \psi_\sigma dy_\sigma + \psi_n dy_n$$

$$= \sum_1^{n-1} \psi_\sigma dx_\sigma + \psi_n dw$$

so

$$dw = -\sum_1^{n-1} \frac{\psi_\sigma}{\psi_n} dx_\sigma + \frac{1}{\psi_n} dx_n . \tag{2.5}$$

(Subscripts with respect to σ denote differentiation with respect to the y-variables.) Second derivatives of w are easily calculated in terms of ψ, so, for example,

$$\begin{aligned}\Delta w &= -\frac{\psi_{nn}}{\psi_n^3} + \sum_{\sigma<n}\left\{-\left(\frac{\psi_\sigma}{\psi_n}\right)_\sigma + \frac{\psi_\sigma}{\psi_n}\left(\frac{\psi_\sigma}{\psi_n}\right)_n\right\} \\ &= F(D^2\psi, D\psi)\,,\end{aligned} \tag{2.6}$$

which is a second order (quasilinear) elliptic equation for ψ. To pursue this example further, suppose we are given

$$\begin{aligned} w \in C^2(\Omega) \cap C^1(\bar{\Omega}) : \Delta w &= \phi(x,w) \quad \text{in} \quad \Omega \\ w_\nu &= f(x) \quad \text{on} \quad \Gamma \\ w &= 0 \end{aligned} \tag{2.7}$$

where ν denotes the exterior normal to Γ. This leads immediately to the problem

$$\begin{aligned} F(D^2\psi, D\psi) &= \phi(y', \psi, y_n) \quad \text{in} \quad U \\ \psi_n &= \frac{1}{f(y',\psi)}\sqrt{1+\Sigma\psi_\sigma^2} \quad \text{on} \quad \Sigma \end{aligned}, \tag{2.8}$$

which is a nonlinear elliptic equation with a nonlinear Neumann boundary condition.

The transformation (2.1), or its inverse (2.3), will not be sufficient for us because it cannot accommodate information from both Ω^+ and Ω^-. So given $w(x) \in C^1(\Omega^+ \cup \Gamma)$ we define

$$y = (x', w(x)) \qquad x \in \Omega^+ \tag{2.1'}$$

$$\psi(y) = x_n \tag{2.2'}$$

as before and

$$\begin{aligned} g^+(y) &= (y', \psi(y)),\ y \in U \cup \Sigma, \ \text{the inverse of (2.1)}, \\ g^-(y) &= (y', \psi(y) - Cy_n),\ y \in U \cup \Sigma,\ C > |\sup \psi_n|\,. \end{aligned} \tag{2.9}$$

The mapping $g^-(y)$ is a generalized reflection since our condition on C implies that

$$g^-(U) \subset \Omega^-\,.$$

Note also that

$$g^+(y', 0) = g^-(y', 0) \in \Gamma, \ (y', 0) \in \Sigma . \tag{2.10}$$

Why not review our conventions about the theory of nonlinear elliptic equations with a rehearsal on the example (2.8)? We must show that every bounded exponential solution of the linearized principal part of (2.8) vanishes identically.

Suppose that f and ϕ are analytic functions of their arguments near $(x, w) = (0, 0)$, and that

$$f(0) < 0 .$$

Assume that $\nu(0) = -e_n = (0, \cdots, 0, -1)$. Then $w_{x_n}(0) = a > 0$ and $w_{x_\sigma}(0) = 0$, $\sigma = 1, \cdots, n-1$, for $a = -f(0) > 0$. Now at $y = 0$,

$$\begin{aligned} L\bar{\psi} &= \frac{d}{d\varepsilon} F(D^2\psi(0) + \varepsilon D^2\bar{\psi}, D\psi(0) + \varepsilon D\bar{\psi})\Big|_{\varepsilon=0} \\ &= -a^3\bar{\psi}_{nn} - a \sum_{\sigma<n} \bar{\psi}_{\sigma\sigma} + \text{lower order terms,} \end{aligned}$$

which is obviously elliptic. The operator

$$L'\bar{\psi} = -a\left(a^2\bar{\psi}_{nn} + \sum_{\sigma<n} \bar{\psi}_{\sigma\sigma}\right) \tag{2.11}$$

is the principal part or principal symbol of L. Also

$$\begin{aligned} B\bar{\psi} &= \frac{d}{d\varepsilon}\left\{\psi_n(0) + \varepsilon\bar{\psi}_n - \frac{1}{f(y', \psi(0) + \varepsilon\bar{\psi})} \sqrt{1 + \Sigma(\psi_\sigma(0) + \varepsilon\bar{\psi}_\sigma)^2}\right\}\Bigg|_{\varepsilon=0} \\ &= \bar{\psi}_n + \text{lower order terms} \end{aligned}$$

so

$$B'\bar{\psi} = \bar{\psi}_n \tag{2.12}$$

is the principal part of B.

Thus we must check that any solution $\bar{\psi}$ of the problem

$$L'\bar{\psi} = 0 \quad \text{in} \quad y_n > 0$$

$$B'\bar{\psi} = 0 \quad \text{on} \quad y_n = 0$$

which is bounded for $y_n > 0$ and has the form

$$\bar{\psi}(y', t) = \zeta(t)e^{iy'\xi'}, \ \xi' \neq 0, (t = y_n)$$

vanishes identically. Since in this case $\zeta(t) = Me^{-\frac{|\xi'|}{a}t}$, $t > 0$ by (2.11), from (2.12) we conclude that $\bar{\psi} \equiv 0$. According to [16], especially Theorems 6.8.2 and 6.8.4, and [1] $\psi(y)$ is real analytic in $U \cup \Sigma$. Thus Γ is an analytic curve (cf. (2.4)).

3. *The hodograph method applied to the liquid edge*

We use the ideas of the previous section to analyze (*). Suppose that $0 \in \Gamma$ and that the normal to Γ at $x = 0$ is parallel to the x_n-axis. Set

$$(3.1) \qquad a_j = \frac{\partial}{\partial x_n} u^j(0), \ j = 1, 2,$$

which are not equal by hypothesis. Suppose that $a_2 > a_1$ so that $u^2 > u^1$ in Ω^+ near $x = 0$. Let $\mu_j = \mu_j(0,0)$, $j = 1,2$, so $\mu_2 - \mu_1 \neq 0$. For simplicity of exposition we suppose that $\mu_j \neq 0$, $j = 1,2$. Note that $\cos \mu_j = (1 + a_j^2)^{-1/2}$. Now define

$$(3.2) \qquad w(x) = u^2(x) - u^1(x), \ x \in \Omega^+,$$

and determine $y(x)$, and $\psi(y)$ by (2.1′), (2.2′). Introduce the functions

$$(3.3) \qquad \begin{aligned} \phi^+(y) &= u^1(x) = u^1(g^+(y)) \\ \phi^-(y) &= u^3(x) = u^3(g^-(y)) \end{aligned}, \quad y \in U,$$

according to (2.9).

Our object is to show that the functions ψ, ϕ^+, ϕ^- are the solution of an elliptic system with coercive boundary conditions on $\Sigma \subset \{y_n = 0\}$. We determine this system and the boundary conditions. With $1 \leqq \sigma \leqq n-1$ and obvious notations, one checks that

$$(3.4)\qquad \begin{aligned}
\nabla w &= \left(-\frac{\psi_\sigma}{\psi_n}, \frac{1}{\psi_n}\right) && (x \in \Omega^+)\\
\nabla u^1 &= \left(\phi_\sigma^+ - \frac{\psi_\sigma}{\psi_n}\phi_n^+, \frac{1}{\psi_n}\phi_n^+\right) && (x \in \Omega^+)\\
\nabla u^2 &= \left(\phi_\sigma^+ - \frac{\psi_\sigma}{\psi_n}(\phi_n^+ + 1), \frac{1}{\psi_n}(\phi_n^+ + 1)\right) && (x \in \Omega^+)\\
\nabla u^3 &= \left(\phi_\sigma^- - \frac{\phi_\sigma}{\psi_n - C}\phi_n^-, \frac{1}{\psi_n - C}\phi_n^-\right) && (x \in \Omega^-)\,.
\end{aligned}$$

The second derivatives of the u^j may be calculated easily observing that

$$(3.5)\qquad \begin{aligned}
\frac{\partial}{\partial x_\sigma} &= \partial_\sigma - \frac{\psi_\sigma}{\psi_n}\partial_n, \quad \frac{\partial}{\partial x_n} = \frac{1}{\psi_n}\partial_n && (x \in \Omega^+)\\
\frac{\partial}{\partial x_\sigma} &= \partial_\sigma - \frac{\psi_\sigma}{\psi_n - C}\partial_n, \quad \frac{\partial}{\partial x_n} = \frac{1}{\psi_n - C}\partial_n && (x \in \Omega^-)
\end{aligned}$$

$\partial_i = \partial/\partial y_i$ and $1 \leqq \sigma \leqq n-1$.

Those of u^1, u^2 involve derivatives up to second order in ψ, ϕ^+ while those of u^3 involve derivatives up to second order in ψ, ϕ^-. Consequently, the equations of (*) give rise to a system

$$(3.6)\qquad \begin{aligned}
F_1(D^2\psi, D\psi, D^2\phi^+, D\phi^+) &= Mu^1 = 0\\
F_2(D^2\psi, D\psi, D^2\phi^+, D\phi^+) &= Mu^2 = 0 \quad \text{in } U\,.\\
F_3(D^2\psi, D\psi, D^2\phi^-, D\phi^-) &= Mu^3 = 0
\end{aligned}$$

Shortly we shall verify that this system is elliptic, but first we consider the intriguing boundary conditions of (*). These we write in the form

$$\Phi_1(D\psi, D\phi^+, D\phi^-) = \nabla u^1 \nabla u^3 + 1 - \cos\mu_1 \sqrt{1+|\nabla u^1|^2}\sqrt{1+|\nabla u^3|^2} = 0\,,$$

$$\Phi_2(D\psi, D\phi^+, D\phi^-) = \nabla u^2 \nabla u^3 + 1 - \cos\mu_2 \sqrt{1+|\nabla u^2|^2}\sqrt{1+|\nabla u^3|^2} = 0\,, \tag{3.7}$$

$$\Phi_3(\phi^+, \phi^-) = \phi^+ - \phi^- = u^1 - u^3 = 0 \text{ on } \Sigma.$$

We illustrate the linearization of these equations by considering Φ_1. So we must calculate

$$B_{11}(0,D)\overline{\psi} + B_{12}(0,D)\overline{\phi^+} + B_{13}(0,D)\overline{\phi^-} =$$

$$\frac{d}{d\varepsilon}\Phi_1(D\psi(0)+\varepsilon D\overline{\psi}, D\phi^+(0)+\varepsilon D\overline{\phi^+}, D\phi^-(0)+\varepsilon D\overline{\phi^-}) = 0\,.$$

(We have neglected the zero order terms which arise through the angle $\mu_1(x,u) = \mu_1(y', \psi, \phi^+)$ because they will be omitted when we consider the principal part.) Denoting the variation of ∇u^j by $\widetilde{\nabla u^j}$, i.e.,

$$\widetilde{\nabla u^1} = \left(\varepsilon\overline{\phi_\sigma^+}, \frac{\phi_n^+(0) + \varepsilon\overline{\phi_n^+}}{\psi_n(0) + \varepsilon\overline{\psi}_n}\right), \text{ etc.,}$$

in the first term of the right-hand side we compute

$$\frac{d}{d\varepsilon}(\widetilde{\nabla u^1}\widetilde{\nabla u^3})\Big|_{\varepsilon=0} = \frac{d}{d\varepsilon}\left\{\frac{\phi_n^+(0) + \varepsilon\overline{\phi_n^+}}{\psi_n(0) + \varepsilon\overline{\psi}_n}\,\frac{\phi_n^-(0) + \varepsilon\overline{\phi_n^-}}{\psi_n(0) + \varepsilon\overline{\psi}_n - C} + \cdots\right\}$$

$$= \left(\left(\frac{d}{d\varepsilon}\widetilde{\nabla u^1}\right)\widetilde{\nabla u^3} + \widetilde{\nabla u^1}\,\frac{d}{d\varepsilon}\widetilde{\nabla u^3}\right)\Big|_{\varepsilon=0}$$

$$= \nabla u^1(0)\,\frac{d}{d\varepsilon}\widetilde{\nabla u^3}\Big|_{\varepsilon=0}$$

$$= \nabla u^1(0)\left(\frac{d}{d\varepsilon}\widetilde{u^3_{x_\sigma}}, \frac{d}{d\varepsilon}\widetilde{u^3_{x_n}}\right)\Big|_{\varepsilon=0}$$

$$= a_1\frac{d}{d\varepsilon}\left(\frac{\phi_n^-(0) + \varepsilon\overline{\phi_n^-}}{\psi_n(0) + \varepsilon\overline{\psi}_n - C}\right)\Big|_{\varepsilon=0}$$

$$= a_1\,\frac{\overline{\phi_n^-}}{\psi_n(0) - C}\,,$$

using here that $\nabla u^3(0) = 0$ and $\nabla u^1(0) = a_1 e_n$. Note that $\psi_n(0) = w_{x_n}(0)^{-1} = (a_2 - a_1)^{-1}$. Analyzing the remaining terms in the same fashion, namely, by the chain rule, we obtain the linearized boundary conditions at $y = 0$

$$(\cos \mu_1)^2 (\overline{\phi_n^{\,+}} - a_1 \overline{\psi}_n) + \frac{1}{C(a_2 - a_1) - 1} \overline{\phi_n^{\,-}} = 0$$

(3.8) $$(\cos \mu_2)^2 (\overline{\phi_n^{\,+}} - a_2 \overline{\psi}_n) + \frac{1}{C(a_2 - a_1) - 1} \overline{\phi_n^{\,-}} = 0 \quad \text{for } y_n = 0$$

$$\frac{a_2}{a_2 - a_1} (\overline{\phi^+} - a_1 \overline{\psi}) - \frac{a_1}{a_2 - a_1} (\overline{\phi^+} - a_2 \overline{\psi}) - \overline{\phi^-} = 0 .$$

In the same manner we may write a set of linearized equations in $y_n > 0$ from (3.6). These may be determined without difficulty and they assume the convenient form

$$\sum_{\sigma < n} (\overline{\phi^+} - a_1 \overline{\psi})_{\sigma\sigma} + (a_2 - a_1)^2 (\cos \mu_1)^2 (\overline{\phi^+} - a_1 \overline{\psi})_{nn} = 0$$

(3.9) $$\sum_{\sigma < n} (\overline{\phi^+} - a_2 \overline{\psi})_{\sigma\sigma} + (a_2 - a_1)^2 (\cos \mu_2)^2 (\overline{\phi^+} - a_2 \overline{\psi})_{nn} = 0 \quad y_n > 0$$

$$\sum_{\sigma < n} \overline{\phi^-_{\sigma\sigma}} + \left(\frac{a_2 - a_1}{1 - C(a_2 - a_1)} \right)^2 \overline{\phi^-_{nn}} = 0 .$$

This system is clearly elliptic for the functions $\overline{\phi^+} - a_1 \overline{\psi}$, $\overline{\phi^+} - a_2 \overline{\psi}$, $\overline{\phi^-}$.

To verify the coerciveness of our system (3.6), (3.7) we must examine bounded exponential solutions of (3.9) in $y_n = t > 0$ which fulfill (3.8). Given $0 \neq \xi' \in R^{n-1}$,

$$\overline{\phi^+} - a_1 \overline{\psi} = C_1 e^{iy'\xi'} e^{-\nu_1 t}$$

$$\overline{\phi^+} - a_2 \overline{\psi} = C_2 e^{iy'\xi'} e^{-\nu_2 t} \qquad t > 0$$

$$\overline{\phi^-} = C_3 e^{iy'\xi'} e^{-\nu_3 t}$$

where

$$\nu_1 : \nu_2 : \nu_3 = \frac{1}{\cos \mu_1} : \frac{1}{\cos \mu_2} : C(a_2 - a_1) - 1 .$$

Substituting this in (3.8) gives the condition for (C_1, C_2, C_3)

$$(3.10)\qquad \begin{pmatrix} \cos\mu_1 & 0 & 1 \\ 0 & \cos\mu_2 & 1 \\ \frac{a_2}{a_2-a_1} & \frac{a_1}{a_2-a_1} & -1 \end{pmatrix} \begin{pmatrix} C_1 \\ C_2 \\ C_3 \end{pmatrix} = \begin{pmatrix} 0 \\ 0 \\ 0 \end{pmatrix}.$$

The determinant δ of this matrix is

$$\begin{aligned} \delta &= -\cos\mu_1 \cos\mu_2 - \frac{a_2\cos\mu_2}{a_2-a_1} + \frac{a_1\cos\mu_1}{a_2-a_1} \\ &= -\cos\mu_1\cos\mu_2 - \frac{1}{(a_2-a_1)}(\sin\mu_2 - \sin\mu_1)\,, \end{aligned}$$

since $a_j = \tan\mu_j$. Now $-\frac{\pi}{2} < \mu_1 < \mu_2 < \frac{\pi}{2}$ implies that $\delta < 0$, so $C_1 = C_2 = C_3 = 0$ is the unique solution of (3.10). Thus ψ, ϕ^+, ϕ^- are analytic in $U \cup \Sigma$ and, recalling (2.4), Γ is also.

In [10], [11] illustrations of this method are given to problems involving higher order contact, in particular the variational inequality of two linear membranes. However the analysis of the free boundary in case of higher order contact for surfaces of assigned mean curvature is incomplete. The existence theorem in this case is due to G. Vergara-Caffarelli [20].

Another free boundary problem which involves minimal surfaces is the obstacle problem of variational inequalities. A treatment of this problem may be found in [12]. The reader is also referred to the papers [3], [8], [15].

4. *An extremal eigenvalue problem*

P. Garabedian and M. Schiffer [4] considered an extremal problem for domains involving the second eigenvalue of $-\Delta$ and the perimeter of the domain. The corresponding problem in R^n leads to a domain Ω and a function $u \in C^1(\bar{\Omega})$ satisfying

$$
\begin{aligned}
&\Delta u + \lambda_2 u = 0 && \text{in } \Omega \\
&u = 0 && \text{on } \partial\Omega \qquad (4.1)\\
&\left(\frac{\partial u}{\partial \nu}\right)^2 = \frac{2\lambda_2}{L} H
\end{aligned}
$$

where $L = \text{meas}_{n-1}(\partial\Omega)$ and H is the mean curvature of $\partial\Omega$.

THEOREM 2. *Let* $\Gamma = \partial\Omega$ *be an* $(n-1)$ *dimensional manifold of* R^n *and suppose that* $u \in C^2(\Omega \cup \Gamma)$ *satisfies (4.1). If* $H > 0$, *then* Γ *is analytic.*

Our interest in (4.1) is that both the boundary condition and the equation obtained after transformation by (2.1) are of second order.

As usual, let $0 \in \Gamma$ and let the positive x_n-axis be normal to Γ and point inwards at $x = 0$. Introduce

$$y = (x', u(x)) \quad \text{and} \quad \psi(y) = x_n$$

by (2.1), (2.2) and suppose that $u_{x_n}(0) = 1$. One easily checks that

$$
\begin{aligned}
&H = \frac{1}{(n-1)(1+|\nabla'\psi|^2)^{3/2}} \sum_{\sigma,\tau} ((1+|\nabla'\psi|^2)\delta_{\sigma\tau} - \psi_\sigma\psi_\tau)\psi_{\sigma\tau}, \\
&|\nabla'\psi|^2 = \sum_{\sigma<n} \psi_\sigma^2, \qquad (4.2)
\end{aligned}
$$

summation over $\sigma, \tau = 1, \cdots, n-1$. Using (2.6), (2.8) we see that

$$
\begin{aligned}
&F(D^2\psi, D\psi) + \lambda y_n = 0 && \text{in } U \subset \{y_n > 0\} \\
&\frac{1}{\psi_n^2}(1+|\nabla'\psi|^2) = \frac{C}{n-1} H && \text{on } S \subset \{y_n = 0\} \qquad (4.3)
\end{aligned}
$$

where $C = 2\lambda_2/L$ and F is defined in (2.6). The linearized problem associated to (4.3) is simply

$$\Delta\bar{\psi} = 0 \quad \text{for} \quad y_n > 0$$

$$\sum_{\sigma<n} \bar{\psi}_{\sigma\sigma} = 0 \quad \text{on} \quad y_n = 0 .$$

For a solution $\bar{\psi}(y', t) = e^{i\xi' y'}\zeta(t)$, $0 \neq \xi' \in R^{n-1}$, we obtain

$$\zeta'' - |\xi'|^2\zeta = 0 \quad \text{for} \quad t > 0 \quad \text{and}$$

$$\zeta(0) = 0$$

$$\zeta(\infty) \ \text{bounded.}$$

Thus $\zeta(t) \equiv 0$. Hence Γ is analytic.

5. *An observation about the prescribed and transversal boundary problems for minimal surfaces*

Let S be a minimal surface in R^N which is partially bounded by an analytic arc Γ, say,

$$\Gamma : u^j = \phi_j(u^1), \ |u^1| < 1 , \ j = 2, \cdots, N$$

where $\phi_j(t)$ are analytic functions for $|t| < 1$, $j = 2, \cdots, N$. Let

$$u(z) = (u^1(z), \cdots, u^N(z)), \ z = x_1 + ix_2 \in G ,$$

$G = \{|z| < 1, \operatorname{Im} z > 0\}$ be a conformal representation of a portion of S in such a way that Γ is the monotone image of $-1 < x_1 < 1$. Assuming that $u^j \in C^{1,\alpha}(\bar{G})$, the isothermal relations

$$\left.\begin{aligned} &\sum_1^N \frac{\partial u^j}{\partial x_1}\frac{\partial u^j}{\partial x_2} = 0 \\ &\sum_1^N \left(\frac{\partial u^j}{\partial x_1}\right)^2 = \sum_1^N \left(\frac{\partial u^j}{\partial x_2}\right)^2 \end{aligned}\right\} \quad \text{in } \bar{G}$$

are valid. These may be rewritten on $-1 < x_1 < 1$, $x_2 = 0$ as

$$\frac{\partial u^1}{\partial x_1}\frac{\partial u^1}{\partial x_2} + \sum_2^N \phi_j'(u^1)\frac{\partial u^j}{\partial x_2}\frac{\partial u^1}{\partial x_1} = 0$$

(5.1) $-1 < x_1 < 1\,,\ x_2 = 0\,.$

$$\left(\frac{\partial u^1}{\partial x_1}\right)^2\left(1 + \sum_2^N \phi_j'(u^1)^2\right) = \sum_1^N \left(\frac{\partial u^j}{\partial x_2}\right)^2$$

Observe that $\partial u^1/\partial x_1 \neq 0$ a.e., $-1 < x_1 < 1$, $x_2 = 0$. For the functions

$$f_j(z) = \frac{\partial u^j}{\partial x_1} - i\,\frac{\partial u^j}{\partial x_2}\,,\quad j = 1, \cdots, N\,,$$

are analytic in G, continuous in $\overline{G}$, and hence

$$0 \neq \sum_1^N |f_j'(z)|^2 = \sum_1^N \left(\left(\frac{\partial u^j}{\partial x_1}\right)^2 + \left(\frac{\partial u^j}{\partial x_2}\right)^2\right)$$

$$= 2\left(1 + \sum_2^N \phi_j'(u^1)^2\right)\left(\frac{\partial u^1}{\partial x_1}\right)^2 \quad \text{a.e., } z = x_1\,,\ -1 < x_1 < 1$$

by Fatou's well-known theorem and (5.1) above. Thus dividing by $\partial u^1/\partial x_1$ in the orthogonality relation of (5.1) we obtain the condition

$$\frac{\partial u^1}{\partial x_2} + \sum_2^N \phi_j'(u^1)\frac{\partial u^j}{\partial x_2} = 0,\quad -1 < x_1 < 1\,,\ x_2 = 0\,.$$

Our surface $u(z) = (u^1(z), \cdots, u^N(z))$ is, therefore, a solution of the system

$$\Delta u^j = 0 \qquad \textit{in } G\,,\ j = 1, \cdots, N\,,$$

(5.2) $$\frac{\partial u^1}{\partial x_2} + \sum_2^N \phi_k'(u^1)\frac{\partial u^k}{\partial x_2} = 0 \qquad \textit{on } -1 < x_1 < 1\,,\ x_2 = 0\,,$$

$$\phi_j(u^1) - u^j = 0 \qquad j = 2, \cdots, N\,.$$

To linearize this system we may assume that $u(0) = 0 \in \Gamma$ and that $\phi'_j(0) = 0$, $j = 2, \cdots, N$. The linearized system (of highest order terms) for (5.2) is just

$$\begin{aligned} &\Delta \bar{u}^j = 0 && \text{in } x_2 > 0,\ 1 \leqq j \leqq N, \\ &\frac{\partial \bar{u}^1}{\partial x_2} = 0 && \\ & && \text{on } x_2 = 0,\ j = 2, \cdots, N, \\ &\bar{u}^j = 0 && \end{aligned} \tag{5.3}$$

which is obviously coercive. Thus $u \in C^{1,\alpha}(\bar{G})$ implies that u is analytic in $G \cup \{-1 < x_1 < 1,\ x_2 = 0\}$.

There are many different proofs that $\Gamma \in C^{m,\alpha}$ implies that $u \in C^{m,\alpha}(\bar{G})$, all many years after Lewy's result of course. We cite [5], [7], [17], [21] as examples.

Given once again a minimal surface S in R^N in conformal representation

$$u(z) = (u^1(z), \cdots, u^N(z)), \quad z \in G,$$

let us suppose that the image L of $-1 < x_1 < 1$, $x_2 = 0$ lies on an analytic $N-1$ manifold (eg)

$$M : \Phi(u^1, \cdots, u^N) = 0, \quad \text{grad } \Phi(u^1, \cdots, u^N) \neq 0.$$

Now assume that the surface S is transversal to M on L, which means that the tangent plane to S at each point of L is perpendicular to any plane in the tangent space of M at that point. Assuming that $u \in C^{1,\alpha}(\bar{G})$ we know that $\partial u/\partial x_1 (x_1)$ lies in the tangent space of M at any point of $L \subset \bar{S} \cap M$, thus the transversality condition may be written

$$\frac{\partial u}{\partial x_2}(x_1) \cdot T = 0 \quad \text{for any} \quad T \in T_{u(x_1)}(M), \tag{5.4}$$

the tangent space of M at $u(x_1)$. Assume that $u(0) = 0 \in M$ and that

$$\text{grad } \Phi(0, \cdots, 0) = e_N,$$

so

$$\Phi_\mu(0, \cdots, 0) = 0 \quad \text{and} \quad \Phi_N(0, \cdots, 0) = 1 , \quad \mu = 1, \cdots, N-1 ,$$

where $\Phi_j(u^1, \cdots, u^N) = \partial\Phi/_{\partial u^j}(u^1, \cdots, u^N)$.

The surface $u(z) = (u^1(z), \cdots, u^N(z))$, $z \in G$, is a solution of

$$\Delta u^j = 0, \qquad in\ G, \ j = 1, \cdots, N ,$$

(5.5)
$$-\Phi_N \frac{\partial u^j}{\partial x_2} + \Phi_j \frac{\partial u^N}{\partial x_2} = 0 \qquad on\ -1 < x_1 < 1, \ x_2 = 0 ,$$

$$\Phi(u^1, \cdots, u^N) = 0 \qquad j = 1, \cdots, N-1 .$$

from (5.4). This gives rise to the linearized system

$$\Delta \bar{u}^j = 0 \qquad \text{for } x_2 > 0, \ j = 1, \cdots, N ,$$

(5.6)
$$-\frac{\partial \bar{u}^j}{\partial x_2} = 0 \qquad \text{on } x_2 = 0, \ j = 1, \cdots, N=1$$

$$\bar{u}^N = 0$$

which is clearly coercive. Thus again we are able to conclude that if $u \in C^{1,\alpha}(\bar{G})$, then u is analytic in $G \cup \{-1 < x_1 < 1, \ x_2 = 0\}$.

We note that (5.4) has a variational formulation, namely, let $F = (F^1, \cdots, F^N)$ be a vector field in R^N which is tangent to M near L with small support and let

$$\zeta^j(z) = F^j(u(z)), \quad j = 1, \cdots, N .$$

Then

$$\int_G \text{grad } u^j \text{ grad } \zeta^j dx = 0 .$$

This property is used to establish the initial regularity of $u(z)$, cf. (Jager [6], Hildebrandt & Jager [5′]).

DAVID KINDERLEHRER
SCHOOL OF MATHEMATICS
UNIVERSITY OF MINNESOTA
MINNEAPOLIS, MINNESOTA 55455

REFERENCES

[1] S. Agmon, A. Douglis, L. Nirenberg, *Estimates near the boundary for solutions of elliptic partial differential equations satisfying general boundary conditions I, II*, Comm. Pure Appl. Math. *12*(1959), 623-727, *17*(1964), 35-92.

[2] N. Aronszajn, *A unique continuation theorem for solutions of elliptic partial differential equations*, J. Math. Pures Appl. *36*(1957), 235-249.

[3] L. Caffarelli, *Compactness methods in free boundary problems*, Comm. in PDE, *5*(1980), 427-448.

[4] P. Garabedian and M. Schiffer, *Variational problems in the theory of elliptic partial differential equations*, J. Rat. Mech. Anal. *2*(1953), 137-171.

[5] S. Hildebrandt, *Boundary behavior of minimal surfaces*, Arch. Rat. Mech. and Anal., *35*(1969), 47-82.

[5′] Hildebrandt, S. and Jager, W., *On the regularity of surfaces with prescribed mean curvature at a free boundary*, Math Z. *118*(1970), 289-308.

[6] W. Jager, *Behavior of minimal surfaces with free boundaries*, Comm. Pure Appl. Math *23*(1970), 803-818.

[7] D. Kinderlehrer, *The boundary regularity of minimal surfaces*, Ann. Sc. N.S. Pisa *23*(1969), 711-744.

[8] ————, *How a minimal surface leaves an obstacle*, Acta. Math. *130*(1973), 221-242.

[9] D. Kinderlehrer, L. Nirenberg, *Regularity in free boundary problems*, Ann. Sc. N.S. Pisa *4*(1977), 373-391.

[10] D. Kinderlehrer, L. Nirenberg, J. Spruck, *Regularity in elliptic free boundary problems I*, J. d'Analyse Math. *34*(1978), 86-119.

[11] ————, *Regularity in elliptic free boundary problems II*, Ann. S.N.S. Pisa *6*(1979), 637-683.

[12] D. Kinderlehrer, G. Stampacchia, Variational Inequalities, Academic Press (New York) (1980).

[13] H. Lewy, *On the boundary behavior of minimal surfaces*, Proc. Nat. Acad. Sci. U.S.A. *37*(1951), 103-110.

[14] ————, *On minimal surfaces with partially free boundary*, Comm. Pure Appl. Math *4*(1952), 1-13.

[15] H. Lewy, G. Stampacchia, *On existence and smoothness of solutions of some noncoercive variational inequalities*, Arch. Rat. Mech. Anal. *41*(1971), 241-253.

[16] C.B. Morrey, Jr., Multiple Integrals in the Calculus of Variations, Springer Verlag (New York, Berlin) (1966).

[17] J.C.C. Nitsche, *The boundary behavior of minimal surfaces*, Inventiones Math. *8*(1969), 313-333.

[18] ———, *The higher regularity of liquid edges in aggregates of minimal surfaces*, Nachr. Akad. Wiss. Göttingen (1977), 75-95.

[19] J. Taylor, *The structure of singularities in soap-bubble-like and soap-film-like minimal surfaces*, Ann. of Math. *103*(1976), 489-539.

[20] G. Vergara-Caffarelli, *Variational inequalities for two surfaces of constant mean curvature*, Arch. Rat. Mech. Anal. *56*(1974), 334-347.

[21] S. Warschawski, *Boundary derivatives of minimal surfaces*, Arch. Rat. Mech. Anal. *38*(1970), 241-256.

SOME PROPERTIES OF CAPILLARY FREE SURFACES

Robert Finn

We outline here briefly some results on properties of equilibrium capillary surface interfaces. The characteristic physical configuration consists of two adjacent fluids A and B in a region bounded at least in part by rigid walls Z (Figure 1). The principle of virtual work leads to a

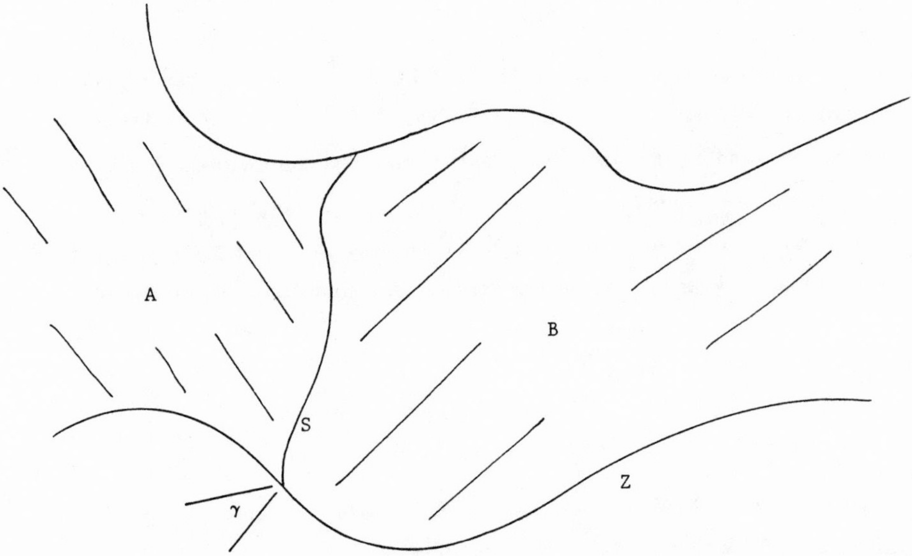

Figure 1

Seminar on Minimal Submanifolds
0-691-08324-X/83/323-16 $1.30/0 (cloth)
0-691-08319-3/83/323-16 $1.30/0 (paperback)

relation of the form

$$H = f(x) + \lambda \tag{1}$$

for the mean curvature H of the surface, the known function $f(x)$ of position x in space being determined by the relative energy densities and physical properties of the fluids, and λ being a Lagrange parameter arising from a volume constraint (if there is no constraint, then $\lambda = 0$). At all points p in the manifold Γ of contact of the surface interface S with the boundary Z, the angle γ between S and Z satisfies

$$\cos \gamma = g(p)$$

where $g(p)$ depends only on the physical properties of the three materials meeting at p. (Thus, if all materials are homogeneous, then $g(p) \equiv$ constant.)

Typical configurations are provided by the familiar rise (or fall) of liquid in (and outside of) a thin glass tube immersed vertically into a reservoir, and by the liquid drop resting on (or pendant from) a horizontal surface. In all these cases the relevant internal energy densities are determined by a gravitational field; in the case S is a graph $u(x)$ over a base plane Ω orthogonal to the gravity field, one finds the equations

$$\begin{aligned} \operatorname{div} Tu &= \kappa u + \lambda \\ Tu &= \frac{1}{\sqrt{1+|Du|^2}}\, Du \end{aligned} \tag{2}$$

where κ is a physical constant, and D denotes the gradient operator. On $\Sigma = \partial\Omega$ we find, for the vertical tube,

$$\nu \cdot Tu = \cos \gamma \tag{3}$$

with ν = unit exterior normal. For the drop resting on a plane (sessile drop) (3) is replaced by

$$\nu \cdot Tu = -\sin \gamma \tag{4}$$

and for the pendent drop we find

(5) $$\nu \cdot Tu = \sin \gamma \, .$$

The constant κ is positive or negative according as the heavier fluid lies below or above the free surface.

Most of the considerations that follow apply without essential change to any situation of co-dimension one, however, for simplicity of exposition only the (physical) case of two-dimensional surfaces in 3-space will be described.

1. *The symmetric case,* $\kappa > 0$

The equations (2) appear first in Laplace [1], who also integrated them approximately in the case Ω is a disk of small radius a, to obtain his celebrated formula

(6) $$u_0 \sim \frac{2 \cos \gamma}{\kappa a} - \frac{a}{\cos \gamma}\left(1 - \frac{2}{3}\,\frac{1 - \sin^3 \gamma}{\cos^2 \gamma}\right) \equiv \mathfrak{L}[a; \gamma]$$

for the height u_0 of S on the axis of symmetry.

The estimate (6) is in fact asymptotically exact as $a \to 0$; this was proved by D. Siegel [2]. An alternative proof, to appear in [3], shows that *the Laplace expression provides a strict lower bound for* u_0,

(7) $$\mathfrak{L}[a; \gamma] < u_0$$

for every radius a.

The proof follows almost immediately from the observation that a lower hemisphere, passing through $(0, u_0)$ and meeting the cylinder walls Z in the given angle γ, bounds with Z and the base plane a larger volume then does S. The volume under S is explicitly obtained by integration of (2) as

(8) $$V = \frac{2\pi}{\kappa} a \cos \gamma \, .$$

The volume under the hemisphere can also be calculated explicitly, and the inequality between these volumes yields (7).

An upper bound for u_0 is obtained by analogous considerations. According to (2) the mean curvature of S increases with height. Thus a lower hemisphere through $(0, u_0)$, having at that point the same mean curvature as S, lies everywhere below S and hence bounds a smaller volume. In this case one finds

$$\text{(9)} \qquad \frac{a^2 u_0}{2} + \frac{a^2}{\kappa u_0} - \frac{8}{3(\kappa u_0)^3}\left[1 - \left(1 - \frac{\kappa^2 a^2 u_0^2}{8}\right)^{3/2}\right] - \frac{a}{\kappa}\cos\gamma < 0\,.$$

The left side of (9) is monotone increasing in u_0, thus *the actual* u_0 *is bounded above by the unique value* u_0^+ *for which equality holds.*

Both the upper and the lower bound are asymptotically equal, hence both are asymptotically exact as $a \to 0$. The estimates are however not accurate for values $\kappa a^2 \gg 0.1$. For such configurations, bounds were obtained in [5, 4, 2] by a direct manipulation of the equations. In terms of the inclination angle ψ of a vertical section of S, we find the system

$$\text{(10)} \qquad \frac{dr}{d\psi} = \frac{r\cos\psi}{\kappa r u - \sin\psi} \qquad \frac{du}{d\psi} = \frac{r\sin\psi}{\kappa r u - \sin\psi}\,.$$

Monotonicity properties of S permit us to write (10) in various ways as a pair of differential inequalities that can be integrated explicitly. One is lead (cf. [4]) to particular estimates that apply in varying situations. We mention here the result

$$\text{(11)} \quad u_0 > \frac{2\cos\gamma}{\kappa a}\,\frac{1+\sqrt{1+a^2}}{1+\sin\gamma+\sqrt{2(1+\sin\gamma)}}\,\exp\left\{2\cos\left(\frac{\pi}{4}-\frac{\gamma}{2}\right) - \sqrt{1+a^2}-1\right\}$$

which applies for all radii a and improves (7) for large a. In the case $\gamma = 0$, (11) is better or worse than (7) according as $\kappa a^2 > 2.889$ or $\kappa a^2 < 2.889$.

The bound from above $(u_0 < u_0^+)$ has been improved by F. Brulois, in a work that is now in preparation. In its present form, Brulois's procedure is based on the analyticity of the solutions.

In his treatise [1], Laplace obtained also an estimate for the height change q of S from the axis to the cylinder wall; his formula is however very complicated. Formal manipulation of (10) yields the simpler and stronger estimate

$$\left(\frac{p+1}{p}\right)^{1/2}\left\{\frac{2}{\kappa}(1-\sin\gamma)+\frac{p+1}{4}u_0^2\right\}^{1/2}-u_0 < q < \frac{\cos\gamma}{\kappa a} + \left\{\frac{2}{\kappa}(1-\sin\gamma)-\frac{\cos^2\gamma}{\kappa^2 a^2}+\frac{u_0^2}{2}\right\}^{1/2}-u_0, \tag{12}$$

with

$$p=\left\{1+\frac{2\kappa a^2}{1+\sin\gamma}\right\}^{1/2}. \tag{13}$$

Both left and right sides of (12) are monotonic in u_0, thus the preceding results yield the desired bounds for q.

A particular consequence of (12) is that $u(r)$ is bounded uniformly throughout its interval $0 \le r < a$ of definition as solution of (2).

The relations (12) hold for all a, and are asymptotically exact in both limiting cases $a \to 0$ and $a \to \infty$. A calculation in [4] for a case with $\kappa a^2 \sim 750$ yielded an accuracy of about 0.6%.

The above results apply also to the problem of the sessile drop, see [4].

2. *The general case,* $\kappa > 0$

The symmetric estimates can be used in conjunction with comparison principles, to obtain information on the behavior of solution surfaces in cylindrical tubes of general section Ω. We indicate here two comparison principles and then, below, some of their consequences.

CP1: *Let* u, v *be solutions of (2) in* Ω, *with* $\kappa > 0$, *and suppose* $\nu\cdot Tu \ge \nu\cdot Tv$ *on* $\Sigma = \partial\Omega$. *Then* $u \ge v$ *in* Ω, *and equality holds at any* $p \in \Omega$ *if and only if it holds at all* $p \in \Omega$.

CP2: *Let* $\Omega^{(1)}$ *be a ball and let* $\Omega^{(2)} \subset \Omega^{(1)}$. *Let* $u^{(1)}, u^{(2)}$ *be solutions in* $\Omega^{(1)}, \Omega^{(2)}$, *with* $\gamma^{(1)} \equiv \text{const.}$, $\nu \cdot Tu^{(2)}]_{\Sigma^{(2)}} \geq \cos \gamma^{(1)}$. *Then* $u^{(2)} > u^{(1)}$ *in* $\Omega^{(2)}$.

In both CP1 and CP2 the data need be prescribed only up to a set of Hausdorff measure zero on Σ, at which points ν need not be defined. This property is crucial to what follows. We point out that CP2 is in general false if $\Omega^{(1)}$ is not a ball, cf. [6, 5] where differing conditions are given. See also Siegel [2], who considered the case $\Omega^{(2)}$ is a ball.

In what follows we write $v(r;\delta;\gamma)$ for the symmetric solution in a ball B_δ of radius δ, corresponding to data γ on ∂B_δ.

DEFINITION. Ω is said to satisfy a *spherical solution condition* (δ,γ) if $\forall x \in \Omega\ \exists B_\delta \ni x$ such that at all points of $\Sigma \cap B_\delta$ (up to a set of Hausdorff measure zero) the surface $v(r;\delta;0)$ meets Σ in an angle not exceeding γ. We write $\Omega \in SSC(\delta;\gamma)$.

If Ω is a bounded $C^{(1)}$ domain, then $\Omega \in SSC(\delta;\gamma)$ for some $\delta > 0$, for any γ in $0 < \gamma \leq \pi$. (See [7, p. 23] for a proof, of a similar property, that applies as well to this case.)

THEOREM. *Let* $\Omega \in SSC(\delta;\gamma)$, *let* $u(x)$ *satisfy (2.3). Then* $\forall x \in \Omega$, $u(x) \leq v(r;\delta;0)$ *corresponding to any* $B_\delta \ni x$ *with the indicated property.*

Proof. We write $\partial(\Omega \cap B_\delta) = \Sigma_\delta \cup \Sigma_\gamma$, where Σ_γ consists of points of Σ interior to B_δ. On Σ_γ there holds by hypothesis $\cos\gamma = \nu \cdot Tu \leq \nu \cdot Tv$ On Σ_δ there holds $\nu \cdot Tv \equiv 1$, $\nu \cdot Tu < 1$. Apply CP1.

An easy corollary is that *if* Σ *is of class* $C^{(2)}$, *then there is a uniform bound in magnitude for all solutions of (2) in* Ω.

For γ in any range $0 \leq \gamma < \gamma_0 < \pi/2$, a bound below (from zero) follows from CP2 whenever Ω is bounded.[1] This bound is independent of smoothness conditions on Σ.

---

[1] The method yields a bound below for any situation in which Ω omits an open set, see [5].

3. *The general case,* $\kappa > 0$ (continued)

The possibility to ignore sets of H-measure zero on Σ has striking consequences. One is a much stronger uniqueness theorem than is possible for, say, the Laplace equation. The simplest nontrivial example of an exceptional set occurs when two straight segments on Σ meet at a point V, forming an interior angle $2\alpha \neq \pi$. The impossibility of prescribing data (3) at V does not affect the CP's. We consider a solution u(x) defined in the region Δ of Figure 2, with boundary data γ at points of Σ distinct from V. For simplicity we suppose in what follows that $\gamma \equiv \text{const.}$, $0 \leq \gamma < \pi/2$.

THEOREM. *If* $\alpha + \gamma \geq \pi/2$, *then*

$$u(x) < \frac{2}{\kappa\delta} + \delta \tag{14}$$

in Δ. *If* $\alpha + \gamma < \pi/2$, *then there holds asymptotically, in terms of polar coordinates based on* V *and the angle bisector,*

$$u(x) \sim \frac{\cos\vartheta - \sqrt{k^2 - \sin^2\vartheta}}{k\kappa r} \tag{15}$$

with $k = \sin\alpha \sec\gamma$.

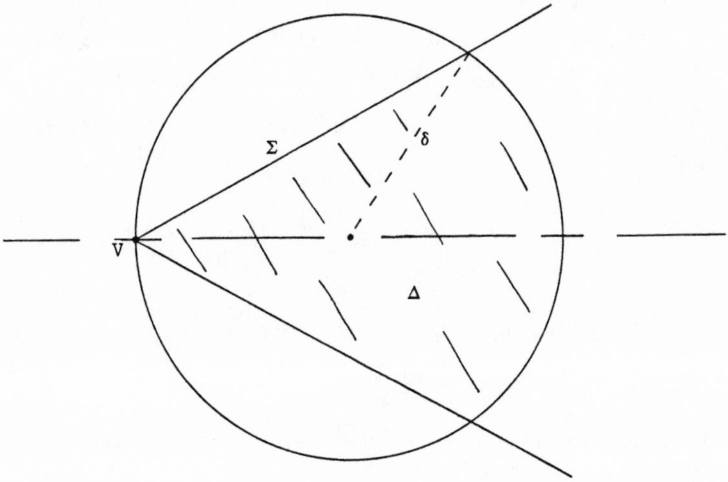

Figure 2

We note the bound (14) is uniform, independent of α or γ in the range considered, whereas in the second case (15), $u(x) \to \infty$ at V. Thus, *the solutions in Δ depend discontinuously on the boundary data.*

The proof of the theorem follows from an extended form of the comparison principles. Details appear in [8].

The behavior of solutions in Δ shaped domains was studied further by Korevaar [9] and by Simon [10]. Simon proved that if $\alpha+\gamma > \pi/2$, $2\alpha < \pi$, then $u(x)$ is differentiable up to V. In entirely independent work, Korevaar showed that if $2\alpha > \pi$, then there exist solutions that are discontinuous at V (all solutions in this case are, according to (14), bounded). Korevaar also obtained general conditions implying continuity of $u(x)$ up to Σ.

If $\kappa > 0$, then under fairly general conditions solutions can be shown to exist. Results under varying hypotheses can be found in [11-23, 7]. From the point of view of the preceding discussion, we note [7] that solutions will in general exist and be uniquely determined, also in the situation of a boundary corner[(2)] for which $\alpha+\gamma < \pi/2$.

4. *The case* $\kappa = 0$

If there is no gravitational field, or if the adjacent media have the same density, then $\kappa = 0$ in (2), and the solutions become surfaces of constant mean curvature. The following result suggests the qualitative change that can occur in behavior of the solution surfaces. Again we suppose $\gamma \equiv \text{const.}$, $0 \le \gamma < \pi/2$.

THEOREM. *If $\kappa = 0$ and $\alpha+\gamma < \pi/2$, then there is no solution of (2) in Δ (Figure 2) for which $\nu \cdot Tu = \cos\gamma$ on $\Sigma \cap B_\delta$.*

Even if Σ is smooth, solutions of (2, 3) need not exist if $\kappa = 0$. A general necessary condition is introduced in [24]; a varied condition was

(2) In the case considered, not only the solution, but also the associated surface energy, will become infinite at V. Under the present hypotheses, which impose no growth restriction at V, uniqueness would fail for the Laplace equation.

used by Giusti [20, 21], who showed it then to be sufficient. The condition as it appears in those references is however not entirely satisfactory, as it requires inspection of every subdomain. The following alternative condition appears in [25].

THEOREM. *If* $\kappa = 0$, *a solution of (2, 3) exists in* Ω *if and only if there is a vector field* $w(x)$ *in* $\bar{\Omega}$, *with* $\operatorname{div} w = \frac{|\Sigma|}{|\Omega|}$, $\nu \cdot w \equiv 1$ *on* Σ, *and* $|w| < \frac{1}{\cos \gamma}$ *in* Ω.

If Ω is a triangle or parallelogram (or parallelopiped), fields can be constructed explicitly, leading to the result that a solution exists if and only if the smallest angle satisfies $\alpha + \gamma \geq \pi/2$.

This result cannot be extended to general polygons. In fact [25], given $\varepsilon > 0$ and any γ in $0 \leq \gamma < \pi/2$, there is a trapezoid for which the smaller angle 2α satisfies $\frac{\pi}{2} - \alpha < \varepsilon$, and for which there exists no solution of (2, 3).

Chen [26] has shown that if a disk of radius $R = \frac{|\Omega|}{|\Sigma|}$ can be rolled around Σ inside Ω, then a solution exists for every $\gamma \geq 0$. The condition is however not necessary.

5. *The pendent drop*

Only the symmetric case has been studied. The drop is known to become unstable with increasing volume, cf. Pitts [27], Wente [28]. From a formal point of view, the equation can be transformed to the parametric system

$$\frac{d\psi}{ds} = -u - \frac{\sin \psi}{r}$$

$$\frac{du}{ds} = \sin \psi \tag{16}$$

$$\frac{dr}{ds} = \cos \psi$$

and the set of all (simply connected) symmetric solution surfaces can be shown to be completely determined by the initial value u_0 on the axis of symmetry. It is shown in [29] that *to each* u_0 *there is a unique solution curve, which has no limit sets or double points, and which extends to infinity as indicated in Figure 3a.* It is also known [30] that *there exists a singular solution in the form*

$$U(r) \sim -\frac{1}{r} + \frac{5}{2} r^3 - \cdots \tag{17}$$

near $r = 0$.

CONJECTURE 1. The solution (17) can be extended as a graph over r for all $r > 0$, and is the unique solution with an isolated singularity as a graph over r.

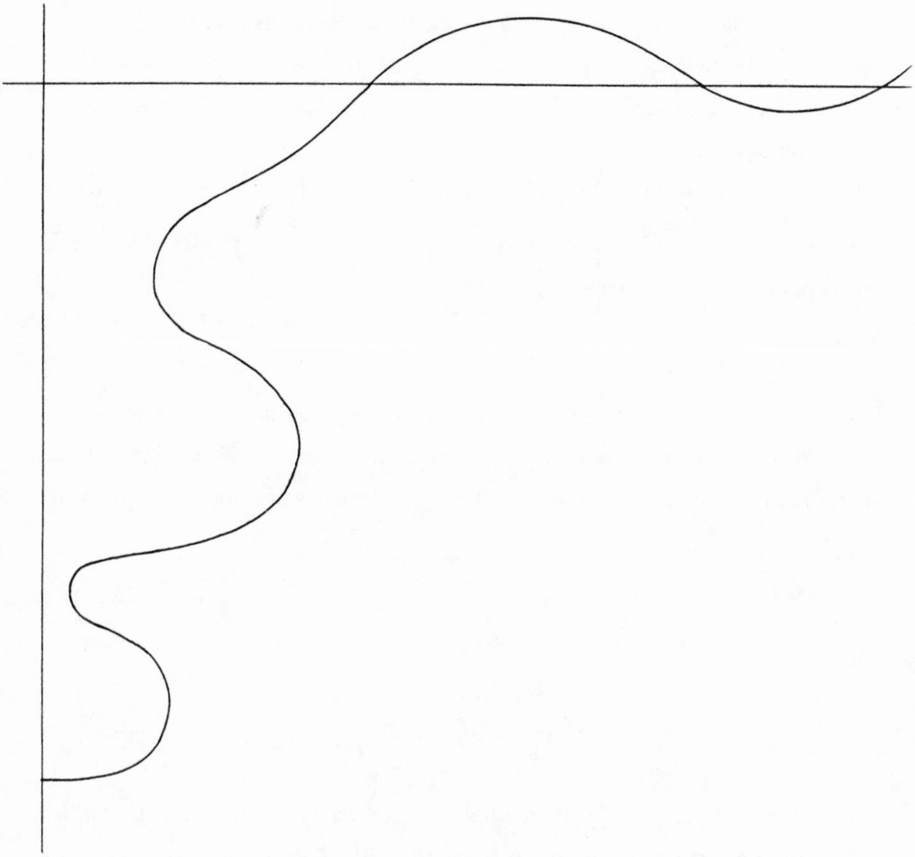

Figure 3a

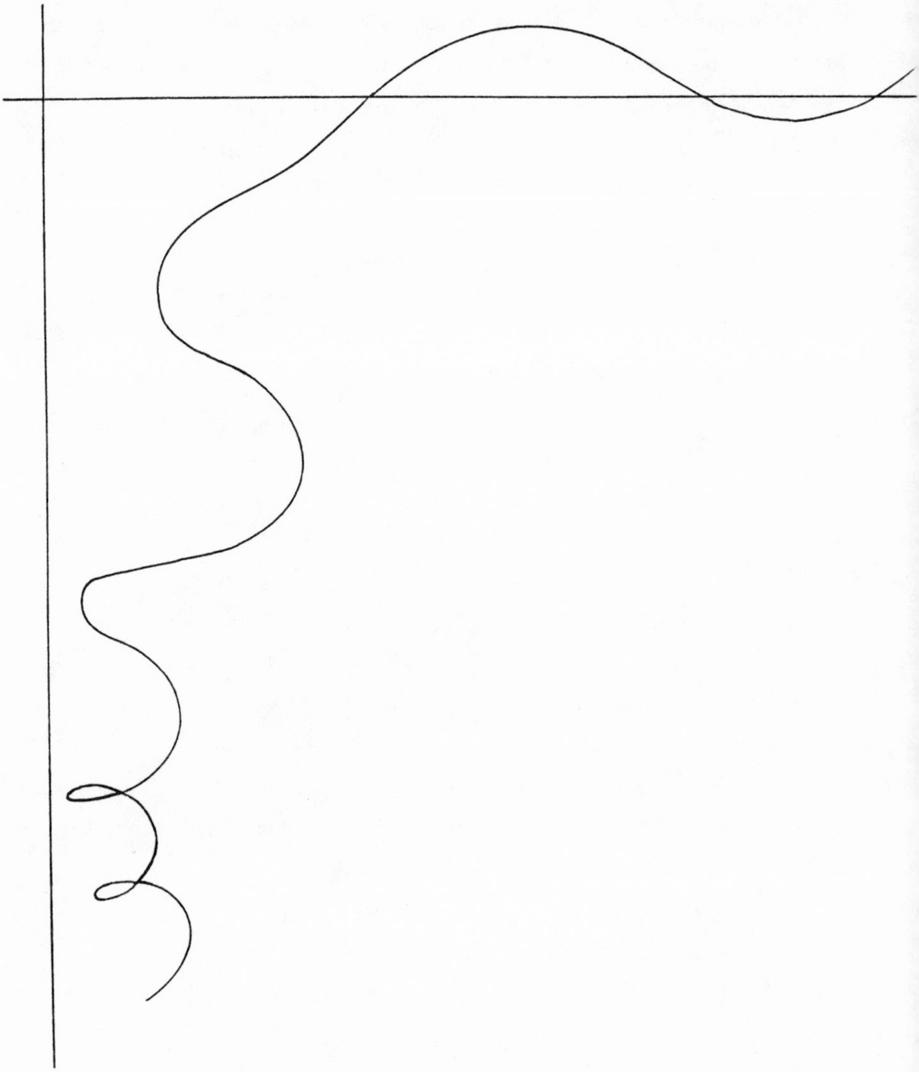

Figure 3b

CONJECTURE 2. On any fixed compact set, the solutions determined by u_0 can be expressed as graphs over r for all sufficiently large u_0, and converge uniformly to $U(r)$.

If a solution determined by u_0 is perturbed slightly, one obtains (in general) a complete curve with double points as in Figure 3b. Letting the perturbation go to zero, one finds an extension of the original curve, with a singularity on the u-axis (Figure 3c); in that limiting sense, the original curve can be regarded as a complete solution with an infinity of double points.

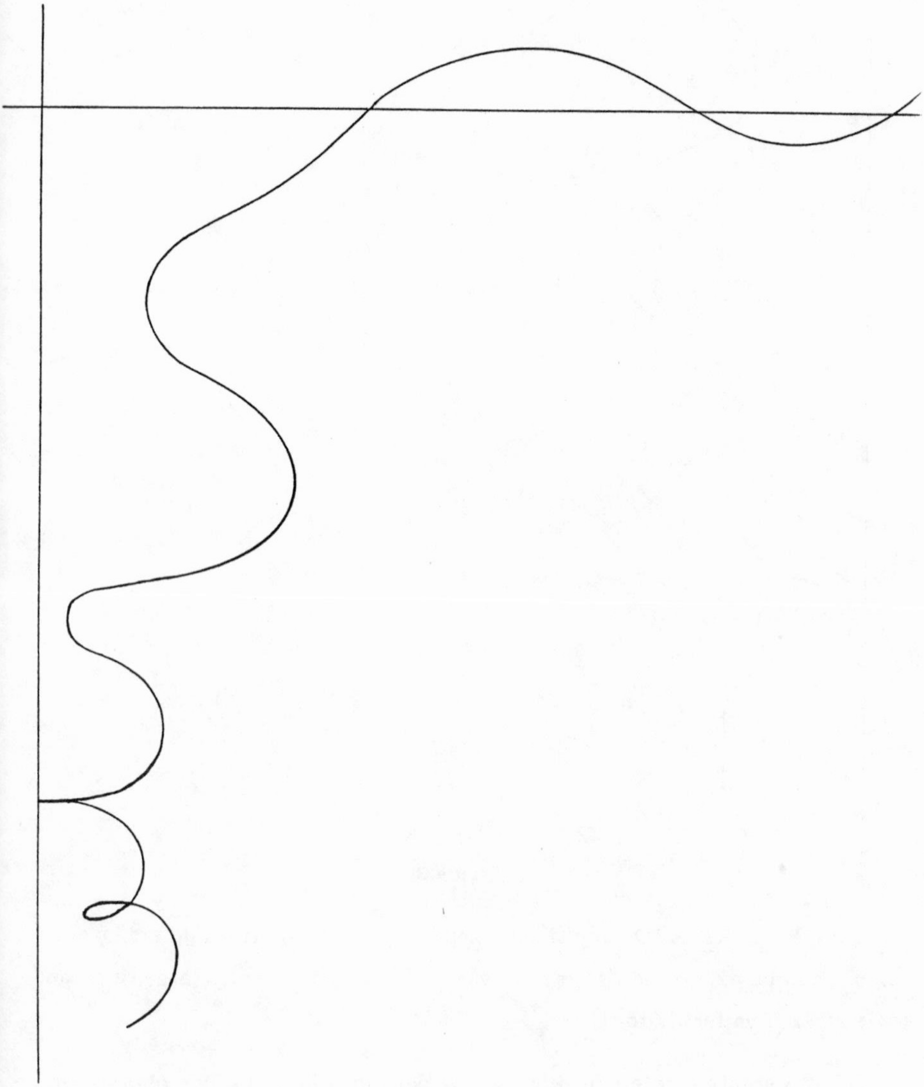

Figure 3c

CONJECTURE 3. In the extended sense just indicated, the singular solution is the unique complete solution of (16) in $r > 0$, without double points or limit sets.

Partial results toward proofs of these conjectures appear in [29, 31].

6. *A gradient estimate*

The following property of surfaces of constant mean curvature was proved [32] by methods in the spirit of the above material, and seems of general interest.

THEOREM. *Let* $u(x)$ *define a surface of constant mean curvature* $H \equiv 1$ *over a disk* $B_R : |x| < R$. *There exists a value* $R_0 = 0.5654062332\cdots$, *and a decreasing function* $C(R)$ *defined in* $R_0 \leq R \leq 1$, *with* $C(R_0) = \infty$, $C(1) = 0$, *such that*

$$|Du(0)| < C(R) . \tag{18}$$

The value R_0 cannot be improved.

A similar result holds in any number n of dimensions (Giusti [21]), but an explicit value for R_0 has been found only for $n = 2$.

7. *Final remarks*

This survey is in no sense complete, nor should it be considered that any specific results not described here were judged of less interest than those that are. Among the omissions we mention contributions by Wente, Gonzalez, Massari, Tamanini, Giusti [33-36] on sessile and pendent drops, also work of Perko [37], of Siegel [2], of Turkington [38], and of Vogel [39] on the wide capillary tube, and on capillary surfaces in exterior domains.

ROBERT FINN
DEPARTMENT OF MATHEMATICS
STANFORD UNIVERSITY
STANFORD, CALIFORNIA 92093

REFERENCES

[1] Laplace, P.S.: Traité de mécanique céleste, Tome quatrième, Supplément au livre X, "Sur l'action capillaire," Gauthier-Villars, Paris 1806. See also the annotated translation by N. Bowditch, vol. IV, Chelsea Publishing Co., Bronx, N. Y., 1966.

[2] Siegel, D.: Height estimates for capillary surfaces, Pacific J. Math., 88 (1980), 471-516.

[3] Finn, R.: On the Laplace formula and the meniscus height for a capillary surface, ZAMM, 61 (1981), 165-173; 175-177.

[4] ________: The sessile liquid drop I: symmetric case, Pacific J. Math., 88 (1980), 541-587.

[5] ________: Some comparison properties and bounds for capillary surfaces, Complex Analysis and its Applications (Russian) Moscow Mathem. Soc., volume dedicated to I. N. Vekua, "Scientific Press," Moscow, 1978.

[6] Concus, P. and R. Finn: On the height of a capillary surface, Math. Z. 147 (1976), 93-95.

[7] Finn, R. and C. Gerhardt: The internal sphere condition and the capillary problem, Ann. Mat. Pura Appl. 112 (1977), 13-31.

[8] Concus, P. and R. Finn: On capillary free surfaces in a gravitational field, Acta Math. 132 (1974), 207-223.

[9] Korevaar, N.: On the behavior of a capillary surface at a re-entrant corner, Pacific J. Math., 88 (1980), 379-386.

[10] Simon, L.: Regularity of capillary surfaces over domains with corners, Pacific J. Math., 88 (1980), 363-378.

[11] Johnson, W. E. and L. M. Perko: Interior and exterior boundary value problems from the theory of the capillary tube, Arch. Rational Mech. Anal. 29 (1968), 125-143.

[12] Emmer, M.: Esistenza, unicità e regolarità delle superfici di equilibrio nei capillari, Ann. Univ. Ferrara Sez. VII, 18 (1973), 79-94.

[13] Pepe, L.: Analiticità delle superfici di equilibrio nei capillari, Symposia Matematica, Academic Press (1974).

[14] Gerhardt, C.: Existence and regularity of capillary surfaces, Boll. Un. Mat. Ital. 10 (1974), 317-335.

[15] Ural'tseva, N. N.: Solution of the capillary problem, Vestnik Leningrad. Univ. 19 (1973), 54-64.

[16] Spruck, J.: On the existence of a capillary surface with a prescribed angle of contact, Comm. Pure Appl. Math. 28 (1975), 189-200.

[17] Simon, L. and J. Spruck: Existence and regularity of a capillary surface with a prescribed contact angle, Arch. Rational Mech. Anal. 61 (1976), 19-34.

[18] Gerhardt, C.: On the capillarity problem with constant volume, Ann. Scuola Norm. Sup. Pisa 2 (1975), 303-320.

[19] ________: Global regularity of the solutions to the capillary problem, Ann. Scuola Norm. Sup. Pisa 3 (1976), 157-175.

[20] Giusti, E.: Boundary value problems for non-parametric surfaces of prescribed mean curvature, Ann. Scuola Norm. Sup. Pisa 3(1976), 501-548.

[21] ________: On the equation of surfaces of prescribed mean curvature. Existence and uniqueness without boundary conditions, Invent. Math. 46(1978), 111-137.

[22] Gerhardt, C.: Boundary value problems for surfaces of prescribed mean curvature, J. Math. Pures Appl. 58(1979), 75-109.

[23] Giusti, E.: Generalized solutions to the mean curvature equation, Pacific J. Math., 88(1980), 297-322.

[24] Concus, P. and R. Finn: On capillary free surfaces in the absence of gravity, Acta Math. 132(1974), 177-198.

[25] Finn, R.: Existence and nonexistence of capillary surfaces, Manuscripta Math. 28(1979), 1-11.

[26] Chen, J.T.: On the existence of capillary free surfaces in the absence of gravity, Pacific J. Math., 88(1980), 323-362.

[27] Pitts, E.: The stability of pendent liquid drops II. The axially symmetric case, J. Fluid Mech. 63(1974), 487-508.

[28] Wente, H.C.: The stability of the axially symmetric pendent drop, Pacific J. Math., 88(1980), 421-470.

[29] Concus, P. and R. Finn: The shape of a pendent liquid drop, Philos. Trans. Roy. Soc. London Ser. A, 292(1979), 307-340.

[30] ________: A singular solution of the capillary equation, I: Existence, Invent. Math. 29(1975), 143-148.

[31] ________: A singular solution of the capillary equation, II: Uniqueness, Invent. Math. 29(1975), 149-160.

[32] Finn, R. and E. Giusti: On nonparametric surfaces of constant mean curvature, Ann. Scuola Norm. Sup. Pisa 4(1977), 13-31.

[33] Wente, H.C.: An existence theorem for surfaces in equilibrium satisfying a volume constraint, Arch. Rational Mech. Anal. 50(1973), 139-158.

[34] Gonzalez, E.H.A.: Sul problema della goccia appoggiata, Rend. Sem. Mat. Univ. Padova 55(1976), 289-302.

[35] Gonzalez, E.H.A., U. Massari and C. Tamanini: Existence and regularity for the problem of a pendent liquid drop, Pacific J. Math., 88(1980), 399-420.

[36] Giusti, E.: The pendent water drop. A direct approach, Boll. Un. Mat. Ital., 17-A(1980), 458-465.

[37] Perko, L.M.: Boundary layer analysis of the wide capillary tube, Arch. Rational Mech. Anal. 45(1972), 120-133.

[38] Turkington, B.: Height estimates for capillary type problems, Pacific J. Math., to appear.

[39] Vogel, T. I.: Symmetric and asymmetric liquid bridges, Doctoral Dissertation, Stanford University. See Indiana Math. J., 31 (1982), 281-288; Ann. Scuola Norm. Sup. Pisa, 9 (1982), 433-442; Pacific J. Math., 102 (1982) in press.

BERNSTEIN CONJECTURE IN HYPERBOLIC GEOMETRY

S. P. Wang and S. Walter Wei

§1. *Introduction*

The Bernstein conjecture in Euclidean space R^n states that any complete minimal graph in R^n is a hyperplane (or equivalently, any entire solution to the equation

$$\sum_{i=1}^{n-1} \frac{\partial}{\partial X_i}\left(\frac{\frac{\partial f}{\partial X_i}}{\sqrt{1+|\Delta f|^2}}\right) = 0$$

is linear). This problem has been completely settled by the efforts of a number of mathematicians. In fact, the conjecture is proven to be true in R^n for $n \leq 8$ by Bernstein [2], De Giorgi [4], Almgren [1], and Simons [11], and false for $n > 8$ by Bombieri, De Giorgi, and Giusti [3].

In recent years, hyperbolic geometry has become one of the most active areas in geometry. To investigate the intrinsic structures and global problems in hyperbolic geometry, it is natural to study the Bernstein conjecture in hyperbolic space that can be expressed in the following fashion:

Any complete, absolutely area-minimizing hypersurface in H^n is totally geodesic for $n > 2$.

The main purpose of this paper is to show the above conjecture to be false in all dimensions (cf. the picture on page 337). In fact, we have

Seminar on Minimal Submanifolds
0-691-08324-X/83/339-20 \$1.50/0 (cloth)
0-691-08319-3/83/339-20 \$1.50/0 (paperback)
For copying information, see copyright page.

constructed a continuum of complete, absolutely area-minimizing hypersurfaces which foliate H^n and only one of them is totally geodesic.

Moreover, by lifting or taking the product structures, we can obtain a continuum of complete, absolutely area-minimizing submanifolds (with arbitrary co-dimensions) which foliate $SO(n_1,1)\times\cdots\times SO(n_\ell,1)\times H^{m_1}\times\cdots\times H^{m_r}$ for appropriate $\ell, r \geq 0$ with $\ell^2+r^2 \neq 0$; $n_i, m_j > 2$ and only one of them is totally geodesic. In particular, the Bernstein conjecture is also false in these spaces and we obtain examples of complete absolutely area-minimizing "smooth" hypersurfaces in $H^n \times H^n$ contrasting with the one with an isolated singularity for $n > 3$ [15].

The technique we employ is to first decompose the hyperbolic space into the product of two hyperbolic spaces with the proper metrics by a technique of Tong and Wang [12]. We then utilize the inherent symmetries of these spaces by employing a non-compact group action. This reduces the complexity of high dimensional global problems to the simplicity of low dimensional corresponding problems in the orbit space in a manner similar to Hsiang and Lawson [7] and Wei [14]. But the method in [7] does not meet the needs of our problem in hyperbolic space because a non-compact group action is being performed. To overcome this difficulty, we bring in uniform discrete subgroups which extend the theorem of Hsiang and Lawson from compact group actions to non-compact group actions. Thus, we vastly simplify the global problem while allowing us to keep the delicate nature of compact group actions. We believe the method employed here can be carried over to a more general setting. The work then proceeds by our studying the geometric and analytic structure of the orbit space. We find a continuum of geodesics foliating the orbit space.

The examples are obtained by lifting this foliation to H^n and verified by constructing a globally defined closed form with comass 1 in H^n (cf. [12]).

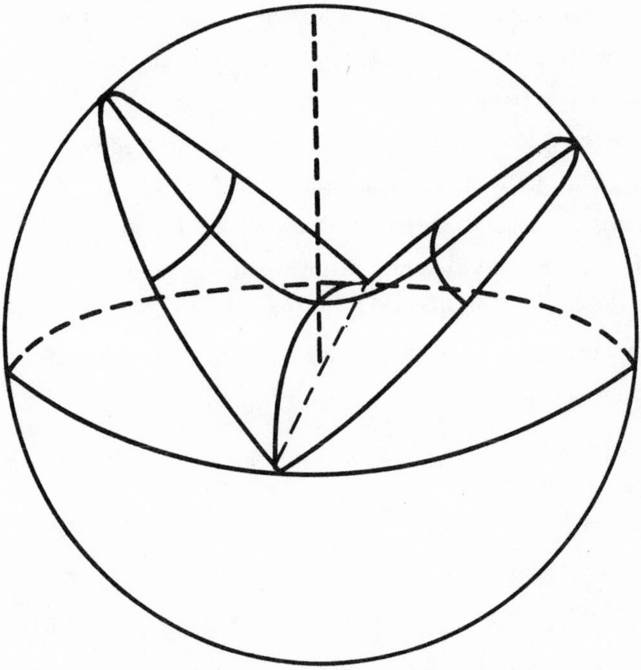

A counter-example to the Bernstein conjecture in H^3.

§2. *Generalization of a theorem of Hsiang and Lawson*

In dealing with minimal varieties in hyperbolic spaces, we fully utilize the inherent symmetries by employing a non-compact group action to transform the problems in hyperbolic space to the reduced problems in the orbit space. For the compact group actions, the reduction lies in the fact that the length of a curve in the orbit space with a proper metric is precisely the area of its preimage, hence the minimality is preserved. But for non-compact group actions, the orbits in general have infinite volume and the method in [7] does not meet the needs of our problem in hyperbolic space. To overcome this difficulty, we extend the theorem of Hsiang and Lawson from compact group action to non-compact group action. In this manner, we can obtain a "nice" orbit space where the global problem is vastly simplified and at the same time we can keep the delicate nature of compact group actions.

Let M be a Riemannian manifold and G a closed subgroup of the isometry group $I(M)$ of M. A submanifold $f: N \to M$ is called G-invariant if there exists a smooth action of G on N such that $gf = fg$ for $g \in G$. By an equivariant variation of a G-invariant submanifold $f: N \to M$, we mean a differentiable variation $f_t: N \to M$, $-\varepsilon < t < \varepsilon$, $f_0 = f$ through submanifolds such that $gf_t = f_t g$ for $g \in G$. Here the action of G on N is independent of t.

LEMMA 2.1. *Let Γ be a discrete subgroup of G and $f: N \to M$ a G-invariant submanifold. Then we have*

(i) *for every compact subset L of N, the set*
$\Gamma(L) = \{\gamma \in \Gamma \mid \gamma L \cap L \neq \emptyset\}$ *is finite,*

(ii) *If Γ is torsion free, Γ acts on N discontinuously, i.e., for any $x_0 \in N$ there is a neighborhood Λ of x_0 such that*

$$\gamma\Lambda \cap \Lambda = \emptyset \quad \text{for all} \quad \gamma \neq e \quad \text{in} \quad \Gamma .$$

Proof. (i) Let $\tilde{L}$ be the image $f(L)$ and $\Gamma(\tilde{L}) = \{\gamma \in \Gamma \mid \gamma\tilde{L} \cap \tilde{L} \neq \emptyset\}$. Clearly $\tilde{L}$ is compact and $\Gamma(L) \subset \Gamma(\tilde{L})$. Since Γ is a discrete subgroup of $I(M)$, $\Gamma(\tilde{L})$ is finite, hence $\Gamma(L)$ is finite.

(ii) Let $\tilde{x}_0$ be the image $f(x_0)$. Γ is torsion free, hence there exists a neighborhood $\tilde{\Omega}$ of $\tilde{x}_0$ in M such that $\gamma\tilde{\Omega} \cap \tilde{\Omega} = \emptyset$ for all $\gamma \neq e$ in Γ. Now take $\Omega = f^{-1}(\Omega)$. Clearly $\tilde{\Omega}$ has the desired property.

Now let Γ be a torsion free discrete subgroup of G. By Lemma 2.1, Γ acts discontinuously on N, and M. Hence we can consider the quotient manifolds $\overline{N} = \Gamma\backslash N$, $\overline{M} = \Gamma\backslash M$. For a Γ-equivariant map $f: N \to M$ (resp. Γ-invariant function $\phi: N \to R$), we denote by $\overline{f}: \overline{N} \to \overline{M}$ (resp. $\overline{\phi}: \overline{N} \to R$) the induced map on $\overline{N}$.

LEMMA 2.2. *Suppose that Γ is a discrete uniform subgroup of G i.e., Γ is discrete and $\Gamma\backslash G$ is compact and $f: N \to M$ a G-invariant submani-*

fold. Then for every $x_0 \in N$, *there exists a smooth* G*-invariant function* $\phi : N \to R$ *such that* $\phi \geq 0$, $\phi(x_0) > 0$ *and the induced map* $\overline{\phi} : \Gamma \backslash N \to R$ *is compactly supported.*

Proof. By Lemma 1, there exists a compact neighborhood Λ of x_0 such that $\gamma\Lambda \cap \Lambda = \emptyset$ for $\gamma \neq e$ in Γ. We choose a smooth function $\theta \in C^\infty(N)$ such that $\theta \geq 0$, $\theta(x_0) > 0$ and θ has support contained in Λ. Introduce a function

$$\psi(x) = \sum_{\gamma \in \Gamma} \theta(\gamma x) . \tag{2.1}$$

Clearly ψ is smooth, $\psi \geq 0$, $\psi(x_0) > 0$ and $\operatorname{supp}(\psi) \subset \Gamma\Lambda$. Moreover, ψ is Γ-invariant, we can consider the integral

$$\phi(x) = \int_{\Gamma \backslash G} \psi(gx)\, dg . \tag{2.2}$$

Obviously, ϕ is smooth, G-invariant, $\phi \geq 0$, $\phi(x_0) > 0$, and $\operatorname{supp}(\phi) \subset$ closure of $G\Lambda$. However, $\Gamma \backslash G$ is compact. There is a compact subset Ω of G with $G = \Gamma\Omega$. It follows that $G\Lambda = \Gamma\Omega\Lambda$ and its image in $\Gamma \backslash N$ is compact, consequently $\operatorname{supp}(\overline{\phi})$ is compact.

Now we are ready to generalize the fundamental theorem in [7].

THEOREM 2.3. *Let* $f : N \to M$ *be a* G*-invariant submanifold of* M. *Suppose that* G *has a torsion free discrete uniform subgroup* Γ. *The following conditions are equivalent:*

(i) $f : N \to M$ *is minimal.*

(ii) *The induced map in the quotient manifold* $\overline{f} : \overline{N} \to \overline{M}$ *is minimal.*

(iii) *The volume of* $\overline{N}$ *is stationary with respect to all compactly supported variations* $\overline{f}_t : \overline{N} \to \overline{M}$ *where* $\overline{f}_t$ *is the induced map of a* G*-equivariant variation* $f_t : N \to M$ *of* f.

Proof. (i) $\Longrightarrow$ (ii) $\Longrightarrow$ (iii) are trivial.

(iii) $\Longrightarrow$ (i). Let K be the mean curvature vector field on N. Since $f: N \to M$ is G-invariant, $g_* K = K$ for $g \in G$. Let ϕ be a smooth, G-invariant function of N such that its induced function $\bar{\phi}$ is compactly supported. Now we introduce a variation f_t given by

$$f_t(x) = \exp_y(t\phi(x)\, K(x)) , \tag{2.3}$$

where $y = f(x)$ and $\exp_y$ is the exponential map on $T_y(M)$. We observe that f_t is G-equivariant. Since $\mathrm{supp}(\bar{\phi})$ is compact, one can choose $\varepsilon > 0$ small enough so that f_t, $-\varepsilon < t < \varepsilon$ are immersions. Without loss of generality, we may assume that the volume of $\bar{N}$ is finite. Now let us consider the variation $\bar{f}_t : \bar{N} \to \bar{M}$. Let ω_t be the volume element of the metric induced by $\bar{f}_t$ and set

$$V(t) = \int_{\bar{N}} \omega_t . \tag{2.4}$$

The variation vector field is simply $\overline{\phi K}$. By the first variation formula, we have

$$\frac{dV}{dt}\Big|_{t=0} = -\int_{\bar{N}} \bar{\phi}\, |\bar{K}|^2\, \omega_0 . \tag{2.5}$$

By Lemma 2, for any $x_0 \in N$, we can always choose ϕ such that $\phi \geq 0$, $\phi(x_0) > 0$ and the induced map $\bar{\phi} : \Gamma \backslash N \to R$ is compactly supported. Then by (2.5) (iii) $\Longrightarrow K = 0$, i.e., $f: N \to M$ is minimal.

§3. *Decomposition of hyperbolic spaces*

In this section, we employ the technique of [12] to decompose the hyperbolic space into the product of two hyperbolic spaces (with proper metrics) so that a non-compact group action is facilitated and we can apply the theory in §2.

Let G be the identity component of $O(n,1)$, $K = G \cap O(n)$ and $H = G/K$. We realize H as the bounded domain

$$H = \{X \in M_{n1}(R) \mid {}^tXX < 1\}. \tag{3.1}$$

For $g \in G$, write

$$g = \begin{pmatrix} A & B \\ C & D \end{pmatrix}$$

with $A \in M_{nn}(R)$, $B \in M_{n1}(R)$, $C \in M_{1n}(R)$ and $D \in R$. The translation of g on H is given by

$$gX = (AX+B)(CX+D)^{-1}, \quad X \in H, \; g \in G. \tag{3.2}$$

Denote by o the image of the identity of G in G/K. G acts transitively on H and the isotropy subgroup of G at o is K. On H, we have a G-invariant metric defined by

$$ds^2 = (1 - {}^tXX)^{-1} d^tX (E - X^tX)^{-1} dX, \tag{3.3}$$

where E is the identity $n \times n$ matrix. Thus H is a symmetric space with a Riemannian metric g. Let $\hat{G}$ and $\hat{K}$ be the Lie algebras of G, K respectively. Denote by $\hat{P}$ the orthogonal complement of $\hat{K}$ in $\hat{G}$ with respect to the Killing form. For $X \in M_{n1}(R)$, we introduce

$$\xi(X) = \begin{pmatrix} 0 & X \\ {}^tX & 0 \end{pmatrix}. \tag{3.4}$$

Then $\hat{P} = \{\xi(X)/X \in M_{n1}(R)\}$. We can identify $\hat{P}$ with the tangent space $T_0(H)$ of H at o. The Riemannian metric becomes

$$g(\xi(X), \xi(Y)) = {}^tXY. \tag{3.5}$$

For $g = \begin{pmatrix} A & B \\ C & D \end{pmatrix} \in G$, $X \in H$, we introduce

$$j(g, X) = CX + D. \tag{3.6}$$

The translation of g on H then takes a convenient matrix multiplication form,

$$g\begin{pmatrix} X \\ 1 \end{pmatrix} = \begin{pmatrix} gX \\ 1 \end{pmatrix} j(g, X) . \tag{3.7}$$

Now let $Q = \begin{pmatrix} E_n & 0 \\ 0 & -1 \end{pmatrix}$. Thus Q defines a bilinear form on R^{n+1}.

DEFINITION 3.1. Let V be a subspace of R^{n+1} such that $Q|V$ is positive definite. Define G_V, H_V by

$$\begin{aligned} G_V &= \{g \in G \mid gv = v \text{ for } v \in V\} , \\ H_V &= \{X \in H \mid {}^t vQ\begin{pmatrix} X \\ 1 \end{pmatrix} = 0 \text{ for } v \in V\} . \end{aligned} \tag{3.8}$$

LEMMA 3.2 ([12]). G_V, H_V *have the following properties*:

(i) *For* $g \in G$, $gH_V = H_{gV}$.

(ii) G_V^o, *the identity component of* G_V, *acts transitively on* H_V.

(iii) H_V *is a totally geodesic sub-symmetric space of* H *and* $\dim H_V = n - \dim V$.

Now, let $e_1, \cdots, e_{n+1}$ be the standard basis of R^{n+1}. Suppose that V is a subspace of R^{n+1} such that $Q|V$ is positive definite and $1 \leq \dim(V) = r < n$. By (i) of Lemma 3.2 and Witt's theorem, we can assume that V is the subspace spanned by $e_{n-r+1}, \cdots, e_n$. For $X \in H$, we decompose X into

$$X = \begin{pmatrix} X_1 \\ X_2 \end{pmatrix} , \tag{3.9}$$

where $X_1 \in M_{n-r,1}(R)$, $X_2 \in M_{r1}(R)$. H_V is then given by

$$H_V = \{X \in H \mid X_2 = 0\}$$

and $g \in G_V$ has a block matrix form

$$g = \begin{pmatrix} A_1 & 0 & B_1 \\ 0 & E_r & 0 \\ C_1 & & D_1 \end{pmatrix}. \tag{3.10}$$

where $A_1 \in M_{n-r,n-r}(R)$, $B_1 \in M_{n-r,1}(R)$, $C_1 \in M_{1,n-r}(R)$, $D_1 \in R$. Denote by H_1 the domain

$$H_1 = \{W \in M_{n-r,1}(R) \mid {}^tWW < 1\}. \tag{3.11}$$

Clearly G_V acts on H_1 and the action of G_V on H takes the form

$$gX = \begin{pmatrix} gX_1 \\ X_2 j^{-1}(g,X) \end{pmatrix}, \quad g \in G_V, \ X \in H. \tag{3.12}$$

where $gX_1 = (A_1X_1 + B_1)(C_1X + D_1)^{-1}$.

Now we discuss the distance function $d(X, H_V)$ from X to H_V. By (2.12), there exists $g \in G_V$ such that gX is of the form

$$X' = gX = \begin{pmatrix} 0 \\ X_2' \end{pmatrix}. \tag{3.13}$$

It is easy to see that

$$\begin{aligned} d(X, H_V) &= d(X', H_V) \\ &= d(0, X_2'), \end{aligned} \tag{3.14}$$

where $d(0, X_2')$ is the distance in the domain

$$H_2 = \{X \in M_{r1}(R) \mid {}^tXX < 1\}. \tag{3.15}$$

LEMMA 3.3. *Let* $X \in H$ *and* $d = d(0,X)$. *Then*

$$(1 - {}^tXX)^{-1} = \cosh^2 d.$$

Proof. There is a unique $Y \in M_{n1}(R)$ such that

$$\exp(\xi(Y)) \cdot o = X . \tag{3.16}$$

The curve $\exp(t\,\xi(Y)) \cdot o$, $0 \le t \le 1$ is a geodesic joining o and X. By (3.5)

$$d^2 = {}^tYY . \tag{3.17}$$

From (3.16) and some computation, we have

$$\frac{Y}{d} \cdot \frac{\sinh d}{\cosh d} = X \tag{3.18}$$

and $\cosh^2 d = (1 - {}^tXX)^{-1}$ follows immediately from (3.18).

DEFINITION 3.4. For $X = \begin{pmatrix} X_1 \\ X_2 \end{pmatrix} \in H$, we introduce two functions $a(X)$, $b(X)$ by

$$\begin{aligned} a(X) &= 1 - {}^tXX , \\ b(X) &= 1 - {}^tX_1X_1 . \end{aligned} \tag{3.19}$$

LEMMA 3.5. $b(X)/a(X)$ *is* G_V*-invariant, moreover,* $\cosh^2 d(X, H_V) = b(X)/a(X)$.

Proof. For $g \in G_V$,

$$\begin{aligned} a(X) &= j(g, X)^2\, a(gX) \quad \text{and} \\ b(X) &= j(g, X)^2\, b(gX) , \end{aligned} \tag{3.20}$$

hence $b(X)/a(X)$ is G_V-invariant. There exists $g \in G_V$ such that $gX = X' = \begin{pmatrix} 0 \\ X_2' \end{pmatrix}$. By (3.14)

$$\begin{aligned} d(X, H_V) &= d(X', H_V) \\ &= d(0, X_2') . \end{aligned}$$

It follows from Lemma 3.3 that $\cosh^2 d(X, H_V) = (1 - {}^tX'_2X'_2)^{-1} = b(X')/a(X') = b(X)/a(X)$.

Now let G_1 be the identity component of G_V, V the subspace spanned by $e_{n-r+1}, \cdots, e_n$, and H_1, H_2 the bounded domain given by (3.11) and (3.15) respectively. In the following, we consider the orbit space H/G_1 of G_1-orbits and the induced metric on $H/G_1 \cdot G_1$ acts on H_1 and H. The following is a G_1-equivariant decomposition of H.

LEMMA 3.6. *Let* $\alpha: H_1 \times H_2 \to H$ *by the map given by*

$$(3.21) \qquad \alpha(X_1, X_2) = \begin{pmatrix} X_1 \\ X_2(1 - {}^tX_1X_1)^{\frac{1}{2}} \end{pmatrix}, \quad X_1 \in H_1, \ X_2 \in H_2 .$$

Then (i) α *is a* G_1*-equivariant diffeomorphism.*

(ii) $\cosh^2 d(\alpha(X_1, X_2), H_1) = (1 - {}^tX_2X_2)^{-1}$.

Proof. (i) First let us consider the map $\beta: G_1 \times H_2 \to H$ defined by

$$(3.22) \qquad \beta(g, X_2) = g\begin{pmatrix} 0 \\ X_2 \end{pmatrix}, \quad g \in G_1, \ X_2 \in H_2 .$$

Clearly β is G_1-equivariant. Let $X = \begin{pmatrix} X_1 \\ X_2 \end{pmatrix} \in H$. By (ii) of Lemma 6, G_1 acts transitively on H_1. Thus there is $g \in G_1$ such that $gX_1 = 0$ and by (3.12)

$$(3.23) \qquad g\begin{pmatrix} X_1 \\ X_2 \end{pmatrix} = \begin{pmatrix} 0 \\ X_2 j^{-1}(g, X) \end{pmatrix} .$$

It is easy to see $X_2 j^{-1}(g, X) \in H_2$ and $\beta(g^{-1}, X_2 j^{-1}(g, X)) = \begin{pmatrix} X_1 \\ X_2 \end{pmatrix}$. Thus β is onto. Let $K_1 = G_1 \cap K$. $K_1 = \left\{ \begin{pmatrix} A & 0 \\ 0 & E_{n-r+1} \end{pmatrix} \middle| A \in SO(r)^o \right\}$. K_1 is the isotropy subgroup of G_1 at o in H_1. It follows that $\beta(g, X_2) =$

$\beta(g', X_2')$ if and only if $g^{-1}g' \in K_1$ and $X_2 = X_2'$. Since $G_1/K = H_1$, β induces a G_1-equivariant diffeomorphism $\alpha : H_1 \times H_2 \to H$. To obtain the explicit formula for α, let us recall that for each $X_1 \in H_1$, there is

$$g_{X_1} = \begin{pmatrix} (E_r - X_1{}^tX_1)^{\frac{-1}{2}} & & X_1(1 - {}^tX_1X_1)^{\frac{-1}{2}} \\ 0 & E_{n-r} & 0 \\ (1 - {}^tX_1X_1)^{-\frac{1}{2}}X_1 & & (1 - {}^tX_1X_1)^{\frac{-1}{2}} \end{pmatrix}$$

satisfying $g_{X_1} \cdot o = X_1$ in H_1 and by

$$(3.23) \qquad g_{X_1}\begin{pmatrix} 0 \\ X_2 \end{pmatrix} = \begin{pmatrix} X_1 \\ X_2(1 - {}^tX_1X_1)^{\frac{1}{2}} \end{pmatrix}.$$

Clearly $\alpha(X_1, X_2) = g_{X_1}\begin{pmatrix} 0 \\ X_2 \end{pmatrix} = \begin{pmatrix} X_1 \\ X_2(1 - {}^tX_1X_1)^{\frac{1}{2}} \end{pmatrix}$ and the assertion (i) of Lemma 3.6 follows. (ii) follows from Lemma 3.5.

LEMMA 3.7. $\alpha^*(ds^2) = ds_2^2 + (1 - {}^tX_2X_2)^{-1}\, ds_1^2$ *where* ds_1^2 *and* ds_2^2 *are the metrics on* H_1, H_2 *respectively.*

Proof. α is G_1-equivariant. Hence it suffices to establish the assertion at ${}^t(0, X_2)$. At $\begin{pmatrix} 0 \\ X_2 \end{pmatrix}$ the metric of H is given by (2.3)

$$(3.24) \qquad ds^2 = (1 - {}^tX_2X_2)^{-1}\, d^tX \begin{pmatrix} E_1 & 0 \\ 0 & E_{n-r} - X_2{}^tX_2 \end{pmatrix}^{-1} dX\,.$$

Let $A \in T_o(H_1)$. $\tau(t) = \begin{pmatrix} tA \\ X_2(1 - t^{2t}AA)^{\frac{1}{2}} \end{pmatrix} = \alpha(tA, X_2)$, so $\dot{\tau}(0) = \begin{pmatrix} A \\ 0 \end{pmatrix}$.

Thus $\alpha_* A = \begin{pmatrix} A \\ 0 \end{pmatrix}$. Let $B \in T_{X_2}(H_2)$. It is easy to see $\alpha_* B = \begin{pmatrix} 0 \\ B \end{pmatrix}$. By (2.24), $\alpha_* A$ and $\alpha_* B$ are orthogonal, and moreover

$$(\alpha^* g)(A, A) = (1 - {}^t X_2 X_2)^{-1}\, {}^t AA = (1 - {}^t X_2 X_2)^{-1} g_1(A, A)\,,$$

$$(\alpha^* g)(B, B) = (1 - {}^t X_2 X_2)^{-1}\, {}^t B(E_{n-r} - X_2 {}^t X_2)^{-1} B = g_2(B, B)\,,$$

where g_1, g_2 are the metrics of H_1, H_2 respective. Therefore $\alpha^*(ds^2) = ds_2^2 + (1 - {}^t X_2 X_2)^{-1} ds_1^2$.

§4. *Non-compact group actions and geometric aspects of the orbit spaces*

In this section, we employ a non-compact group action and study the geometric and analytic structure of the orbit space. As a result, we obtain a continuum of geodesics foliating the orbit space.

From the preceding section, we now study the case $r = n-2$, which implies $\dim H_2 = 2$ by Lemma 3.2.

PROPOSITION 4.1. *Let G_1 act on H by the translation (2.2). The orbit space H/G_1 is homeomorphic to H_2 and for any discrete torsion free uniform subgroup Γ of G_1, α (in Lemma 3.5) induces a differentiable bundle*

$$\Gamma\backslash H_1 \longrightarrow \Gamma\backslash H \xrightarrow{\pi'} H_2\,. \tag{4.1}$$

Proof. Immediate from Lemma 3.6.

Now let us define a volume function $V : H_2 \to \mathbf{R}$. For $X_2 \in H_2$, let $V(X_2) = (n-2)$-dimensional volume of $\pi^{-1}(X_2)$. By Lemma 3.7, $V(X_2) = c(1 - {}^t X_2 X_2)^{\frac{-(n-2)}{2}}$, where $c = \mathrm{vol}(\Gamma\backslash H_1)$. For simplicity, we take $c = 1$. We introduce a metric

$$\rho = (1 - {}^t X_2 X_2)^{-(n-2)} ds_2^2 \tag{4.2}$$

on H_2 where ds_2^2 is the usual metric on H_2. Then the arc length of a curve in H_2 w.r.t. the metric ρ is equal to the area of its inverse image in $\Gamma\backslash H$.

On H_2, let us use the polar coordinates (t, θ) where t = distance from X_2 to 0. Then we know $ds_2^2 = dt^2 + \sinh^2 t\, d\theta^2$ and by Lemma 3.3, $(1 - {}^tX_2X_2)^{-1} = \cosh^2 t$. Thus ρ becomes

$$\rho = \cosh^{2(n-2)} t (dt^2 + \sinh^2 t\, d\theta^2)\,. \tag{4.3}$$

By a straightforward computation, we obtain

PROPOSITION 4.2. *H_2 with metric ρ is complete and the Gaussian curvature K is given by*

$$K = \frac{-1}{\cosh^{2(n-2)} t} \{(2n-3) - (n-2) \tanh^2 t\}\,. \tag{4.4}$$

By solving an ordinary differential equation, we can characterize geodesics in the following manner:

LEMMA 4.3. *The pregeodesics of (H_2, ρ) are given by the following curves:*

(i) *t-parameter curves, i.e., geodesic lines in (H_2, ds_2^2) passing through o.*

(ii) *curves of the form $(t, \theta(t))$ with*

$$\theta(t) = \frac{\pm c\, dt}{\sinh t \sqrt{\cosh^{2(n-2)} t \sinh^2 t - c^2}} \tag{4.5}$$

$(c \neq 0)$.

REMARK. The only geodesics of (H_2, ρ) passing through o are those geodesics of (H_2, ds_2^2) passing through o, for curves in (ii) do not pass through o.

LEMMA 4.4. *Let* τ *be a geodesic of* (H_2, ρ) *and* $\tilde{\tau}$ *the preimage of* τ *in* H. $\tilde{\tau}$ *is a totally geodesic submanifold if and only if* τ *passes through* 0.

Proof. $\Longrightarrow$) Observe that $\tilde{\tau}$ is G_1-invariant. If $\tilde{\tau}$ is a totally geodesic submanifold, $\tilde{\tau}$ is isometric to H^{n-1} ((n–1)-dimensional hyperbolic space) and $G_1 \subset I(\tilde{\tau})$. Let $G_0 = I(\tilde{\tau})$, and

$$A = \left\{ \begin{pmatrix} \cosh s & 0 & \sinh s \\ & E_{n-1} & \\ \sinh s & 0 & \cosh s \end{pmatrix} s \in R \right\} \subset G_1 .$$

Then A is a connected, maximal abelian subgroup of G_0 such that Ad(A) consists of R-diagonalizable matrices. It follows that in $\tilde{\tau}$, there exists $X \in \tilde{\tau}$ such that AX is a geodesic. However, A0 is the only A-invariant geodesic, consequently, τ passes through 0.

$\Longleftarrow$) By Lemma 14, τ is just a t-parameter curve, more precisely τ is a line in H_2 passing through 0. It follows that

$$\tilde{\tau} = \left\{ X = \begin{pmatrix} X_1 \\ X_2 \end{pmatrix} \in H \mid X_2 \text{ lies in a line in } H^2 \text{ passing through } 0 \right\}.$$

We can find $k \in K$ so that

$$k\tilde{\tau} = \left\{ X = \begin{pmatrix} X_1 \\ \vdots \\ X_n \end{pmatrix} \in H \mid X_n = 0 \right\} .$$

Clearly $k\tilde{\tau}$ is a totally geodesic submanifold of H.

In H_2, we take a fixed geodesic λ passing through 0. Let $T(\lambda)^{\perp}$ be the normal bundle of λ in H_2. Consider the exponential map $\mathrm{Exp} : T(\lambda)^{\perp} \to H_2$. Since H_2 is simply connected and (H_2, ρ) has negative curvature, Exp is a diffeomorphism. Clearly λ is diffeomorphic to

R. Let $T(\lambda)^{\perp} \xrightarrow{p} \lambda$ be the projection map and $\lambda \xrightarrow{a} R$ a diffeomorphism. Then we have a C^∞ function $\beta: H_2 \to R$ where $\beta = a \circ p \circ \mathrm{Exp}^{-1}$. By our construction the level curves $\beta = c$ are geodesics in (H_2, ρ) and these geodesics fill out H_2 (please see the diagram below). Moreover, there is only one level curve passing through 0.

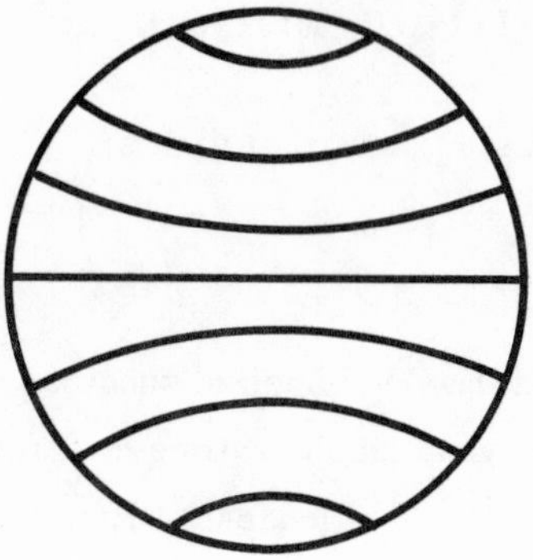

§5. *Counter-examples to the Bernstein conjecture in* H^n *and other homogeneous spaces*

The main work in this section is to show the lift of geodesics in §4 are absolutely area-minimizing leaves foliating H^n by constructing a globally defined closed form with comass 1.

Use the same notations as before. Now let $\pi: H \to H_2 = H/G_1$ be the projection map and $h = \beta \circ \pi: H \to R$.

LEMMA 5.1. *The level sets of* h *are minimal surfaces of* H, *thus they foliate* H.

Proof. Observe that G_1 contains a uniform discrete subgroup Γ. Consider $\overline{H} = H/\Gamma$ and $\overline{\pi}: \overline{H} \to H/G_1$. Let $\overline{h} = \beta \circ \overline{\pi}$. Since the level curves $\beta = c$ are geodesics in H_2, by Theorem 2.3, the level surfaces $\overline{h} = c$ are minimal in $\overline{H}$; hence the level surfaces $h = c$ are minimal in H.

REMARK. By the construction of h, it is easy to see $|\nabla h(x)| \neq 0$ for every $x \in H$.

PROPOSITION 5.2. *Let* M *be an orientable* n*-dimensional Riemannian manifold and* $h: M \to R$ *a smooth function with* $|\nabla h(x)| \neq 0$ *for every* $x \in M$. *Suppose that each level surface is minimal in* M. *Then it is homologically area minimizing.*

Proof. Let X be the unit vector field normal to level surfaces in the direction of increasing the value of h. Since $|\nabla h(x)| \neq 0$ for every $x \in M$; X is globally defined on M. Now we introduce an $(n-1)$-form defined on M

$$\phi(X_1, \cdots, X_{n-1}) = \omega(X, X_1, \cdots, X_{n-1}) . \tag{5.1}$$

where ω is the volume form of M.

Clearly ϕ has mass 1 on the level surfaces and has mass < 1 otherwise. We assert that $d\phi = 0$. Let $x_o \in M$. Here we can choose local coordinates $(x_1, x_2, \cdots, x_n)$ such that $x_0 = (0, \cdots, 0)$, $x_1 = c$ are level surfaces $\frac{\partial}{\partial x_1} \perp$ level surfaces, and $\left(\frac{\partial}{\partial x_1}\right)_o, \cdots, \left(\frac{\partial}{\partial x_n}\right)_o$ are orthonormal (at x_o). Let $X_i = \frac{\partial}{\partial x_i}$, $1 \leq i \leq n$. Then

$$\begin{aligned}(d\phi)(X_1, X_2, \cdots, X_n) &= \frac{1}{n} \sum_{i=2}^{n} (-1)^{i-1} X_i \phi(X_1, \cdots, \hat{X}_i, \cdots, X_n) \\ &\quad + \frac{1}{n} X_1 \phi(X_2, \cdots, X_n) \\ &= \frac{1}{n} X_1 \omega(X, X_2, \cdots, X_n) \\ &= \frac{1}{n} \omega(\nabla_{X_1} X, X_2, \cdots, X_n) \\ &\quad + \frac{1}{n} \sum_{2 \leq i \leq n} \omega(X, X_2, \cdots, \nabla_{X_1} X_i, \cdots, X_n)\end{aligned} \tag{5.2}$$

where ∇ is the Riemannian connection on M. Furthermore, $g(X, X) = 1$, implies $\nabla_{X_1} X$ is orthogonal to X, thus $\omega(\nabla_{X_1} X, X_2, \cdots, X_n) = 0$ at x_o.

Since $g(X_1, X_i) = 0$ is

$$g(\nabla_{X_i} X_1, X_i) + g(X_1, \nabla_{X_i} X_i) = 0 \tag{5.3}$$

and since $\nabla_{X_1} X_i = \nabla_{X_i} X_1$ we have at the point x_o,

$$\begin{aligned} &\omega(X, X_2, \cdots, \nabla_{X_1} X_i, \cdots, X_n) \\ &= \omega(X, X_2, \cdots, \nabla_{X_i} X_1, \cdots, X_n) \\ &= \omega(X, X_2, \cdots, X_n)\, g(\nabla_{X_i} X_1, X_i) \\ &= -\omega(X, X_2, \cdots, X_n)\, g(X_1, \nabla_{X_i} X_i) . \end{aligned} \tag{5.4}$$

Therefore, at x_o

$$\begin{aligned} &d\phi(X, X_2, \cdots, X_n) \\ &= -\frac{1}{n}\, g(X, K(x_o)) \\ &= 0 \end{aligned}$$

where $K(x_o)$ is the mean curvature of $h = h(x_o)$ at $x_o \cdot d\phi = 0$, the assertion now follows from Stokes' theorem.

Combining the results of Lemma 5.1 and Proposition 5.2, we can settle the Bernstein conjecture in H^n in all dimensions $(n > 2)$.

THEOREM 5.3. *The level sets* S_i *of* h *are absolutely area-minimizing imbedded hypersurfaces in* H^n $(n > 2)$. *Furthermore, they are of topological type* R^{n-1} *but only one of them is totally geodesic which passes through* o.

REMARK 1. We can describe the area-minimizing hypersurface S_i explicitly as follows:

Let $X_2 = \begin{pmatrix} y_1 \\ y_2 \end{pmatrix}$ and $r = \sqrt{y_1^2 + y_2^2}$. Then $(1 - y_1^2 - y_2^2)^{-1} = \cosh^2 t$. $(1-r^2)^{-1} = \cosh t$, and $X_2 = \begin{pmatrix} r\cos\theta \\ r\sin\theta \end{pmatrix}$. The pre-geodesics in H_2, ρ

can be given by $\begin{pmatrix} r\cos\theta(r) \\ r\sin\theta(r) \end{pmatrix}$ with

$$\theta(r) = \int \frac{\pm c\, dr}{1-(1-r^2)^{\frac{1}{2}}\sqrt{\frac{r^2}{(1-r^2)^{n-1}}-c^2}} \,. \tag{5.5}$$

The set

$$F = \left\{ \begin{pmatrix} X_1 \\ r(1-{}^tX_1X_1)^{\frac{1}{2}}\cos\theta(r) \\ \pm r(1-{}^tX_1X_1)^{\frac{1}{2}}\sin\theta(r) \end{pmatrix} \,\middle|\, X_1 \in H_1\,,\ r_0 \le r < 1 \right\}$$

$\frac{r_0^2}{(1-r_0^2)^{n-1}} = c^2$ and $\theta(r)$ given by (5.5) is an area-minimizing surface of H. The boundary F_∞ of S_i at ∞ is the set

$$F_\infty = \left\{ \begin{pmatrix} X_1 \\ (1-{}^tX_1X_1)^{\frac{1}{2}}\cos\theta(1) \\ \pm(1-{}^tX_1X_1)^{\frac{1}{2}}\sin\theta(1) \end{pmatrix} \,\middle|\, {}^tX_1X_1 \le 1 \right\}.$$

(cf. the picture on page 337).

REMARK 2. Since $H^n = SO(n,1)/SO(n)$, by lifting or taking the product structures, we can obtain a continuum of complete, absolutely area-minimizing submanifolds (with arbitrary co-dimensions) which foliate $SO(n_1,1)\times\cdots\times SO(n_\ell,1)\times H^{m_1}\times\cdots\times H^{m_r}$ for proper $\ell, r \ge 0$ with $\ell^2+r^2 \ne 0$; $n_i, m_j > 2$ and only one of them is totally geodesic. In particular, the Bernstein conjecture is also false in these spaces.

S. P. WANG
SCHOOL OF MATHEMATICS
INSTITUTE FOR ADVANCED STUDY
PRINCETON, NEW JERSEY 08540

S. WALTER WEI
DEPARTMENT OF MATHEMATICS
UNIVERSITY OF HAWAII
HONOLULU, HAWAII 96822

REFERENCES

[1] Almgren, F. J., Jr. Some interior regularity theorems for minimal surfaces and an extension of Berstein Theorem. Ann. Math. 84 (1966), 277-292.

[2] Bernstein, S. Sur un théorème de Géométrie et ses applications aux équations aux dérivées partielles du type elliptique, Comm. de la Soc. Math. de Kharkov (2 éme sér) 5 (1915-1917), 38-45.

[3] Bombieri, E., de Giorgi, E., Giutti, E. Minimal cones and the Berstein problem. Inventiones Math. 7 (1969), 243-269.

[4] de Giorgi, E. Una extensione del theoreme di Berstein. Ann. Scuola Norm. Sup. Pisa 19 (1965), 79-85.

[5] Fleming, W. On the oriented plateau problem. Rend. Circ. Mat. Palermo 11 (1962), 1-22.

[6] Harvey, R., Lawson, B. Calibrated foliations.

[7] Hsiang, W. Y., Lawson, H. B. Minimal Submanifolds of Low Cohomogeneity. J. Diff. Geom. 5 (1970), 1-37.

[8] Kobayashi, S., Nomizu, K. Foundations of Differential Geometry. Vol. I and II. Interscience, NY 1965 and 1968.

[9] Lawson, H. B. The equivariant Plateau problem and interior regularity. Trans. A.M.S. 173 (1972), 231-249.

[10] O'Neill, B. Elementary Differential Geometry, Academic Press, 1969.

[11] Simons, J. Minimal Varieties in Riemannian manifolds. Ann. of Math. 88 (1968), 62-105.

[12] Tong, Y. L., Wang, S. P. Poincaré dual of some cycles in quotient of bounded domains.

[13] Wei, S. W. Remarks on functions of Least Gradient.

[14] ———. Minimality, Stability and plateau's problem. Berkeley Thesis, 1980.

[15] ———. Plateau problems in Symmetric Spaces.

Library of Congress Cataloging in Publication Data
Main entry under title:

Seminar on minimal submanifolds.

(Annals of mathematics studies ; 103)
Bibliography: p.
1. Submanifolds, Minimal—Addresses, essays, lectures. I. Bombieri, Enrico, 1940- . II. Series.
QA649.S45 1983 516.3′6 82-61356
ISBN 0-691-08324-X
ISBN 0-691-08319-3 (pbk.)

Enrico Bombieri is Professor of Mathematics at the Institute for Advanced Study in Princeton, New Jersey.